Applied Iterative Methods

Applied Iterative Methods

Charles L. Byrne

University of Massachusetts Lowell

CRC Press
Taylor & Francis Group
Boca Raton London New York

CRC Press is an imprint of the
Taylor & Francis Group, an **informa** business

AN A K PETERS BOOK

CRC Press
Taylor & Francis Group
6000 Broken Sound Parkway NW, Suite 300
Boca Raton, FL 33487-2742

First issued in hardback 2020

© 2008 by Taylor & Francis Group, LLC

First issued in paperback 2021

CRC Press is an imprint of the Taylor & Francis Group, an Informa business

No claim to original U.S. Government works

ISBN 13: 978-0-367-44616-1 (pbk)
ISBN 13: 978-1-56881-342-4 (hbk)

Library of Congress Cataloging-in-Publication Data

Byrne, C. L. (Charles L.), 1947-
    Applied iterative methods / Charles L. Byrne.
        p. cm.
    Includes bibliographical references and index.
    ISBN-13: 978-1-56881-342-4 (alk. paper)
    ISBN-10: 1-56881-271-X (alk. paper)
    1. Iterative methods (Mathematics) 2. Algorithms. I. Title.

    QA297.8.B97 2007
    518'.26--dc22

                                                        2007027595

Cover image: Radar image of Death Valley, California, and the different surface types in the area. NASA Planetary Photojournal.

Visit the Taylor & Francis Web site at
http://www.taylorandfrancis.com

and the CRC Press Web site at
http://www.crcpress.com

To Eileen

Contents

Preface

Much has been written on the theory and applications of iterative algorithms, so any book on the subject must be but a glimpse. The topics included here are those most familiar to me, and not necessarily those most familiar to others. Well-known algorithms that have been exhaustively discussed in other books, such as Dantzig's simplex method, are mentioned here only in passing, with more attention given to methods, like the expectation maximization maximum likelihood algorithm, that are popular within a confined group, but perhaps less familiar to those outside the group. Over the past two or three decades, I have had the opportunity to work on the application of mathematical methods to problems arising in acoustic signal processing, optical imaging and medical tomography. Many of the problems and algorithms I discuss here are ones I became familiar with during this work. It is the interplay between problems and algorithms, how problems can lead to algorithms, old algorithms and proofs lead to new ones by analogy, and algorithms are applied to new problems, that fascinates me, and provides the main theme for this book.

This book is aimed at a fairly broad audience of scientists and engineers. With few exceptions, the problems and algorithms discussed here are presented in the context of operators on finite-dimensional Euclidean space, although extension to infinite-dimensional spaces is often possible.

This book is not a textbook, but rather a collection of essays on iterative algorithms and their uses. I have used earlier versions of this book as the text in a graduate course on numerical linear algebra, concentrating more on specific algorithms, somewhat on the applications, and less on the general unifying framework of operators and their properties. I have also used substantial portions of the book in a graduate class on the mathematics of medical image reconstruction, with emphasis on likelihood maximization methods and Fourier inversion. Certain topics in the book will be appropriate for an undergraduate class, but generally the book is aimed at a graduate-level audience.

Some of the chapters end with a section devoted to exercises. In addition, throughout the book there are a number of lemmas given without proof, with the tacit understanding that the proofs are left as additional exercises for the reader.

Most of the referenced papers that I have authored or co-authored are available at http://faculty.uml.edu/cbyrne/cbyrne.html.

Acknowledgments

Over the years, I have had the good fortune to work with many people who have contributed to my growth as a mathematician. I particularly wish to thank Francis Sullivan, Ray Fitzgerald, Mike Fiddy, Alan Steele, Bill Penney, Mike King, Steve Glick, and Yair Censor.

Glossary of Symbols

R^N	The space of N-dimensional real vectors
C^N	The space of N-dimensional complex vectors
$\mathcal{J}(g)(x)$	Jacobian matrix of g at x
A^T	Transpose of the matrix A
$A^\dagger$	Conjugate transpose of the matrix A
$\mathcal{X}$	Either R^N or C^N
$\|x\|_2$	Euclidean norm of the vector x
$\overline{\alpha}$	Complex conjugate of the scalar α
$\langle x, y \rangle$	Inner product of x and y
$x \cdot y$	$y^T x$ in R^N and $y^\dagger x$ in C^N
$H = H(a, \gamma)$	$\{x \mid \langle x, a \rangle = \gamma\}$
$P_C x$	Orthogonal projection of x onto the set C
$P_C^f x$	Bregman projection of x onto the set C
I	The identity matrix
$S^\perp$	The set of vectors orthogonal to all vectors in S
$CS(A)$	Span of the columns of A
$NS(A)$	Null space of A
s_j	$\sum_{i=1}^{I} A_{ij}$
$\{\mathcal{S}, d\}$	A metric space
$\|x\|$	A norm of x
$\rho(S)$	Spectral radius of the matrix S
$\lambda_{\max}(S)$	Largest eigenvalue of Hermitian S
$Re(z)$	Real part of the complex number z
$Im(z)$	Imaginary part of the complex number z
$\text{Fix}(T)$	The fixed points of T
$KL(x, z)$	Kullback-Leibler distance from x to z
α_+	$\max\{\alpha, o\}$
$\text{prox}_{\gamma f}(\cdot)$	Proximity operator
$\text{dom}(f)$	Effective domain of f
$\iota_C(x)$	Indicator function of the set C
$\text{int} D$	Interior of the set D
$\text{bdry} D$	Boundary of the set D

$\mathrm{Ext}(C)$	Extreme points of the set C
$\partial f(x)$	Subdifferential of f at x
m_f	Moreau envelope
f^*	Conjugate function of f
$Q_i x$	Weighted KL projection of x

Glossary of Abbreviations

ABMART	MART with lower and upper bound vectors a and b (u and v in this book)
ABEMML	EMML with lower and upper bound vectors a and b (u and v in this book)
AMS	Agmon-Motzkin-Schoenberg Algorithm
ART	Algebraic reconstruction technique
BI-ART	Block-iterative ART
BI-EMML	Block-iterative EMML
BI-SMART	Block-iterative SMART
BLUE	Best linear unbiased estimator
BPC	Bregman paracontraction
CAT	Computer-assisted tomography
CFP	Convex feasibility problem
CGM	Conjugate gradient method
CSP	Cyclic subgradient projection method
DART	Double ART
DFT	Discrete Fourier transform
DPDFT	Discretized PDFT
EKN	Elsner-Koltracht-Neumann Theorem 5.21
EM-MART	Row-action version of EMML
EMML	Expectation maximization maximum likelihood method
ET	Emission tomography
EUD	Equivalent uniform dosage
GS	Gauss-Seidel method
HLWB	Halpern-Lions-Wittmann-Bauschke Algorithm
IMRT	Intensity-modulated radiation therapy
IPA	Interior-point algorithm
IPDFT	Indirect prior DFT method
ism	Inverse strongly monotone operator
JOR	Jacobi over-relaxation
KL	Kullback-Leibler distance
KM	Krasnoselskii-Mann Theorem 5.16
LC	Limit cycle

LOR	Line of response
LP	Linear programming
LS-ART	Least-squares ART
MAP	Maximum a posteriori method
MART	Multiplicative algebraic reconstruction technique
ML	Maximum likelihood
MRI	Magnetic-resonance imaging
MSGP	Multiple-distance SGP method
MSSFP	Multiset split feasibility problem
MTF	Modulation transfer function
NR	Newton-Raphson method
OAR	Organs at risk
OSEM	Ordered subset EM method
OSL	One step late
OSSMART	Ordered subset SMART
OTF	Optical transfer function
PET	Positron emission tomography
pdf	Probability density function
PDFT	Prior DFT method
POCS	Projection onto convex sets method
PTV	Planned target volume
RAMLA	Row-action maximum likelihood algorithm
RBI	Rescaled block-iterative
RBI-ART	Rescaled block-iterative ART
RBI-EMML	Rescaled block-iterative EMML
RBI-SMART	Rescaled block-iterative SMART
REM-MART	Rescaled EM-MART
RF	Radio frequency
SAR	Synthetic-aperture radar
SART	Simultaneous ART
SFP	Split feasibility problem
SGP	Successive generalized projection method
SIMOP	Simultaneous orthogonal projection method
SMART	Simultaneous multiplicative ART
SOP	Successive orthogonal projection method
SOR	Successive over-relaxation
SPECT	Single-photon emission computed tomography
SVD	Singular-value decomposition of a matrix
UV	Ultraviolet

I | Preliminaries

VALENTINE: What she's doing is, every time she works out a value for y, she's using *that* as her next value for x. And so on. Like a feedback. She's feeding the solution back into the equation, and then solving it again. Iteration, you see. . . . This thing works for any phenomenon which eats its own numbers. . . .

HANNAH: What I don't understand is . . . why nobody did this feedback thing before—it's not like relativity, you don't have to be Einstein.

VALENTINE: You couldn't see to look before. The electronic calculator was what the telescope was for Galileo.

HANNAH: Calculator?

VALENTINE: There wasn't enough time before. There weren't enough *pencils*! . . . Now she'd only have to press a button, the same button over and over. Iteration. . . . with only a *pencil* the calculations would take me the rest of my life to do again—thousands of pages—tens of thousands! And so boring! . . .

HANNAH: Do you mean that was the only problem? Enough time? And paper? And the boredom? . . .

VALENTINE: Well, the other thing is, you'd have to be insane.

Tom Stoppard, *Arcadia,* Act 1, Scene 4

1 | Introduction

A typical iterative algorithm (the name comes from the Latin word *iterum*, meaning "again") involves a relatively simple calculation, performed repeatedly. An iterative method produces a sequence of approximate answers that, in the best case, converges to the solution of the problem. The idea of using iterative procedures for solving problems is an ancient one. Archimedes' use of the areas of inscribed and circumscribed regular polygons to estimate the area of a circle is a famous instance of an iterative procedure, as is his method of exhaustion for finding the area of a section of a parabola.

The well-known formula for solving a quadratic equation produces the answer in a finite number of calculations; it is a non-iterative method, if we are willing to accept a square-root symbol in our answer. Similarly, Gauss elimination gives the solution to a system of linear equations, if there is one, in a finite number of steps; it, too, is a non-iterative method. The bisection method for root-finding is an iterative method. Some iterative sequences arise not from algorithms but from discrete models of continuous systems. The study of dynamical systems provides several interesting examples.

1.1 Dynamical Systems

The characters in Stoppard's play are discussing the apparent anticipation, by a (fictional) teenage girl in 1809, of the essential role of iterative algorithms in chaos theory and fractal geometry.

To illustrate the role of iteration in chaos theory, consider the simplest differential equation describing population dynamics:

$$p'(t) = ap(t),$$

3

with exponential solutions. More realistic models impose limits to growth, and may take the form

$$p'(t) = a(L - p(t))p(t),$$

where L is an asymptotic limit for $p(t)$. Discrete versions of the limited-population problem then have the form

$$x_{k+1} - x_k = a(L - x_k)x_k,$$

which, for $z_k = \frac{a}{1+aL}x_k$, can be written as

$$z_{k+1} = r(1 - z_k)z_k; \tag{1.1}$$

we shall assume that $r = 1 + aL > 1$. With $Tz = r(1 - z)z = f(z)$ and $z_{k+1} = Tz_k$, we are interested in the behavior of the sequence, as a function of r.

The operator T has a fixed point at $z_* = 0$, for every value of r, and another fixed point at $z_* = 1 - \frac{1}{r}$, if $r > 1$. From the Mean-Value Theorem we know that

$$z_* - z_{k+1} = f(z_*) - f(z_k) = f'(c_k)(z_* - z_k),$$

for some c_k between z_* and z_k. If z_k is sufficiently close to z_*, then c_k will be even closer to z_* and $f'(c_k)$ can be approximated by $f'(z_*)$.

A fixed point z_* of $f(z)$ is said to be *stable* if $|f'(z_*)| < 1$. The fixed point $z_* = 0$ is stable if $r < 1$, while $z_* = 1 - \frac{1}{r}$ is stable if $1 < r < 3$. When z_* is a stable fixed point, and z_k is sufficiently close to z_*, we have

$$|z_* - z_{k+1}| < |z_* - z_k|,$$

so we get closer to z_* with each iterative step. Such a fixed point is *attractive*. In fact, if $r = 2$, $z_* = 1 - \frac{1}{r} = \frac{1}{2}$ is *superstable* and convergence is quite rapid, since $f'(\frac{1}{2}) = 0$. What happens for $r > 3$ is more interesting.

Using the change of variable $x = -rz + \frac{r}{2}$, the iteration in Equation (1.1) becomes

$$x_{k+1} = x_k^2 + (\frac{r}{2} - \frac{r^2}{4}). \tag{1.2}$$

For $r > 1$, the fixed points become $x_* = \frac{r}{2}$ and $x_* = 1 - \frac{r}{2}$.

For $r = 4$ there is a starting point x_0 for which the iterates are periodic of period three, which implies, according to the results of Li and Yorke, that there are periodic orbits of all periods [122]. Using Equation (1.2), the iteration for $r = 4$ can be written as

$$x_{k+1} = x_k^2 - 2.$$

In [23] Burger and Starbird illustrate the sensitivity of this iterative scheme to the choice of x^0 by comparing, for $k = 1, ..., 50$, the computed values of x_k for $x_0 = 0.5$ with those for $x_0 = 0.50001$. For $r > 4$ the set of starting points in $[0, 1]$ for which the sequence of iterates never leaves $[0, 1]$ is a Cantor set, which is a fractal. The book by Devaney [80] gives a rigorous treatment of these topics; Young's book [159] contains a more elementary discussion of some of the same notions.

1.1.1 The Newton-Raphson Algorithm

The well-known Newton-Raphson (NR) iterative algorithm is used to find a root of a function $g : R \to R$.

Algorithm 1.1 (Newton-Raphson). *Let* $x^0 \in R$ *be arbitrary. Having calculated* x^k, *let*

$$x^{k+1} = x^k - g(x^k)/g'(x^k).$$

The operator T is now the ordinary function

$$Tx = x - g(x)/g'(x).$$

If g is a vector-valued function, $g : R^J \to R^J$, then $g(x)$ has the form $g(x) = (g_1(x), ..., g_J(x))^T$, where $g_j : R^J \to R$ are the component functions of $g(x)$. The NR algorithm is then as follows:

Algorithm 1.2 (Newton-Raphson). *Let* $x^0 \in R^J$ *be arbitrary. Having calculated* x^k, *let*

$$x^{k+1} = x^k - [\mathcal{J}(g)(x^k)]^{-1}g(x^k).$$

Here $\mathcal{J}(g)(x)$ is the Jacobian matrix of first partial derivatives of the component functions of g; that is, its entries are $\frac{\partial g_m}{\partial x_j}(x)$. The operator T is now

$$Tx = x - [\mathcal{J}(g)(x)]^{-1}g(x).$$

Convergence of the NR algorithm is not guaranteed and depends on the starting point being sufficiently close to a solution. When it does converge, however, it does so fairly rapidly. In both the scalar and vector cases, the limit is a fixed point of T, and therefore a root of $g(x)$.

1.1.2 Newton-Raphson and Chaos

It is interesting to consider how the behavior of the NR iteration depends on the starting point.

A simple case. The complex-valued function $f(z) = z^2 - 1$ of the complex variable z has two roots, $z = 1$ and $z = -1$. The NR method for finding a root now has the iterative step

$$z_{k+1} = Tz_k = \frac{z_k}{2} + \frac{1}{2z_k}.$$

If z_0 is selected closer to $z = 1$ than to $z = -1$, then the iterative sequence converges to $z = 1$; similarly, if z_0 is closer to $z = -1$, the limit is $z = -1$. If z_0 is on the vertical axis of points with real part equal to zero, then the sequence does not converge, and is not even defined for $z_0 = 0$. This axis separates the two *basins of attraction* of the algorithm.

A not-so-simple case. Now consider the function $f(z) = z^3 - 1$, which has the three roots $z = 1$, $z = \omega = e^{2\pi i/3}$, and $z = \omega^2 = e^{4\pi i/3}$. The NR method for finding a root now has the iterative step

$$z_{k+1} = Tz_k = \frac{2z_k}{3} + \frac{1}{3z_k^2}.$$

Where are the *basins of attraction* now? Is the complex plane divided up as three people would divide a pizza, into three wedge-shaped slices, each containing one of the roots? Far from it. In fact, it can be shown that, if the sequence starting at $z_0 = a$ converges to $z = 1$ and the sequence starting at $z_0 = b$ converges to ω, then there is a starting point $z_0 = c$, closer to a than b is, whose sequence converges to ω^2. For more details and beautiful colored pictures illustrating this remarkable behavior, see Schroeder's delightful book [142].

1.1.3   The Sir Pinski Game

In [142] Schroeder discusses several iterative sequences that lead to fractals or chaotic behavior. The *Sir Pinski Game* has the following rules. Let P_0 be a point chosen arbitrarily within the interior of the equilateral triangle with vertices $(1, 0, 0), (0, 1, 0)$, and $(0, 0, 1)$. Let V be the vertex closest to P_0 and P_1 chosen so that P_0 is the midpoint of the line segment VP_1. Repeat the process, with P_1 in place of P_0. The game is lost when P_n falls outside the original triangle. The objective of the game is to select P_0 that will allow the player to win the game. Where are these winning points?

The *inverse Sir Pinski Game* is similar. Select any point P_0 in the plane of the equilateral triangle, let V be the most distance vertex, and P_1

the midpoint of the line segment $P_0 V$. Replace P_0 with P_1 and repeat the procedure. The resulting sequence of points is convergent. Which points are limit points of sequences obtained in this way?

1.1.4   The Chaos Game

Schroeder also mentions Barnsley's *Chaos Game*. Select P_0 inside the equilateral triangle. Roll a fair die and let $V = (1,0,0)$ if 1 or 2 is rolled, $V = (0,1,0)$ if 3 or 4 is rolled, and $V = (0,0,1)$ if 5 or 6 is rolled. Let P_1 again be the midpoint of VP_0. Replace P_0 with P_1 and repeat the procedure. Which points are limits of such sequences of points?

1.2   Iterative Root-Finding

A good example of an iterative algorithm is the *bisection method* for finding a root of a real-valued continuous function $f(x)$ of the real variable x: begin with an interval $[a, b]$ such that $f(a)f(b) < 0$ and then replace one of the endpoints with the average $\frac{a+b}{2}$, maintaining the negative product. The length of each interval so constructed is half the length of the previous interval and each interval contains a root. In the limit, the two sequences defined by the left endpoints and right endpoints converge to the same root. The bisection approach can be used to calculate $\sqrt{2}$, by finding a positive root of the function $f(x) = x^2 - 2$, or to solve the equation $\tan x = x$, by finding a root of the function $f(x) = x - \tan x$. It can also be used to optimize a function $F(x)$, by finding the roots of its derivative, $f(x) = F'(x)$.

Iterative algorithms are used to solve problems for which there is no noniterative solution method, such as the two just mentioned, as well as problems for which noniterative methods are impractical, such as using Gauss elimination to solve a system of thousands of linear equations in thousands of unknowns. If our goal is to minimize $F(x)$, we may choose an iterative algorithm, such as steepest descent, that generates an iterative sequence $\{x^k\}$, $k = 0, 1, ...$, that, in the best cases, at least, will converge to a minimizer of $F(x)$.

1.2.1   Computer Division

Iterative algorithms that use only addition, subtraction, and multiplication can be designed to perform division. The solution of the equation $ax = b$ is $x = b/a$. We can solve this equation by minimizing the function $f(x) = \frac{1}{2}(ax - b)^2$, using the iterative sequence

$$x_{k+1} = x_k + (b - ax_k).$$

Since, in the binary system, multiplication by powers of two is easily implemented as a shift, we assume that the equation has been rescaled so that $\frac{1}{2} \le a < 1$. Then, the operator $Tx = x + (b - ax)$ is an *affine linear* function and a *strict contraction*:

$$|Tx - Ty| = (1 - a)|x - y|,$$

for any x and y. Convergence of the iterative sequence to a fixed point of T is a consequence of the Banach-Picard Theorem [82], also called the Contraction Mapping Theorem. A fixed point of T satisfies $Tx = x$, or $x = x + (b - ax)$, so must be the quotient, $x = b/a$. As we shall see later, the iterative sequence is also an instance of the Landweber method for solving systems of linear equations.

1.3   Iterative Fixed-Point Algorithms

Iterative algorithms are often formulated as *fixed-point* methods: the equation $f(x) = 0$ is equivalent to $x = f(x) + x = g(x)$, so we may try to find a fixed point of $g(x)$, that is, an x for which $g(x) = x$. The iterative algorithms we discuss take the form $x^{k+1} = Tx^k$, where T is some (usually nonlinear) continuous operator on the space R^J of J-dimensional real vectors, or C^J, the space of J-dimensional complex vectors. If the sequence $\{T^k x^0\}$ converges to x^*, then $Tx^* = x^*$, that is, x^* is a *fixed point* of T. In order to discuss convergence of a sequence of vectors, we need a measure of distance between vectors. A vector norm is one such distance, but there are other useful distances, such as the Kullback-Leibler, or cross-entropy distance.

1.4   Convergence Theorems

To be sure that the sequence $\{T^k x^0\}$ converges, we need to know that T has fixed points, but we need more than that. Most of the operators T that we shall encounter fall into two broad classes, those that are *averaged, nonexpansive* with respect to the Euclidean vector norm, and those that are *paracontractive* with respect to some vector norm. Convergence for the first class of operators is a consequence of the Krasnoselskii-Mann (KM) Theorem 5.16, and the Elsner-Koltracht-Neumann (EKN) Theorem 5.21 establishes convergence for the second class. The definitions of these classes are derived from basic properties of orthogonal projection operators, which are members of both classes.

1.5 Positivity Constraints

In many remote-sensing applications, the object sought is naturally represented as a nonnegative function or a vector with nonnegative entries. For such problems, we can incorporate nonnegativity in the algorithms through the use of projections with respect to entropy-based distances. These algorithms are often developed by analogy with those methods using orthogonal projections. As we shall see, this analogy can often be further exploited to derive convergence theorems. The cross-entropy, or Kullback-Leibler (KL), distance is just one example of a Bregman distance. The KL distance exhibits several convenient properties that are reminiscent of the Euclidean distance, making it a useful tool in extending linear algorithms for general vector variables to nonlinear algorithms for positively constrained variables. The notion of an operator being paracontractive, with respect to a norm, can be extended to being paracontractive, with respect to a Bregman distance. Bregman projections onto convex sets are paracontractive in this generalized sense, as are many of the operators of interest. The EKN Theorem and many of its corollaries can be extended to operators that are paracontractive, with respect to Bregman distances.

1.6 Fundamental Concepts

Although the object function of interest is often a function of one or more continuous variables, it may be necessary to discretize the problem, and to represent that function as a finite-dimensional real or complex vector, that is, as a member of R^J or C^J. When we impose a norm on the spaces R^J and C^J we make them *metric spaces*. The basic properties of such metric spaces are important in our analysis of the behavior of iterative algorithms. It is often the case that the data we have measured is related to the object function in a linear manner. Consequently, the estimation procedure will involve solving a system of linear equations, sometimes inconsistent, and usually quite large. There may also be constraints, such as positivity, imposed on the solution. For these reasons, the fundamentals of linear algebra, including matrix norms and eigenvalues, will also play an important role.

We begin with a discussion of the basic properties of finite-dimensional spaces and the fundamentals of linear algebra. Then we turn to an overview of operators and the mathematical problems and algorithms to be treated in more detail in subsequent chapters.

2 | Background

In this book our focus is on applications that employ iterative algorithms, and on the mathematics behind those algorithms. In this chapter, we sketch briefly some of the applications we investigate in more detail subsequently.

2.1 Iterative Algorithms and their Applications

Iterative algorithms are playing an increasingly prominent role in a variety of applications where it is necessary to solve large systems of linear equations, often with side constraints on the unknowns, or to optimize functions of many variables. Such mathematical problems arise in many remote-sensing applications, such as sonar, radar, radio astronomy, optical and hyperspectral imaging, transmission and emission tomography, magnetic-resonance imaging, radiation therapy, and so on. We shall be particularly interested in algorithms that are used to solve such inverse problems.

In the chapters that follow, we present several iterative algorithms and discuss their theoretical properties. In each case, we attempt to motivate the algorithm with a brief discussion of applications in which the algorithm is used. More detailed treatment of the particular applications is left to later chapters. Whenever it seems helpful to do so, we include, usually in separate chapters, background material to clarify certain points of the theoretical discussion. For this purpose, the reader will find early chapters on basic notions concerning the analysis and geometry of finite-dimensional Euclidean space, and linear algebra and matrices, as well as later ones on convex functions and optimization, alternatives to the Euclidean distance, the Fourier transform, and so on.

One theme of the book is to find unity underlying the different algorithms. Many of the algorithms discussed here are special cases of general iterative schemes involving one of two types of operators; in such cases, convergence will follow from the theorems pertaining to the general case. Some algorithms and their convergence proofs have been discovered by analogy with other algorithms. We shall see this, in particular, when we move from algorithms for solving general systems of linear equations, to solving those with positivity requirements or other constraints, and from gradient-descent optimization to related methods involving positive-valued variables.

It usually comes as a bit of a shock to mathematicians, accustomed to focusing on convergence and limits of iterative sequences, when they discover that, in practice, iterative algorithms may only be incremented one or two times. The need to produce usable reconstructions in a short time is a constraint that is not easily incorporated into the mathematical treatment of iterative methods, and one that makes the entire enterprise partly experimental.

2.2   A Basic Inverse Problem

A basic inverse problem is to estimate or reconstruct an object function from a finite number of measurements pertaining to that function. When the object function being estimated is a distribution of something, it is natural to display the estimate in the form of an image. For this reason, we often speak of these problems as *image reconstruction* problems. For example, in passive sonar we estimate the distribution of sources of acoustic energy in the ocean, based on readings taken at some finite number of hydrophones. In medical emission tomography we estimate the spatial (and, perhaps, temporal) distribution of radionuclides within the patient's body, based on photon counts at detectors exterior to the patient. These problems are highly under-determined; even in the absence of noise, the data are insufficient to specify a single solution. It is common, therefore, to seek an estimate that optimizes some second function, subject to data constraints and other prior information about the object function being estimated. The second function may measure the distance from the estimate to a prior estimate of the object function, or the statistical likelihood, or the energy in the estimate, or its entropy, and so on. Typically, such optimization problems can be solved only with iterative algorithms.

2.3   Some Applications

A main theme of this book is the interplay between problems and algorithms. Each application presents a unique set of desiderata and requirements. We know, more or less, how the data we have measured relates to the information we seek, and usually have a decent idea of what an acceptable solution looks like. Sometimes, general-purpose methods are satisfactory, while often algorithms tailored to the problem at hand perform better. In this section we describe some of the applications to be treated in more detail later in the book.

2.3.1   Transmission Tomography

In transmission tomography, radiation, usually x-ray, is transmitted along many lines through the object of interest and the initial and final intensities are measured. The intensity drop associated with a given line indicates the amount of attenuation the ray encountered as it passed along the line. It is this distribution of attenuating matter within the patient, described by a function of two or three spatial variables, that is the object of interest. Unexpected absence of attenuation can indicate a break in a bone, for example. The data are usually modeled as line integrals of that function. The *Radon transform* is the function that associates with each line its line integral.

If we had the line integrals for every line, then we could use that data to determine the Fourier transform of the attenuation function. In practice, of course, we have finitely many noisy line-integral values, so finding the attenuation function using Fourier methods is approximate. Both iterative and noniterative methods are used to obtain the final estimate.

The estimated attenuation function will ultimately be reduced to a finite array of numbers. This discretization can be performed at the end, or can be made part of the problem model from the start. In the latter case, the attenuation function is assumed to be constant over small pixels or voxels; these constants are the object of interest now. The problem has been reduced to solving a large system of linear equations, possibly subject to nonnegativity or other constraints.

If the physical nature of the radiation is described using a statistical model, then the pixel values can be viewed as parameters to be estimated. The well-known maximum likelihood parameter estimation method can then be employed to obtain these pixel values. This involves a large-scale optimization of the likelihood function.

Because components of the x-ray beam that have higher energies can be attenuated less, the loss of low-energy components as the beam proceeds along a line, which is known as beam hardening, may need to be considered. The Laplace transform connects the energy spectrum of the x-ray beam to the amount of attenuation produced by a given thickness of material. Measurements of the attenuation as a function of thickness can be used to determine the energy spectrum by Laplace-transform inversion.

In similar fashion, measurements of the intensities of ultraviolet light from the sun that has been scattered by the ozone layer are related by Laplace transformation to the density of the ozone as a function of altitude and can be used to estimate that density.

2.3.2 Emission Tomography

In emission tomography, a carefully designed chemical tagged with a radioisotope is introduced into the body of the patient. The chemical is selected to accumulate in a specific organ or region of the body, such as the brain, or the heart wall. On the basis of emissions from the radioisotope that are detected outside the body, the distribution of the chemical within the body is estimated. Unexpected absence of the radionuclide from a given region, or a higher than expected concentration, can indicate a medical problem.

There are two basic types of emission tomography: single-photon emission computed tomography (SPECT); and positron emission tomography (PET). In SPECT the radioisotope emits a single photon, while in PET a positron is emitted, which shortly meets an electron and the resulting annihilation produces two gamma-ray photons travelling in essentially opposite directions.

In both SPECT and PET the data can be viewed as integrals along a line through the body. However, more sophisticated models that more accurately describe the physics of the situation are preferred. The photons that travel through the body toward the external detectors are sometimes absorbed by the body itself and not detected. The probability of being detected depends on the attenuation presented by the body. This attenuation, while not the object of interest now, is an important part of the physical model and needs to be included in the reconstruction method. The randomness inherent in emission can also be included, leading once again to probabilistic models and a maximum likelihood approach to reconstruction.

Although in both transmission and emission medical tomography the dosage to the patient is restricted, thereby decreasing the signal-to-noise ratio, the amount of data is still considerable and the need to produce the reconstructed image in a few minutes paramount. Much work has gone into methods for accelerating the iterative reconstruction algorithms.

2.3.3 Array Processing

The term *array processing* refers to those applications, such as sonar, radar and astronomy, in which the data are measurements of a propagating spatio-temporal field, taken in the farfield, using an array of sensors. Plane-wave solutions to the wave equation are used to model the situation, with the result that the data so obtained are usually related by Fourier transformation to the distribution of interest. In some cases, the array is too short to provide the desired resolution and line-integral models for the data can be used. Reconstruction then proceeds as in tomography.

The data are finite, while the object of interest is often, at least initially, viewed as a function of continuous variables; therefore, even in the absence of noise, no unique solution is specified by the data. Solutions can be obtained by minimizing a cost function, such as a norm, or entropy, subject to the data constraints.

2.3.4 Optical Imaging and the Phase Problem

In certain applications of optical imaging, such as imaging through a turbulent atmospheric layer, only the magnitude of the Fourier transform data is available and the phase information is lost. The problem of reconstructing the image from magnitude-only Fourier data can be solved using algorithms that iteratively estimate the missing phases.

2.3.5 Magnetic-Resonance Imaging

When the body is placed inside a strong magnetic field, a small fraction of the spinning protons in, say, the hydrogen nuclei in water, are induced to align their spin axes with the external field. When a second magnetic field perturbs the spin axes, the precession results in a detectable signal, providing information about the spatial density of the water molecules. This is magnetic-resonance imaging (MRI).

The detected signals are related to the distribution of interest by means of the Fourier transform. Which values of the Fourier transform we obtain depend on the particular magnetic fields activated. In some approaches, the data are line integrals, as in tomography. Both iterative and noniterative methods can be used to obtain the reconstruction.

2.3.6 Intensity-Modulated Radiation Therapy

The problem in intensity-modulated radiation therapy (IMRT) is to determine the various intensities of radiation to apply to the patient so as to deliver the desired minimum dosage to the tumor, while not exceeding the acceptable dosage to nearby parts of the body. Mathematically, the problem is one of solving a system of linear equations, subject to inequality constraints involving convex sets. Iterative algorithms developed to solve the convex feasibility and split feasibility problems can be applied to solve the IMRT problem.

2.3.7   Hyperspectral Imaging

In hyperspectral imaging the problem is to obtain subpixel resolution in
radar imaging through the use of multifrequency data. The problem is
a *mixture problem*, which can be solved using methods for reconstructing
from Fourier data, along with iterative procedures for solving large systems of linear equations, subject to positivity constraints. Similar mixture
problems arise in determining photon-count statistics in optics.

2.3.8   Discrimination and Classification

A fundamental problem in pattern recognition is to determine to which of
several classes an object belongs, based on measured data pertaining to that
object. Commonly, the measurements are numerical and the classification
or discrimination task can be performed by calculating some (possibly linear) function of the measurements. Such linear discriminants are obtained
by finding vectors that satisfy a certain system of linear inequalities. This
problem can be solved using various iterative methods for convex feasibility.

2.4   The Urn Model for Remote Sensing

There seems to be a tradition in physics of using simple models or examples involving urns and marbles to illustrate important principles. In
keeping with that tradition, we present here such a model, to illustrate
various aspects of remote sensing. We begin with the model itself, and
then give several examples to show how the model illustrates randomness
in tomography.

Although remote-sensing problems differ from one another in many re-
spects, they often share a fundamental aspect that can best be illustrated
by a simple model involving urns containing colored marbles.

2.4.1   The Model

Suppose that we have J urns numbered $j = 1, ..., J$, each containing marbles of various colors. Suppose that there are I colors, numbered $i = 1, ..., I$.
Suppose also that there is a box containing N small pieces of paper, and
on each piece is written the number of one of the J urns. Assume that N
is much larger than J. Assume that I know the precise contents of each
urn. My objective is to determine the precise contents of the box, that
is, to estimate the number of pieces of paper corresponding to each of the
numbers $j = 1, ..., J$.

Out of my view, my assistant removes one piece of paper from the box,
takes one marble from the indicated urn, announces to me the color of the

marble, and then replaces both the piece of paper and the marble. This action is repeated many times, at the end of which I have a long list of colors. This list is my data, from which I must determine the contents of the box.

This is a form of remote sensing; what we have access to is related to, but not equal to, what we are interested in. Sometimes such data is called "incomplete data," in contrast to the "complete data," which would be the list of the actual urn numbers drawn from the box.

If all the marbles of one color are in a single urn, the problem is trivial; when I hear a color, I know immediately which urn contained that marble. My list of colors is then a list of urn numbers; I have the complete data now. My estimate of the number of pieces of paper containing the urn number j is then simply N times the proportion of draws that resulted in urn j being selected.

At the other extreme, suppose two urns had identical contents. Then I could not distinguish one urn from the other and would be unable to estimate more than the total number of pieces of paper containing either of the two urn numbers. Generally, the more the contents of the urns differ, the easier the task of estimating the contents of the box. In remote sensing applications, these issues affect our ability to resolve individual components contributing to the data.

To introduce some mathematics, let us denote by x_j the proportion of the pieces of paper that have the number j written on them. Let P_{ij} be the proportion of the marbles in urn j that have the color i. Let y_i be the proportion of times the color i occurs on the list of colors. The expected proportion of times i occurs on the list is $E(y_i) = \sum_{j=1}^{J} P_{ij}x_j = (Px)_i$, where P is the I by J matrix with entries P_{ij} and x is the J by 1 column vector with entries x_j. A reasonable way to estimate x is to replace $E(y_i)$ with the actual y_i and solve the system of linear equations $y_i = \sum_{j=1}^{J} P_{ij}x_j$, $i = 1, ..., I$. Of course, we require that the x_j be nonnegative and sum to one, so special algorithms, such as the EMML algorithm to be introduced later, may be needed to find such solutions.

2.4.2   The Case of SPECT

In the SPECT case, let there be J pixels or voxels, numbered $j = 1, ..., J$ and I detectors, numbered $i = 1, ..., I$. Let P_{ij} be the probability that a photon emitted at pixel j will be detected at detector i; we assume these probabilities are known to us. Let y_i be the proportion of the total photon count that was recorded at the ith detector. Denote by x_j the (unknown) proportion of the total photon count that was emitted from pixel j. Selecting an urn randomly is analogous to selecting which pixel will be the next to emit a photon. Learning the color of the marble is analogous to learning where the photon was detected; for simplicity we are assuming that all emitted photons are detected, but this is not essential. The data we have, the counts at each detector, constitute the "incomplete data" ; the "complete data" would be the counts of emissions from each of the J pixels.

We can determine the x_j by finding nonnegative solutions of the system $y_i = \sum_{j=1}^{J} P_{ij} x_j$; this is what the various iterative algorithms to be introduced later, such as MART, EMML, and RBI-EMML, seek to do.

2.4.3   The Case of PET

In the PET case, let there be J pixels or voxels, numbered $j = 1, ..., J$ and I lines of response (LOR), numbered $i = 1, ..., I$. Let P_{ij} be the probability that a positron emitted at pixel j will result in a coincidence detection associated with LOR i; we assume these probabilities are known to us. Let y_i be the proportion of the total detections that was associated with the ith LOR. Denote by x_j the (unknown) proportion of the total count that was due to a positron emitted from pixel j. Selecting an urn randomly is analogous to selecting which pixel will be the next to emit a positron. Learning the color of the marble is analogous to learning which LOR was detected; again, for simplicity we are assuming that all emitted positrons are detected, but this is not essential. As in the SPECT case, we can determine the x_j by finding nonnegative solutions of the system $y_i = \sum_{j=1}^{J} P_{ij} x_j$.

2.4.4   The Case of Transmission Tomography

Assume that x-ray beams are sent along I line segments, numbered $i = 1, ..., I$, and that the initial strength of each beam is known. By measuring the final strength, we determine the drop in intensity due to absorption along the ith line segment. Associated with each line segment we then have the proportion of transmitted photons that were absorbed, but we do not know where along the line segment the absorption took place. The proportion of absorbed photons for each line is our data, and corresponds to the proportion of each color in the list. The rate of change of the intensity of the x-ray beam as it passes through the jth pixel is proportional to the intensity itself, to P_{ij}, the length of the ith segment that is within the jth pixel, and to x_j, the amount of attenuating material present in the jth pixel. Therefore, the intensity of the x-ray beam leaving the jth pixel is the product of the intensity of the beam upon entering the jth pixel and the decay term, $e^{-P_{ij} x_j}$.

The "complete data" is the proportion of photons entering the jth pixel that were absorbed within it; the "incomplete data" is the proportion of photons sent along each line segment that were absorbed. Selecting the jth urn is analogous to having an absorption occurring at the jth pixel. Knowing that an absorption has occurred along the ith line segment does tell us that an absorption occurred at one of the pixels that intersections that line segment, but that is analogous to knowing that there are certain urns that are the only ones that contain the ith color.

The (measured) intensity of the beam at the end of the ith line segment is $e^{-(Px)_i}$ times the (known) intensity of the beam when it began its journey along the ith line segment. Taking logs, we obtain a system of linear equations which we can solve for the x_j.

3 | Basic Concepts

In iterative methods, we begin with an initial vector, say x^0, and, for each nonnegative integer k, we calculate the next vector, x^{k+1}, from the current vector x^k. The limit of such a sequence of vectors $\{x^k\}$, when the limit exists, is the desired solution to our problem. The fundamental tools we need to understand iterative algorithms are the geometric concepts of distance between vectors and mutual orthogonality of vectors, the algebraic concept of transformation or operator on vectors, and the vector-space notions of subspaces and convex sets.

3.1   The Geometry of Euclidean Space

We denote by R^J the real Euclidean space consisting of all J-dimensional column vectors $x = (x_1, ..., x_J)^T$ with real entries x_j; here the superscript T denotes the transpose of the 1 by J matrix (or, row vector) $(x_1, ..., x_J)$. We denote by C^J the collection of all J-dimensional column vectors $x = (x_1, ..., x_J)^\dagger$ with complex entries x_j; here the superscript $\dagger$ denotes the conjugate transpose of the 1 by J matrix (or, row vector) $(x_1, ..., x_J)$. When discussing matters that apply to both R^J and C^J we denote the underlying space simply as $\mathcal{X}$.

3.1.1   Inner Products

For $x = (x_1, ..., x_J)^T$ and $y = (y_1, ..., y_J)^T$ in R^J, the dot product $x \cdot y$ is defined to be

$$x \cdot y = \sum_{j=1}^{J} x_j y_j.$$

Note that we can write

$$x \cdot y = y^T x = x^T y,$$

where juxtaposition indicates matrix multiplication. The 2-norm, or *Euclidean norm*, or *Euclidean length*, of x is

$$||x||_2 = \sqrt{x \cdot x} = \sqrt{x^T x}.$$

The *Euclidean distance* between two vectors x and y in R^J is $||x - y||_2$. As we discuss in the chapter on metric spaces, there are other norms on $\mathcal{X}$; nevertheless, in this chapter we focus on the 2-norm of x.

For $x = (x_1, ..., x_J)^T$ and $y = (y_1, ..., y_J)^T$ in C^J, the dot product $x \cdot y$ is defined to be

$$x \cdot y = \sum_{j=1}^{J} x_j \overline{y_j}.$$

Note that we can write

$$x \cdot y = y^\dagger x.$$

The norm, or Euclidean length, of x is

$$||x||_2 = \sqrt{x \cdot x} = \sqrt{x^\dagger x}.$$

As in the real case, the distance between vectors x and y is $||x - y||_2$.

Both of the spaces R^J and C^J, along with their dot products, are examples of finite-dimensional Hilbert space. Much of what follows in this chapter applies to both R^J and C^J. In such cases, we shall simply refer to the underlying space as $\mathcal{X}$.

Definition 3.1. Let V be a real or complex vector space. The scalar-valued function $\langle u, v \rangle$ is called an inner product on V if the following four properties hold, for all u, w, and v in V, and scalars c:

$$\langle u + w, v \rangle = \langle u, v \rangle + \langle w, v \rangle;$$

$$\langle cu, v \rangle = c \langle u, v \rangle;$$

$$\langle v, u \rangle = \overline{\langle u, v \rangle};$$

and

$$\langle u, u \rangle \geq 0, \tag{3.1}$$

with equality in Inequality (3.1) if and only if $u = 0$.

The usual real or complex dot product of vectors are examples of inner products. The properties of an inner product are precisely the ones needed to prove Cauchy's Inequality, so that inequality holds for any inner product, as we shall see shortly. We shall favor the dot product notation $u \cdot v$ for the inner product of vectors, although we shall occasionally use the matrix multiplication form, $v^\dagger u$ or the inner product notation $\langle u, v \rangle$.

3.1.2 Cauchy's Inequality

Cauchy's Inequality, also called the Cauchy-Schwarz Inequality, tells us that

$$|\langle x, y \rangle| \leq ||x||_2 ||y||_2,$$

with equality if and only if $y = \alpha x$, for some scalar α. The Cauchy-Schwarz Inequality holds for any inner product.

Proof (of Cauchy's Inequality): To prove Cauchy's Inequality, we write

$$\langle x, y \rangle = |\langle x, y \rangle| e^{i\theta}.$$

Let t be a real variable and consider

$$0 \leq ||e^{-i\theta} x - ty||_2^2 = \langle e^{-i\theta} x - ty, e^{-i\theta} x - ty \rangle$$

$$= ||x||_2^2 - t[\langle e^{-i\theta} x, y \rangle + \langle y, e^{-i\theta} x \rangle] + t^2 ||y||_2^2$$

$$= ||x||_2^2 - t[\langle e^{-i\theta} x, y \rangle + \overline{\langle e^{-i\theta} x, y \rangle}] + t^2 ||y||_2^2$$

$$= ||x||_2^2 - 2\text{Re}(te^{-i\theta} \langle x, y \rangle) + t^2 ||y||_2^2$$

$$= ||x||_2^2 - 2\text{Re}(t|\langle x, y \rangle|) + t^2 ||y||_2^2$$

$$= ||x||_2^2 - 2t|\langle x, y \rangle| + t^2 ||y||_2^2.$$

This is a nonnegative quadratic polynomial in the variable t, so it cannot have two distinct real roots. Therefore, the discriminant is nonpositive, that is,

$$4|\langle x, y \rangle|^2 - 4||y||_2^2 ||x||_2^2 \leq 0,$$

and so

$$|\langle x, y \rangle|^2 \leq ||x||_2^2 ||y||_2^2.$$

This is the desired inequality. $\qquad\square$

A simple application of Cauchy's Inequality gives us

$$||x + y||_2 \le ||x||_2 + ||y||_2; \tag{3.2}$$

this is called the *triangle inequality*. We say that the vectors x and y are *mutually orthogonal* if $\langle x, y \rangle = 0$.

The *parallelogram law* is an easy consequence of the definition of the 2-norm:

$$||x + y||_2^2 + ||x - y||_2^2 = 2||x||_2^2 + 2||y||_2^2. \tag{3.3}$$

It is important to remember that Cauchy's Inequality and the parallelogram law hold only for the 2-norm.

3.2   Hyperplanes in Euclidean Space

For a fixed column vector a with Euclidean length one and a fixed scalar γ the *hyperplane* determined by a and γ is the set $H(a, \gamma) = \{z | \langle a, z \rangle = \gamma\}$.

For an arbitrary vector x in $\mathcal{X}$ and arbitrary hyperplane $H = H(a, \gamma)$, the *orthogonal projection* of x onto H is the member $z = P_H x$ of H that is closest to x. For $H = H(a, \gamma)$, $z = P_H x$ is the vector

$$z = P_H x = x + (\gamma - \langle a, x \rangle)a. \tag{3.4}$$

Definition 3.2. A subset H of $\mathcal{X}$ is a *subspace* if, for every x and y in H and scalars α and β, the linear combination $\alpha x + \beta y$ is again in H.

For $\gamma = 0$, the hyperplane $H = H(a, 0)$ is also a subspace of $\mathcal{X}$; in particular, the zero vector 0 is in $H(a, 0)$.

3.3   Convex Sets in Euclidean Space

The notion of a convex set will play an important role in our discussions.

Definition 3.3. A subset C of $\mathcal{X}$ is said to be *convex* if, for every pair of members x and y of C, and for every α in the open interval $(0, 1)$, the vector $\alpha x + (1 - \alpha)y$ is also in C.

For example, the unit ball U in $\mathcal{X}$, consisting of all x with $||x||_2 \le 1$, is convex, while the surface of the ball, the set of all x with $||x||_2 = 1$, is not convex.

Definition 3.4. A subset B of $\mathcal{X}$ is *closed* if, whenever x^k is in B for each nonnegative integer k and $||x - x^k|| \to 0$, as $k \to +\infty$, then x is in B.

For example, $B = [0, 1]$ is closed as a subset of R, but $B = (0, 1)$ is not.

Definition 3.5. We say that $d \geq 0$ is the *distance from the point x to the set B* if, for every $\epsilon > 0$, there is b_ϵ in B, with $||x - b_\epsilon||_2 < d + \epsilon$, and no b in B with $||x - b||_2 < d$.

It follows easily that, if B is closed and $d = 0$, then x is in B.

The following proposition is fundamental in the study of convexity and can be found in most books on the subject; see, for example, the text by Goebel and Reich [92].

Proposition 3.6. *Given any nonempty closed convex set C and an arbitrary vector x in $\mathcal{X}$, there is a unique member of C closest to x, denoted $P_C x$, the orthogonal (or metric) projection of x onto C.*

Proof: If x is in C, then $P_C x = x$, so assume that x is not in C. Then $d > 0$, where d is the distance from x to C. For each positive integer n, select c_n in C with $||x - c_n||_2 < d + \frac{1}{n}$, and $||x - c_n||_2 < ||x - c_{n-1}||_2$. Then the sequence $\{c_n\}$ is bounded; let c^* be any cluster point. It follows easily that $||x - c^*||_2 = d$ and that c^* is in C. If there is any other member c of C with $||x - c||_2 = d$, then, by the parallelogram law, we would have $||x - (c^* + c)/2||_2 < d$, which is a contradiction. Therefore, c^* is $P_C x$. $\square$

For example, if $C = U$, the unit ball, then $P_C x = x/||x||_2$, for all x such that $||x||_2 > 1$, and $P_C x = x$ otherwise. If C is R_+^J, the nonnegative cone of R^J, consisting of all vectors x with $x_j \geq 0$, for each j, then $P_C x = x_+$, the vector whose entries are $\max(x_j, 0)$.

3.4 Basic Linear Algebra

In this section we discuss systems of linear equations, Gaussian elimination, basic and nonbasic variables, the fundamental subspaces of linear algebra and eigenvalues and norms of square matrices.

3.4.1 Bases

The notions of a basis and of linear independence are fundamental in linear algebra.

Definition 3.7. A collection of vectors $\{u^1, ..., u^N\}$ in $\mathcal{X}$ is *linearly independent* if there is no choice of scalars $\alpha_1, ..., \alpha_N$, not all zero, such that

$$0 = \alpha_1 u^1 + ... + \alpha_N u^N.$$

Definition 3.8. The *span* of a collection of vectors $\{u^1, ..., u^N\}$ in $\mathcal{X}$ is the set of all vectors x that can be written as linear combinations of the u^n; that is, for which there are scalars $c_1, ..., c_N$, such that

$$x = c_1 u^1 + ... + c_N u^N.$$

Definition 3.9. A collection of vectors $\{u^1, ..., u^N\}$ in $\mathcal{X}$ is called a *basis* for a subspace S if the collection is linearly independent and S is their span.

Definition 3.10. A collection of vectors $\{u^1, ..., u^N\}$ in $\mathcal{X}$ is called *orthonormal* if $||u^n||_2 = 1$, for all n, and $\langle u^m, u^n \rangle = 0$, for $m \neq n$.

3.4.2 Systems of Linear Equations

Consider the system of three linear equations in five unknowns given by

$$
\begin{array}{rcl}
x_1 +2x_2 \quad\quad\quad +2x_4 \;+\; x_5 &=& 0, \\
-x_1 \;-\; x_2 \;+\; x_3 \;+\; x_4 \quad\quad\quad &=& 0, \\
x_1 +2x_2 \;-3x_3 \;-\; x_4 \;-2x_5 &=& 0.
\end{array}
$$

This system can be written in matrix form as $Ax = 0$, with A the coefficient matrix

$$
A = \begin{bmatrix} 1 & 2 & 0 & 2 & 1 \\ -1 & -1 & 1 & 1 & 0 \\ 1 & 2 & -3 & -1 & -2 \end{bmatrix},
$$

and $x = (x_1, x_2, x_3, x_4, x_5)^T$. Applying Gaussian elimination to this system, we obtain a second, simpler, system with the same solutions:

$$
\begin{array}{rcl}
x_1 \quad\quad\quad\quad -2x_4 \;+x_5 &=& 0, \\
x_2 \quad\quad +2x_4 \quad\quad &=& 0, \\
x_3 \;+x_4 \;+x_5 &=& 0.
\end{array}
$$

From this simpler system we see that the variables x_4 and x_5 can be freely chosen, with the other three variables then determined by this system of equations. The variables x_4 and x_5 are then independent, the others dependent. The variables x_1, x_2, and x_3 are then called *basic variables*. To obtain a basis of solutions we can let $x_4 = 1$ and $x_5 = 0$, obtaining the solution $x = (2, -2, -1, 1, 0)^T$, and then choose $x_4 = 0$ and $x_5 = 1$ to get the solution $x = (-1, 0, -1, 0, 1)^T$. Every solution to $Ax = 0$ is then a linear combination of these two solutions. Notice that which variables are basic and which are nonbasic is somewhat arbitrary, in that we could have chosen as the nonbasic variables any two whose columns are independent.

Having decided that x_4 and x_5 are the nonbasic variables, we can write the original matrix A as $A = \begin{bmatrix} B & N \end{bmatrix}$, where B is the square invertible matrix

$$
B = \begin{bmatrix} 1 & 2 & 0 \\ -1 & -1 & 1 \\ 1 & 2 & -3 \end{bmatrix},
$$

and N is the matrix

$$N = \begin{bmatrix} 2 & 1 \\ 1 & 0 \\ -1 & -2 \end{bmatrix},$$

With $x_B = (x_1, x_2, x_3)^T$ and $x_N = (x_4, x_5)^T$ we can write

$$Ax = Bx_B + Nx_N = 0,$$

so that

$$x_B = -B^{-1}Nx_N. \tag{3.5}$$

3.4.3 Real and Complex Systems of Linear Equations

A system $Ax = b$ of linear equations is called a *complex system*, or a *real system* if the entries of A, x and b are complex, or real, respectively. Any complex system can be converted to a real system in the following way. A complex matrix A can be written as $A = A_1 + iA_2$, where A_1 and A_2 are real matrices and $i = \sqrt{-1}$. Similarly, $x = x^1 + ix^2$ and $b = b^1 + ib^2$, where x^1, x^2, b^1, and b^2 are real vectors. Denote by $\tilde{A}$ the real matrix

$$\tilde{A} = \begin{bmatrix} A_1 & -A_2 \\ A_2 & A_1 \end{bmatrix},$$

by $\tilde{x}$ the real vector

$$\tilde{x} = \begin{bmatrix} x^1 \\ x^2 \end{bmatrix},$$

and by $\tilde{b}$ the real vector

$$\tilde{b} = \begin{bmatrix} b^1 \\ b^2 \end{bmatrix}.$$

Then x satisfies the system $Ax = b$ if and only if $\tilde{x}$ satisfies the system $\tilde{A}\tilde{x} = \tilde{b}$.

Definition 3.11. A square matrix A is symmetric if $A^T = A$ and Hermitian if $A^\dagger = A$.

Definition 3.12. A nonzero vector x is said to be an eigenvector of the square matrix A if there is a scalar λ such that $Ax = \lambda x$. Then λ is said to be an eigenvalue of A.

If x is an eigenvector of A with eigenvalue λ, then the matrix $A - \lambda I$ has no inverse, so its determinant is zero; here I is the identity matrix with ones on the main diagonal and zeros elsewhere. Solving for the roots of the determinant is one way to calculate the eigenvalues of A. For example, the eigenvalues of the Hermitian matrix

$$B = \begin{bmatrix} 1 & 2+i \\ 2-i & 1 \end{bmatrix}$$

are $\lambda = 1 + \sqrt{5}$ and $\lambda = 1 - \sqrt{5}$, with corresponding eigenvectors $u = (\sqrt{5}, 2 - i)^T$ and $v = (\sqrt{5}, i - 2)^T$, respectively. Then $\breve{B}$ has the same eigenvalues, but both with multiplicity two. Finally, the associated eigenvectors of $\breve{B}$ are

$$\begin{bmatrix} u^1 \\ u^2 \end{bmatrix},$$

and

$$\begin{bmatrix} -u^2 \\ u^1 \end{bmatrix},$$

for $\lambda = 1 + \sqrt{5}$, and

$$\begin{bmatrix} v^1 \\ v^2 \end{bmatrix},$$

and

$$\begin{bmatrix} -v^2 \\ v^1 \end{bmatrix},$$

for $\lambda = 1 - \sqrt{5}$.

Definition 3.13. The complex square matrix N is nonexpansive (with respect to the Euclidean norm) if $||Nx||_2 \leq ||x||_2$, for all x.

Lemma 3.14. *The matrix N is nonexpansive if and only if $\tilde{N}$ is nonexpansive.*

Definition 3.15. The complex square matrix A is averaged if there is a nonexpansive N and scalar α in the interval $(0, 1)$, with $A = (1 - \alpha)I + \alpha N$.

Lemma 3.16. *The matrix A is averaged if and only if $\tilde{A}$ is averaged.*

3.4.4   The Fundamental Subspaces

We begin with some notation. Let S be a subspace of $\mathcal{X}$. We denote by $S^{\perp}$ the set of vectors u that are orthogonal to every member of S; that is,

$$S^{\perp} = \{u | u^{\dagger}s = 0, \text{for every } s \in S\}.$$

Definition 3.17. Let A be an I by J matrix. Then $CS(A)$, the column space of A, is the subspace of C^I consisting of all the linear combinations of the columns of A; we also say that $CS(A)$ is the *range* of A. The null space of $A^{\dagger}$, denoted $NS(A^{\dagger})$, is the subspace of C^I containing all the vectors w for which $A^{\dagger}w = 0$.

The subspaces $CS(A), NS(A^{\dagger}), CS(A^{\dagger})$ and $NS(A)$ play a prominent role in linear algebra and, for that reason, are called the *four fundamental subspaces.*

Let Q be a I by I matrix. We denote by $Q(S)$ the set

$$Q(S) = \{t | \text{there exists } s \in S \text{ with } t = Qs\}$$

and by $Q^{-1}(S)$ the set

$$Q^{-1}(S) = \{u | Qu \in S\}.$$

Note that the set $Q^{-1}(S)$ is defined whether or not the matrix Q is invertible.

We assume, now, that the matrix Q is Hermitian and invertible and that the matrix $A^{\dagger}A$ is invertible. Note that the matrix $A^{\dagger}Q^{-1}A$ need not be invertible under these assumptions. We shall denote by S an arbitrary subspace of R^J.

Lemma 3.18. *For a given set S, $Q(S) = S$ if and only if $Q(S^{\perp}) = S^{\perp}$.*

Proof: Use Exercise 3.9. □

Lemma 3.19. *If $Q(CS(A)) = CS(A)$ then $A^{\dagger}Q^{-1}A$ is invertible.*

Proof: Show that $A^{\dagger}Q^{-1}Ax = 0$ if and only if $x = 0$. Recall that $Q^{-1}Ax \in CS(A)$, by Exercise 3.8. Then use Exercise 3.6. □

3.5   Linear and Nonlinear Operators

In our study of iterative algorithms we shall be concerned with sequences of vectors $\{x^k | k = 0, 1, ...\}$. The core of an iterative algorithm is the transition from the current vector x^k to the next one x^{k+1}. To understand the algorithm, we must understand the operation (or operator) T by which x^k is transformed into $x^{k+1} = Tx^k$. An *operator* is any function T defined on $\mathcal{X}$ with values again in $\mathcal{X}$.

3.5.1   Linear and Affine Linear Operators

For example, if $\mathcal{X} = C^J$ and A is a J by J complex matrix, then we can define an operator T by setting $Tx = Ax$, for each x in C^J; here Ax denotes the multiplication of the matrix A and the column vector x.

Definition 3.20. An operator T is said to be a linear operator if

$$T(\alpha x + \beta y) = \alpha Tx + \beta Ty,$$

for each pair of vectors x and y and each pair of scalars α and β.

Any operator T that comes from matrix multiplication, that is, for which $Tx = Ax$, is linear.

Lemma 3.21. *For $H = H(a,\gamma)$, $H_0 = H(a,0)$, and any x and y in $\mathcal{X}$, we have*

$$P_H(x + y) = P_H x + P_H y - P_H 0,$$

so that

$$P_{H_0}(x + y) = P_{H_0} x + P_{H_0} y,$$

that is, the operator P_{H_0} is an additive operator. In addition,

$$P_{H_0}(\alpha x) = \alpha P_{H_0} x,$$

so that P_{H_0} is a linear operator.

Definition 3.22. If A is a square matrix and d is a fixed nonzero vector in $\mathcal{X}$, the operator defined by $Tx = Ax + d$ is an affine linear operator.

Lemma 3.23. *For any hyperplane $H = H(a,\gamma)$ and $H_0 = H(a,0)$,*

$$P_H x = P_{H_0} x + P_H 0,$$

so P_H is an affine linear operator.

Lemma 3.24. *For $i = 1,...,I$ let H_i be the hyperplane $H_i = H(a^i,\gamma_i)$, $H_{i0} = H(a^i,0)$, and P_i and P_{i0} the orthogonal projections onto H_i and H_{i0}, respectively. Let T be the operator $T = P_I P_{I-1} \cdots P_2 P_1$. Then $Tx = Bx + d$, for some square matrix B and vector d; that is, T is an affine linear operator.*

3.5.2 Orthogonal Projection onto Convex Sets

For an arbitrary nonempty closed convex set C in $\mathcal{X}$, the orthogonal projection $T = P_C$ is a nonlinear operator, unless, of course, C is a subspace. We may not be able to describe $P_C x$ explicitly, but we do know a useful property of $P_C x$.

Proposition 3.25. *For a given x, a vector z in C is $P_C x$ if and only if*

$$Re(\langle c - z, z - x \rangle) \geq 0,$$

for all c in the set C.

Proof: For simplicity, we consider only the real case, $\mathcal{X} = R^J$. Let c be arbitrary in C and α in $(0, 1)$. Then

$$||x - P_C x||_2^2 \leq ||x - (1 - \alpha)P_C x - \alpha c||_2^2$$
$$= ||x - P_C x + \alpha(P_C x - c)||_2^2$$
$$= ||x - P_C x||_2^2 - 2\alpha\langle x - P_C x, c - P_C x \rangle + \alpha^2||P_C x - c||_2^2.$$

Therefore,

$$-2\alpha\langle x - P_C x, c - P_C x \rangle + \alpha^2||P_C x - c||_2^2 \geq 0,$$

so that

$$2\langle x - P_C x, c - P_C x \rangle \leq \alpha||P_C x - c||_2^2.$$

Taking the limit, as $\alpha \to 0$, we conclude that

$$\langle c - P_C x, P_C x - x \rangle \geq 0.$$

If z is a member of C that also has the property

$$\langle c - z, z - x \rangle \geq 0,$$

for all c in C, then we have both

$$\langle z - P_C x, P_C x - x \rangle \geq 0,$$

and

$$\langle z - P_C x, x - z \rangle \geq 0.$$

Adding on both sides of these two inequalities lead to

$$\langle z - P_C x, P_C x - z \rangle \geq 0.$$

But,

$$\langle z - P_C x, P_C x - z \rangle = -||z - P_C x||_2^2,$$

so it must be the case that $z = P_C x$. This completes the proof. $\qquad \square$

Corollary 3.26. *Let S be any subspace of $\mathcal{X}$. Then, for any x in $\mathcal{X}$ and s in S, we have*

$$\langle P_S x - x, s \rangle = 0.$$

Proof: Since S is a subspace, $s + P_S x$ is again in S, for all s, as is cs, for every scalar c. $\qquad\square$

Corollary 3.27. *Let S be any subspace of $\mathcal{X}$, d a fixed vector, and V the linear manifold $V = S + d = \{v = s + d | s \in S\}$, obtained by translating the members of S by the vector d. Then, for every x in $\mathcal{X}$ and every v in V, we have*

$$\langle P_V x - x, v - P_V x \rangle = 0.$$

Proof: Since v and $P_V x$ are in V, they have the form $v = s + d$, and $P_V x = \hat{s} + d$, for some s and $\hat{s}$ in S. Then $v - P_V x = s - \hat{s}$. $\qquad\square$

Corollary 3.28. *Let H be the hyperplane $H(a, \gamma)$. Then, for every x, and every h in H, we have*

$$\langle P_H x - x, h - P_H x \rangle = 0.$$

Corollary 3.29. *Let S be a subspace of $\mathcal{X}$. Then, every x in $\mathcal{X}$ can be written as $x = s + u$, for a unique s in S and a unique u in $S^\perp$.*

Proof: The vector $P_S x - x$ is in $S^\perp$. $\qquad\square$

Corollary 3.30. *Let S be a subspace of $\mathcal{X}$. Then $(S^\perp)^\perp = S$.*

Proof: Every x in $\mathcal{X}$ has the form $x = s + u$, with s in S and u in $S^\perp$. Suppose x is in $(S^\perp)^\perp$. Then $u = 0$. $\qquad\square$

3.5.3 Gradient Operators

Another important example of a nonlinear operator is the gradient of a real-valued function of several variables. Let $f(x) = f(x_i, ..., x_J)$ be a real number for each vector x in R^J. The *gradient* of f at the point x is the vector whose entries are the partial derivatives of f; that is,

$$\nabla f(x) = (\frac{\partial f}{\partial x_1}(x), ..., \frac{\partial f}{\partial x_J}(x))^T.$$

The operator $Tx = \nabla f(x)$ is linear only if the function $f(x)$ is quadratic; that is, $f(x) = x^T A x$ for some square matrix x, in which case the gradient of f is $\nabla f(x) = \frac{1}{2}(A + A^T)x$.

If u is any vector in $\mathcal{X}$ with $||u||_2 = 1$, then u is said to be a *direction vector*. Let $f : R^J \to R$. The *directional derivative* of f, at the point x, in the direction of u, is

$$D_u f(x) = \lim_{t \to 0}(1/t)(f(x + tu) - f(x)),$$

if this limit exists. If the partial derivatives of f are continuous, then

$$D_u f(x) = u_1 \frac{\partial f}{\partial x_1}(x) + \dots + u_J \frac{\partial f}{\partial x_J}(x).$$

It follows from the Cauchy Inequality that $|D_u f(x)| \leq ||\nabla f(x)||_2$, with equality if and only if u is parallel to the gradient vector, $\nabla f(x)$. The gradient points in the direction of the greatest increase in $f(x)$.

3.6   Exercises

3.1. Show that the vector a is orthogonal to the hyperplane $H = H(u, \gamma)$; that is, if u and v are in H, then a is orthogonal to $u - v$.

3.2. Show that B is Hermitian if and only if the real matrix $\tilde{B}$ is symmetric.

3.3. Let B be Hermitian. For any $x = x^1 + ix^2$, define $\tilde{x}' = (-x^2, x^1)^T$. Show that the following are equivalent:

1. $Bx = \lambda x$;

2. $\tilde{B}\tilde{x} = \lambda \tilde{x}$;

3. $\tilde{B}\tilde{x}' = \lambda \tilde{x}'$.

3.4. Show that $B^\dagger Bx = c$ if and only if $\tilde{B}^T \tilde{B}\tilde{x} = \tilde{c}$.

3.5. Show that $CS(A)^\perp = NS(A^\dagger)$. *Hint:* If $v \in CS(A)^\perp$, then $v^\dagger Ax = 0$ for all x, including $x = A^\dagger v$.

3.6. Show that $CS(A) \cap NS(A^\dagger) = \{0\}$. *Hint:* If $y = Ax \in NS(A^\dagger)$, consider $||y||_2^2 = y^\dagger y$.

3.7. Show that $Ax = b$ has solutions if and only if the associated Björck-Elfving equations $AA^\dagger z = b$ has solutions.

3.8. Let S be any subspace of C^I. Show that if Q is invertible and $Q(S) = S$ then $Q^{-1}(S) = S$. *Hint:* If $Qt = Qs$, then $t = s$.

3.9. Let Q be Hermitian. Show that $Q(S)^\perp = Q^{-1}(S^\perp)$ for every subspace S. If Q is also invertible, then $Q^{-1}(S)^\perp = Q(S^\perp)$. Find an example of a noninvertible Hermitian Q for which $Q^{-1}(S)^\perp$ and $Q(S^\perp)$ are different.

3.10. Show that we can write P_{H_0} as a matrix multiplication:

$$P_{H_0} x = (I - aa^\dagger)x.$$

3.11. Prove Lemma 3.24. *Hint*: Use the previous exercise and the fact that P_{i0} is linear to show that

$$B = (I - a^I(a^I)^\dagger) \cdots (I - a^1(a^1)^\dagger).$$

3.12. Let A be a complex I by J matrix with $I < J$, b a fixed vector in C^I, and S the linear manifold of C^J consisting of all vectors x with $Ax = b$. Denote by $P_S z$ the orthogonal projection of vector z onto S. Assume that A has rank I, so that the matrix $AA^\dagger$ is invertible. Show that

$$P_S z = (I - A^\dagger(AA^\dagger)^{-1}A)z + A^\dagger(AA^\dagger)^{-1}b.$$

Hint: Note that, if $z = 0$, then $P_S z$ is the minimum-norm solution of the system $Ax = b$.

3.13. Let C be a fixed, nonempty, closed convex subset of $\mathcal{X}$, and x not in C. Where are the vectors z for which $P_C z = P_C x$? Prove your conjecture.

4 | Metric Spaces and Norms

As we have seen, the inner product on $\mathcal{X} = R^J$ or $\mathcal{X} = C^J$ can be used to define the Euclidean norm $||x||_2$ of a vector x, which, in turn, provides a *metric*, or a measure of distance between two vectors, $d(x, y) = ||x - y||_2$. The notions of metric and norm are actually more general notions, with no necessary connection to the inner product.

4.1 Metric Spaces

We begin with the basic definitions.

Definition 4.1. Let S be a nonempty set. We say that the function $d : S \times S \to [0, +\infty)$ is a *metric* if the following hold:

$$d(s, t) \geq 0,$$

for all s and t in S;

$$d(s, t) = 0$$

if and only if $s = t$;

$$d(s, t) = d(t, s)$$

for all s and t in S; and, for all s, t, and u in S,

$$d(s, t) \leq d(s, u) + d(u, t).$$

The pair $\{S, d\}$ is a metric space.

The last inequality is the triangle inequality for this metric.

4.2   Analysis in Metric Space

Analysis is concerned with issues of convergence and limits.

Definition 4.2. A sequence $\{s^k\}$ in the metric space $(\mathcal{S}, d)$ is said to have limit s^* if

$$\lim_{k \to +\infty} d(s^k, s^*) = 0.$$

Any sequence with a limit is said to be *convergent*.

A sequence can have at most one limit.

Definition 4.3. The sequence $\{s^k\}$ is said to be a Cauchy sequence if, for any $\epsilon > 0$, there is a positive integer m, such that, for any nonnegative integer n,

$$d(s^m, s^{m+n}) \le \epsilon.$$

Every convergent sequence is a Cauchy sequence.

Definition 4.4. The metric space $(\mathcal{S}, d)$ is said to be *complete* if every Cauchy sequence is a convergent sequence.

The finite-dimensional spaces R^J and C^J are complete metric spaces, with respect to the usual Euclidean distance.

Definition 4.5. An infinite sequence $\{s^k\}$ in $\mathcal{S}$ is said to be bounded if there is an element a and a positive constant $b > 0$ such that $d(a, s^k) \le b$, for all k.

Definition 4.6. A subset K of the metric space is said to be *closed* if, for every convergent sequence $\{s^k\}$ of elements in K, the limit point is again in K. The closure of a set K is the smallest closed set containing K.

For example, in $\mathcal{X} = R$, the set $K = (0, 1]$ is not closed, because it does not contain the point $s = 0$, which is the limit of the sequence $\{s^k = \frac{1}{k}\}$; the set $K = [0, 1]$ is closed and is the *closure* of the set $(0, 1]$, that is, it is the smallest closed set containing $(0, 1]$.

Definition 4.7. For any bounded sequence $\{x^k\}$ in $\mathcal{X}$, there is at least one subsequence, often denoted $\{x^{k_n}\}$, that is convergent; the notation implies that the positive integers k_n are ordered, so that $k_1 < k_2 < \dots$. The limit of such a subsequence is then said to be a cluster point of the original sequence.

When we investigate iterative algorithms, we will want to know if the sequence $\{x^k\}$ generated by the algorithm converges. As a first step, we will usually ask if the sequence is bounded? If it is bounded, then it will have at least one cluster point. We then try to discover if that cluster point is really the limit of the sequence. We turn now to metrics that come from norms.

4.3 Norms

The metric spaces that interest us most are those for which the metric comes from a norm, which is a measure of the length of a vector.

Definition 4.8. We say that $|| \cdot ||$ is a norm on $\mathcal{X}$ if

$$||x|| \geq 0$$

for all x,

$$||x|| = 0$$

if and only if $x = 0$,

$$||\gamma x|| = |\gamma| \, ||x||$$

for all x and scalars γ, and

$$||x + y|| \leq ||x|| + ||y||$$

for all vectors x and y.

Lemma 4.9. *The function* $d(x, y) = ||x - y||$ *defines a metric on* $\mathcal{X}$.

It can be shown that R^J and C^J are complete for any metric arising from a norm.

4.3.1 Some Common Norms on $\mathcal{X}$

We consider now the most common norms on the space $\mathcal{X}$.

The 1-norm. The 1-norm on $\mathcal{X}$ is defined by

$$||x||_1 = \sum_{j=1}^{J} |x_j|.$$

The ∞-norm. The ∞-norm on $\mathcal{X}$ is defined by

$$||x||_\infty = \max\{|x_j| \, | j = 1, ..., J\}.$$

The 2-norm. The 2-norm, also called the Euclidean norm, is the most commonly used norm on $\mathcal{X}$. It is the one that comes from the inner product:

$$||x||_2 = \sqrt{\langle x, x \rangle} = \sqrt{x^\dagger x}.$$

Weighted 2-norms. Let A be an invertible matrix and $Q = A^\dagger A$. Define

$$||x||_Q = ||Ax||_2 = \sqrt{x^\dagger Q x},$$

for all vectors x. If Q is the diagonal matrix with diagonal entries $Q_{jj} > 0$, then

$$||x||_Q = \sqrt{\sum_{j=1}^{J} Q_{jj} |x_j|^2};$$

for that reason we speak of $||x||_Q$ as the Q-*weighted* 2-norm of x.

4.4 Eigenvalues and Eigenvectors

Let S be a complex, square matrix. We say that λ is an eigenvalue of S if λ is a root of the complex polynomial $\det(\lambda I - S)$. Therefore, each S has as many (possibly complex) eigenvalues as it has rows or columns, although some of the eigenvalues may be repeated.

An equivalent definition is that λ is an eigenvalue of S if there is a nonzero vector x with $Sx = \lambda x$, in which case the vector x is called an *eigenvector* of S. From this definition, we see that the matrix S is invertible if and only if zero is not one of its eigenvalues. The *spectral radius* of S, denoted $\rho(S)$, is the maximum of $|\lambda|$, over all eigenvalues λ of S.

If S is an I by I Hermitian matrix with (necessarily real) eigenvalues

$$\lambda_1 \geq \lambda_2 \geq \cdots \geq \lambda_I,$$

and associated (column) eigenvectors $\{u_i | i = 1, ..., I\}$ (which we may assume are mutually orthogonal), then S can be written as

$$S = \lambda_1 u_1 u_1^\dagger + \cdots + \lambda_I u_I u_I^\dagger.$$

This is the *eigenvalue/eigenvector decomposition* of S. The Hermitian matrix S is invertible if and only if all of its eigenvalues are nonzero, in which case we can write the inverse of S as

$$S^{-1} = \lambda_1^{-1} u_1 u_1^\dagger + \cdots + \lambda_I^{-1} u_I u_I^\dagger.$$

Definition 4.10. A Hermitian matrix S is positive-definite if each of its eigenvalues is positive.

It follows from the eigenvector decomposition of S that $S = QQ^\dagger$ for the Hermitian, positive-definite matrix

$$Q = \sqrt{\lambda_1} u_1 u_1^\dagger + \cdots + \sqrt{\lambda_I} u_I u_I^\dagger;$$

Q is called the *Hermitian square root* of S.

4.4.1 The Singular-Value Decomposition

The eigenvector/eigenvalue decomposition applies only to square matrices. The singular-value decomposition is similar, but applies to any matrix.

Definition 4.11. Let A be an I by J complex matrix. The rank of A is the number of linearly independent rows, which always equals the number of linearly independent columns. The matrix A is said to have full rank if its rank is the smaller of I and J.

Let $I \leq J$. Let $B = AA^\dagger$ and $C = A^\dagger A$. Let $\lambda_i \geq 0$, for $i = 1, ..., I$, be the eigenvalues of B, and let $\{u^1, ..., u^I\}$ be associated orthonormal eigenvectors of B. Assume that $\lambda_i > 0$ for $i = 1, ..., N \leq I$, and, if $N < I$, $\lambda_i = 0$, for $i = N + 1, ..., I$; if $N = I$, then the matrix A has full rank. For $i = 1, ..., N$, let $v^i = \lambda_i^{-1/2} A^\dagger u^i$. It is easily shown that the collection $\{v^1, ..., v^N\}$ is orthonormal. Let $\{v^{N+1}, ..., v^J\}$ be selected so that $\{v^1, ..., v^J\}$ is orthonormal. Then the sets $\{u^1, ..., u^N\}$, $\{u^{N+1}, ..., u^I\}$, $\{v^1, ..., v^N\}$, and $\{v^{N+1}, ..., v^J\}$ are orthonormal bases for the subspaces $CS(A)$, $NS(A^\dagger)$, $CS(A^\dagger)$, and $NS(A)$, respectively.

Definition 4.12. We have

$$A = \sum_{i=1}^{N} \sqrt{\lambda_i} u^i (v^i)^\dagger,$$

which is the *singular-value decomposition* (SVD) of the matrix A.

The SVD of the matrix $A^\dagger$ is then

$$A^\dagger = \sum_{i=1}^{N} \sqrt{\lambda_i} v^i (u^i)^\dagger.$$

Definition 4.13. The pseudo-inverse of the matrix A is the J by I matrix

$$A^\sharp = \sum_{i=1}^{N} \lambda_i^{-1/2} v^i (u^i)^\dagger.$$

Lemma 4.14. *For any matrix A, we have*

$$(A^\dagger)^\sharp = (A^\sharp)^\dagger.$$

For A that has full rank, if $N = I \leq J$, then

$$A^\sharp = A^\dagger B^{-1}$$

and

$$(A^\dagger)^\sharp = B^{-1} A.$$

4.4.2   An Upper Bound for the Singular Values of A

Several of the iterative algorithms we shall encounter later involve a positive parameter γ that can be no larger than $2/\lambda_{max}$, where λ_{max} is the largest eigenvalue of the matrix $A^{\dagger}A$, which is also the square of the largest singular value of A itself. In order for these iterations to converge quickly, it is necessary that the parameter be chosen reasonably large, which implies that we must have a good estimate of λ_{max}. When A is not too large, finding λ_{max} poses no significant problem, but, for many of our applications, A is large. Even calculating $A^{\dagger}A$, not to mention finding eigenvalues, is expensive in those cases. We would like a good estimate of λ_{max} that can be obtained from A itself. The upper bounds for λ_{max} we present here apply to any matrix A, but will be particularly helpful when A is sparse, that is, most of its entries are zero.

The normalized case. We assume now that the matrix A has been normalized so that each of its rows has Euclidean length one. Denote by s_j the number of nonzero entries in the jth column of A, and let s be the maximum of the s_j. Our first result is the following:

Theorem 4.15. *For normalized A, λ_{max}, the largest eigenvalue of the matrix $A^{\dagger}A$, does not exceed s.*

Proof: For notational simplicity, we consider only the case of real matrices and vectors. Let $A^T A v = c v$ for some nonzero vector v. We show that $c \leq s$. We have $A A^T A v = c A v$ and so $w^T A A^T w = v^T A^T A A^T A v = c v^T A^T A v = c w^T w$, for $w = A v$. Then, with $e_{ij} = 1$ if $A_{ij} \neq 0$ and $e_{ij} = 0$ otherwise, we have

$$(\sum_{i=1}^{I} A_{ij} w_i)^2 \; = \; (\sum_{i=1}^{I} A_{ij} e_{ij} w_i)^2$$

$$\leq \; (\sum_{i=1}^{I} A_{ij}^2 w_i^2)(\sum_{i=1}^{I} e_{ij}^2)$$

$$= \; (\sum_{i=1}^{I} A_{ij}^2 w_i^2) s_j$$

$$\leq \; (\sum_{i=1}^{I} A_{ij}^2 w_i^2) s.$$

Therefore,

$$w^T A A^T w = \sum_{j=1}^{J}(\sum_{i=1}^{I} A_{ij} w_i)^2 \leq \sum_{j=1}^{J}(\sum_{i=1}^{I} A_{ij}^2 w_i^2) s,$$

and

$$w^T AA^T w = c \sum_{i=1}^{I} w_i^2 = c \sum_{i=1}^{I} w_i^2 (\sum_{j=1}^{J} A_{ij}^2)$$

$$= c \sum_{i=1}^{I} \sum_{j=1}^{J} w_i^2 A_{ij}^2.$$

The result follows immediately. $\square$

When A is normalized, the trace of AA^T, that is, the sum of its diagonal entries, is I. Since the trace is also the sum of the eigenvalues of both AA^T and $A^T A$, we have $\lambda_{max} \leq I$. When A is sparse, s is much smaller than I, so provides a much tighter upper bound for λ_{max}.

The general case. A similar upper bound for λ_{max} is given for the case in which A is not normalized.

Theorem 4.16. *For each $i = 1, ..., I$ let $\nu_i = \sum_{j=1}^{J} |A_{ij}|^2 > 0$. For each $j = 1, ..., J$, let $\sigma_j = \sum_{i=1}^{I} e_{ij} \nu_i$, where $e_{ij} = 1$ if $A_{ij} \neq 0$ and $e_{ij} = 0$ otherwise. Let σ denote the maximum of the σ_j. Then the eigenvalues of the matrix $A^\dagger A$ do not exceed σ.*

The proof of Theorem 4.16 is similar to that of Theorem 4.15.

Upper bounds for ϵ-sparse matrices. If A is not sparse, but most of its entries have magnitude not exceeding $\epsilon > 0$ we say that A is ϵ-sparse. We can extend the results for the sparse case to the ϵ-sparse case.

Given a matrix A, define the entries of the matrix B to be $B_{ij} = A_{ij}$ if $|A_{ij}| > \epsilon$, and $B_{ij} = 0$, otherwise. Let $C = A - B$; then $|C_{ij}| \leq \epsilon$, for all i and j. If A is ϵ-sparse, then B is sparse. The 2-norm of the matrix A, written $||A||_2$, is defined to be the square root of the largest eigenvalue of the matrix $A^\dagger A$, that is, $||A||_2 = \sqrt{\lambda_{\max}}$. From Theorem 4.16 we know that $||B||_2 \leq \sigma$. The trace of the matrix $C^\dagger C$ does not exceed $IJ\epsilon^2$. Therefore

$$\sqrt{\lambda_{\max}} = ||A||_2 = ||B + C||_2 \leq ||B||_2 + ||C||_2 \leq \sqrt{\sigma} + \sqrt{IJ}\epsilon, \qquad (4.1)$$

so that

$$\lambda_{\max} \leq \sigma + 2\sqrt{\sigma IJ}\epsilon + IJ\epsilon^2. \qquad (4.2)$$

Simulation studies have shown that these upper bounds become tighter as the size of the matrix A increases. In hundreds of runs, with I and J in the hundreds, we found that the relative error of the upper bound was around one percent [50].

4.5   Matrix Norms

Any matrix can be turned into a vector by vectorization. Therefore, we can define a norm for any matrix by simply vectorizing and taking a norm of the resulting vector. Such norms for matrices may not be compatible with the role of a matrix as representing a linear transformation.

4.5.1   Induced Matrix Norms

One way to obtain a compatible norm for matrices is through the use of an induced matrix norm.

Definition 4.17. Let $||x||$ be any norm on C^J, not necessarily the Euclidean norm, $||b||$ any norm on C^I, and A a rectangular I by J matrix. The induced matrix norm of A, simply denoted $||A||$, derived from these two vectors norms, is the smallest positive constant c such that

$$||Ax|| \leq c||x||,$$

for all x in C^J. This induced norm can be written as

$$||A|| = \max_{x \neq 0}\{||Ax||/||x||\}.$$

We study induced matrix norms in order to measure the distance $||Ax - Az||$, relative to the distance $||x - z||$:

$$||Ax - Az|| \leq ||A|| \, ||x - z||,$$

for all vectors x and z and $||A||$ is the smallest number for which this statement can be made.

4.5.2   Condition Number of a Square Matrix

Let S be a square, invertible matrix and z the solution to $Sz = h$. We are concerned with the extent to which the solution changes as the right side, h, changes. Denote by δ_h a small perturbation of h, and by δ_z the solution of $S\delta_z = \delta_h$. Then $S(z + \delta_z) = h + \delta_h$. Applying the compatibility condition $||Ax|| \leq ||A|| \, ||x||$, we get

$$||\delta_z|| \leq ||S^{-1}|| \, ||\delta_h||$$

and

$$||z|| \geq ||h||/||S||.$$

Therefore,

$$\frac{||\delta_z||}{||z||} \leq ||S|| \, ||S^{-1}||\frac{||\delta_h||}{||h||}. \qquad (4.3)$$

Definition 4.18. The quantity $c = ||S||||S^{-1}||$ is the condition number of S, with respect to the given matrix norm.

Note that $c \geq 1$: for any nonzero z, we have

$$||S^{-1}|| \geq ||S^{-1}z||/||z|| = ||S^{-1}z||/||SS^{-1}z|| \geq 1/||S||.$$

When S is Hermitian and positive-definite, the condition number of S, with respect to the matrix norm induced by the Euclidean vector norm, is

$$c = \lambda_{\max}(S)/\lambda_{\min}(S),$$

the ratio of the largest to the smallest eigenvalues of S.

4.5.3   Some Examples of Induced Matrix Norms

If we choose the two vector norms carefully, then we can get an explicit description of $||A||$, but, in general, we cannot.

For example, let $||x|| = ||x||_1$ and $||Ax|| = ||Ax||_1$ be the 1-norms of the vectors x and Ax, where

$$||x||_1 = \sum_{j=1}^{J} |x_j|.$$

Lemma 4.19. *The 1-norm of A, induced by the 1-norms of vectors in C^J and C^I, is*

$$||A||_1 = \max \{\sum_{i=1}^{I} |A_{ij}|, j = 1, 2, ..., J\}.$$

Proof: Use basic properties of the absolute value to show that

$$||Ax||_1 \leq \sum_{j=1}^{J}(\sum_{i=1}^{I} |A_{ij}|)|x_j|.$$

Then let $j = m$ be the index for which the maximum column sum is reached and select $x_j = 0$, for $j \neq m$, and $x_m = 1$.   □

The *infinity norm* of the vector x is

$$||x||_\infty = \max \{|x_j|, j = 1, 2, ..., J\}.$$

Lemma 4.20. *The infinity norm of the matrix A, induced by the infinity norms of vectors in C^J and C^I, is*

$$||A||_\infty = \max\{\sum_{j=1}^{J} |A_{ij}|, i = 1, 2, ..., I\}.$$

The proof is similar to that of the previous lemma.

Lemma 4.21. *Let M be an invertible matrix and $||x||$ any vector norm. Define*

$$||x||_M = ||Mx||.$$

Then, for any square matrix S, the matrix norm

$$||S||_M = \max_{x \neq 0}\{||Sx||_M/||x||_M\}$$

is

$$||S||_M = ||MSM^{-1}||.$$

In [4] this result is used to prove the following lemma:

Lemma 4.22. *Let S be any square matrix and let $\epsilon > 0$ be given. Then there is an invertible matrix M such that*

$$||S||_M \leq \rho(S) + \epsilon.$$

4.5.4   The Euclidean Norm of a Square Matrix

We shall be particularly interested in the Euclidean norm (or 2-norm) of the square matrix A, denoted by $||A||_2$, which is the induced matrix norm derived from the Euclidean vector norms.

From the definition of the Euclidean norm of A, we know that

$$||A||_2 = \max\{||Ax||_2/||x||_2\},$$

with the maximum over all nonzero vectors x. Since

$$||Ax||_2^2 = x^\dagger A^\dagger A x,$$

we have

$$||A||_2 = \sqrt{\max\{\frac{x^\dagger A^\dagger A x}{x^\dagger x}\}}, \tag{4.4}$$

over all nonzero vectors x.

Proposition 4.23. *The Euclidean norm of a square matrix is*

$$||A||_2 = \sqrt{\rho(A^\dagger A)};$$

that is, the term inside the square-root in Equation (4.4) is the largest eigenvalue of the matrix $A^\dagger A$.

Proof: Let

$$\lambda_1 \geq \lambda_2 \geq ... \geq \lambda_J \geq 0$$

and let $\{u^j, j = 1, ..., J\}$ be mutually orthogonal eigenvectors of $A^\dagger A$ with $||u^j||_2 = 1$. Then, for any x, we have

$$x = \sum_{j=1}^{J} [(u^j)^\dagger x] u^j,$$

while

$$A^\dagger A x = \sum_{j=1}^{J} [(u^j)^\dagger x] A^\dagger A u^j = \sum_{j=1}^{J} \lambda_j [(u^j)^\dagger x] u^j.$$

It follows that

$$||x||_2^2 = x^\dagger x = \sum_{j=1}^{J} |(u^j)^\dagger x|^2 \tag{4.5}$$

and

$$||Ax||_2^2 = x^\dagger A^\dagger A x = \sum_{j=1}^{J} \lambda_j |(u^j)^\dagger x|^2.$$

Maximizing $||Ax||_2^2/||x||_2^2$ over $x \neq 0$ is equivalent to maximizing $||Ax||_2^2$, subject to $||x||_2^2 = 1$. The right side of Equation (4.6) is then a convex combination of the λ_j, which will have its maximum when only the coefficient of λ_1 is nonzero. $\square$

If S is not Hermitian, then the Euclidean norm of S cannot be calculated directly from the eigenvalues of S. Take, for example, the square, non-Hermitian matrix

$$S = \begin{bmatrix} i & 2 \\ 0 & i \end{bmatrix},$$

having eigenvalues $\lambda = i$ and $\lambda = i$. The eigenvalues of the Hermitian matrix

$$S^{\dagger}S = \begin{bmatrix} 1 & -2i \\ 2i & 5 \end{bmatrix}$$

are $\lambda = 3 + 2\sqrt{2}$ and $\lambda = 3 - 2\sqrt{2}$. Therefore, the Euclidean norm of S is

$$||S||_2 = \sqrt{3 + 2\sqrt{2}}.$$

4.5.5 Diagonalizable Matrices

Definition 4.24. A square matrix S is diagonalizable if $\mathcal{X}$ has a basis of eigenvectors of S.

In the case in which S is diagonalizable, with V be a square matrix whose columns are linearly independent eigenvectors of S and L the diagonal matrix having the eigenvalues of S along its main diagonal, we have $SV = VL$, or $V^{-1}SV = L$. Let $T = V^{-1}$ and define $||x||_T = ||Tx||_2$, the Euclidean norm of Tx. Then the induced matrix norm of S is $||S||_T = \rho(S)$. We see from this that, for any diagonalizable matrix S, in particular, for any Hermitian matrix, there is a vector norm such that the induced matrix norm of S is $\rho(S)$. In the Hermitian case we know that, if the eigenvector columns of V are scaled to have length one, then $V^{-1} = V^{\dagger}$ and $||Tx||_2 = ||V^{\dagger}x||_2 = ||x||_2$, so that the required vector norm is just the Euclidean norm, and $||S||_T$ is just $||S||_2$, which we know to be $\rho(S)$.

4.5.6 Gerschgorin's Theorem

Gerschgorin's theorem gives us a way to estimate the eigenvalues of an arbitrary square matrix A.

Theorem 4.25. *Let A be J by J. For $j = 1, ..., J$, let C_j be the circle in the complex plane with center A_{jj} and radius $r_j = \sum_{m \neq j} |A_{jm}|$. Then every eigenvalue of A lies within one of the C_j.*

Proof: Let λ be an eigenvalue of A, with associated eigenvector u. Let u_j be the entry of the vector u having the largest absolute value. From $Au = \lambda u$, we have

$$(\lambda - A_{jj})u_j = \sum_{m \neq j} A_{jm}u_m,$$

so that

$$|\lambda - A_{jj}| \leq \sum_{m \neq j} |A_{jm}||u_m|/|u_j| \leq r_j.$$

This completes the proof. $\square$

4.5.7 Strictly Diagonally Dominant Matrices

Definition 4.26. A square I by I matrix S is said to be strictly diagonally dominant if, for each $i = 1, ..., I$,

$$|S_{ii}| > r_i = \sum_{m \neq i} |S_{im}|.$$

When the matrix S is strictly diagonally dominant, all the eigenvalues of S lie within the union of the spheres with centers S_{ii} and radii S_{ii}. With D the diagonal component of S, the matrix $D^{-1}S$ then has all its eigenvalues within the circle of radius one, centered at $(1, 0)$. Then $\rho(I - D^{-1}S) < 1$. We use this result in our discussion of the Jacobi splitting method.

4.6 Exercises

4.1. Show that every convergent sequence is a Cauchy sequence.

4.2. Let S be the set of rational numbers, with $d(s, t) = |s - t|$. Show that (S, d) is a metric space, but not a complete metric space.

4.3. Show that any convergent sequence in a metric space is bounded. Find a bounded sequence of real numbers that is not convergent.

4.4. Show that, if $\{s^k\}$ is bounded, then, for any element c in the metric space, there is a constant $r > 0$, with $d(c, s^k) \leq r$, for all k.

4.5. Show that your bounded, but not convergent, sequence found in Exercise 4.3 has a cluster point.

4.6. Show that, if x is a cluster point of the sequence $\{x^k\}$, and if $d(x, x^k) \geq d(x, x^{k+1})$, for all k, then x is the limit of the sequence.

4.7. Show that the 1-norm is a norm.

4.8. Show that the ∞-norm is a norm.

4.9. Show that the 2-norm is a norm. *Hint:* For the triangle inequality, use Cauchy's Inequality.

4.10. Show that the Q-weighted 2-norm is a norm.

4.11. Show that $\rho(S^2) = \rho(S)^2$.

4.12. Show that, if S is Hermitian, then every eigenvalue of S is real. *Hint*: Suppose that $Sx = \lambda x$. Then consider $x^\dagger Sx$.

4.13. Use the SVD of A to obtain the *eigenvalue/eigenvector decompositions* of B and C:

$$B = \sum_{i=1}^{N} \lambda_i u^i (u^i)^\dagger,$$

and

$$C = \sum_{i=1}^{N} \lambda_i v^i (v^i)^\dagger.$$

4.14. Show that, for any square matrix S and any induced matrix norm $||S||$, we have $||S|| \geq \rho(S)$. Consequently, for any induced matrix norm $||S||$,

$$||S|| \geq |\lambda|,$$

for every eigenvalue λ of S. So we know that

$$\rho(S) \leq ||S||,$$

for every induced matrix norm, but, according to Lemma 4.22, we also have

$$||S||_M \leq \rho(S) + \epsilon.$$

4.15. Show that, if $\rho(S) < 1$, then there is a vector norm on $\mathcal{X}$ for which the induced matrix norm of S is less than one.

4.16. Show that, if S is Hermitian, then $||S||_2 = \rho(S)$. *Hint*: Use Exercise 4.11.

II | Overview

5 | Operators

In a broad sense, all iterative algorithms generate a sequence $\{x^k\}$ of vectors. The sequence may converge for any starting vector x^0, or may converge only if x^0 is sufficiently close to a solution. The limit, when it exists, may depend on x^0, and may, or may not, solve the original problem. Convergence to the limit may be slow and the algorithm may need to be accelerated. The algorithm may involve measured data. The limit may be sensitive to noise in the data and the algorithm may need to be regularized to lessen this sensitivity. The algorithm may be quite general, applying to all problems in a broad class, or it may be tailored to the problem at hand. Each step of the algorithm may be costly, but only a few steps generally needed to produce a suitable approximate answer, or, each step may be easily performed, but many such steps needed. Although convergence of an algorithm is important, theoretically, sometimes in practice only a few iterative steps are used.

5.1 Operators

For most of the iterative algorithms we shall consider, the iterative step is

$$x^{k+1} = Tx^k,$$

for some operator T. If T is a continuous operator (and it usually is), and the sequence $\{T^k x^0\}$ converges to $\hat{x}$, then $T\hat{x} = \hat{x}$, that is, $\hat{x}$ is a *fixed point* of the operator T. We denote by $\text{Fix}(T)$ the set of fixed points of T. The convergence of the iterative sequence $\{T^k x^0\}$ will depend on the properties of the operator T.

Our approach here will be to identify several classes of operators for which the iterative sequence is known to converge, to examine the convergence theorems that apply to each class, to describe several applied problems that can be solved by iterative means, to present iterative algorithms for solving these problems, and to establish that the operator involved in each of these algorithms is a member of one of the designated classes.

5.2   Two Useful Identities

The identities in the next two lemmas relate an arbitrary operator T to its complement, $G = I - T$, where I denotes the identity operator. They will allow us to transform properties of T into properties of G that may be easier to work with.

Lemma 5.1. *Let T be an arbitrary operator T on $\mathcal{X}$ and $G = I - T$. Then*

$$||x - y||_2^2 - ||Tx - Ty||_2^2 = 2\mathrm{Re}(\langle Gx - Gy, x - y \rangle) - ||Gx - Gy||_2^2. \quad (5.1)$$

The proof is a simple calculation.

Lemma 5.2. *Let T be an arbitrary operator T on $\mathcal{X}$ and $G = I - T$. Then*

$$\mathrm{Re}(\langle Tx - Ty, x - y \rangle) - ||Tx - Ty||_2^2 = \mathrm{Re}(\langle Gx - Gy, x - y \rangle) \\ - ||Gx - Gy||_2^2.$$

Proof: Use the previous lemma.   □

5.3   Strict Contractions

The strict contraction operators are perhaps the best known class of operators associated with iterative algorithms.

Definition 5.3. An operator T on $\mathcal{X}$ is Lipschitz continuous, with respect to a vector norm $|| \cdot ||$, or L-Lipschitz, if there is a positive constant L such that

$$||Tx - Ty|| \le L||x - y||,$$

for all x and y in $\mathcal{X}$.

Definition 5.4. An operator T on $\mathcal{X}$ is a strict contraction, with respect to a vector norm $|| \cdot ||$, if there is $r \in (0, 1)$ such that

$$||Tx - Ty|| \le r||x - y||,$$

for all vectors x and y.

For strict contractions, we have the Banach-Picard Theorem [82]:

Theorem 5.5. *Let T be a strict contraction. Then, there is a unique fixed point of T and, for any starting vector x^0, the sequence $\{T^k x^0\}$ converges to the fixed point.*

The key step in the proof is to show that $\{x^k\}$ is a Cauchy sequence, therefore, it has a limit.

Lemma 5.6. *Let T be an affine operator, that is, T has the form $Tx = Bx + d$, where B is a linear operator, and d is a fixed vector. Then T is a strict contraction if and only if $||B||$, the induced matrix norm of B, is less than one.*

The spectral radius of B, written $\rho(B)$, is the maximum of $|\lambda|$, over all eigenvalues λ of B. Since $\rho(B) \leq ||B||$ for every norm on B induced by a vector norm, B being a strict contraction implies that $\rho(B) < 1$. When B is Hermitian, the matrix norm of B induced by the Euclidean vector norm is $||B||_2 = \rho(B)$, so if $\rho(B) < 1$, then B is a strict contraction with respect to the Euclidean norm.

When B is not Hermitian, it is not as easy to determine if the affine operator T is a strict contraction with respect to a given norm. Instead, we often tailor the norm to the operator T. Suppose that B is a diagonalizable matrix, that is, there is a basis for $\mathcal{X}$ consisting of eigenvectors of B. Let $\{u^1, ..., u^J\}$ be such a basis, and let $Bu^j = \lambda_j u^j$, for each $j = 1, ..., J$. For each x in $\mathcal{X}$, there are unique coefficients a_j so that

$$x = \sum_{j=1}^{J} a_j u^j.$$

Then let

$$||x|| = \sum_{j=1}^{J} |a_j|. \tag{5.2}$$

Lemma 5.7. *The expression $|| \cdot ||$ in Equation (5.2) defines a norm on $\mathcal{X}$. If $\rho(B) < 1$, then the affine operator T is a strict contraction, with respect to this norm.*

According to Lemma 4.22, for any square matrix B and any $\epsilon > 0$, there is a vector norm for which the induced matrix norm satisfies $||B|| \leq \rho(B) + \epsilon$. Therefore, if B is an arbitrary square matrix with $\rho(B) < 1$, there is a vector norm with respect to which B is a strict contraction.

In many of the applications of interest to us, there will be multiple fixed points of T. Therefore, T will not be a strict contraction for any vector norm, and the Banach-Picard fixed-point theorem will not apply. We need to consider other classes of operators. These classes of operators will emerge as we investigate the properties of orthogonal projection operators.

5.4   Orthogonal Projection Operators

If C is a closed, nonempty convex set in $\mathcal{X}$, and x is any vector, then, as we have seen, there is a unique point $P_C x$ in C closest to x, in the sense of the Euclidean distance. This point is called the orthogonal projection of x onto C. If C is a subspace, then we can get an explicit description of $P_C x$ in terms of x; for general convex sets C, however, we will not be able to express $P_C x$ explicitly, and certain approximations will be needed. Orthogonal projection operators are central to our discussion, and, in this overview, we focus on problems involving convex sets, algorithms involving orthogonal projection onto convex sets, and classes of operators derived from properties of orthogonal projection operators.

5.4.1   Properties of the Operator P_C

Although we usually do not have an explicit expression for $P_C x$, we can, however, characterize $P_C x$ as the unique member of C for which

$$Re(\langle P_C x - x, c - P_C x\rangle) \geq 0, \tag{5.3}$$

for all c in C; see Proposition 3.25.

P_C is nonexpansive.   Recall that an operator T is nonexpansive, with respect to a given norm, if, for all x and y, we have

$$||Tx - Ty|| \leq ||x - y||. \tag{5.4}$$

Lemma 5.8. *The orthogonal projection operator $T = P_C$ is nonexpansive, with respect to the Euclidean norm, that is,*

$$||P_C x - P_C y||_2 \leq ||x - y||_2,$$

for all x and y.

Proof: Use Inequality (5.3) to get

$$Re(\langle P_C y - P_C x, P_C x - x\rangle) \geq 0$$

and

$$Re(\langle P_C x - P_C y, P_C y - y\rangle) \geq 0.$$

Add the two inequalities to obtain

$$\text{Re}(\langle P_C x - P_C y, x - y \rangle) \geq ||P_C x - P_C y||_2^2, \tag{5.5}$$

and use Cauchy's Inequality.						□

Because the operator P_C has multiple fixed points, P_C cannot be a strict contraction, unless the set C is a singleton set.

P_C is firmly nonexpansive. We now refine the condition on the operator.

Definition 5.9. An operator T is said to be firmly nonexpansive if

$$\text{Re}(\langle Tx - Ty, x - y \rangle) \geq ||Tx - Ty||_2^2, \tag{5.6}$$

for all x and y in $\mathcal{X}$.

Lemma 5.10. *An operator T is firmly nonexpansive if and only if $G = I - T$ is firmly nonexpansive.*

Proof: Use the identity in Equation (5.2).						□

From Equation (5.5), we see that the operator $T = P_C$ is not simply nonexpansive, but firmly nonexpansive, as well. A good source for more material on these topics is the book by Goebel and Reich [92].

The search for other properties of P_C. The class of nonexpansive operators is too large for our purposes; the operator $Tx = -x$ is nonexpansive, but the sequence $\{T^k x^0\}$ does not converge, in general, even though a fixed point, $x = 0$, exists. The class of firmly nonexpansive operators is too small for our purposes. Although the convergence of the iterative sequence $\{T^k x^0\}$ to a fixed point does hold for firmly nonexpansive T, whenever fixed points exist, the product of two or more firmly nonexpansive operators need not be firmly nonexpansive; that is, the class of firmly nonexpansive operators is not *closed to finite products*. This poses a problem, since, as we shall see, products of orthogonal projection operators arise in several of the algorithms we wish to consider. We need a class of operators smaller than the nonexpansive ones, but larger than the firmly nonexpansive ones, closed to finite products, and for which the sequence of iterates $\{T^k x^0\}$ will converge, for any x^0, whenever fixed points exist. The class we shall consider is the class of *averaged* operators.

5.5   Averaged Operators

The term "averaged operator" appears in the work of Baillon, Bruck, and Reich [5, 21]. There are several ways to define averaged operators. One way is in terms of the complement operator.

Definition 5.11. An operator G on $\mathcal{X}$ is called ν-inverse strongly monotone (ν-ism) [93] (also called co-coercive in [67]) if there is $\nu > 0$ such that

$$\mathrm{Re}(\langle Gx - Gy, x - y \rangle) \geq \nu \|Gx - Gy\|_2^2. \tag{5.7}$$

Lemma 5.12. *An operator T is nonexpansive if and only if its complement $G = I - T$ is $\frac{1}{2}$-ism, and T is firmly nonexpansive if and only if G is 1-ism, and if and only if G is firmly nonexpansive. Also, T is nonexpansive if and only if $F = (I + T)/2$ is firmly nonexpansive. If G is ν-ism and $\gamma > 0$, then the operator γG is $\frac{\nu}{\gamma}$-ism.*

Definition 5.13. An operator T is called averaged if $G = I - T$ is ν-ism for some $\nu > \frac{1}{2}$. If G is $\frac{1}{2\alpha}$-ism, for some $\alpha \in (0, 1)$, then we say that T is α-averaged.

It follows that every averaged operator is nonexpansive, with respect to the Euclidean norm, and every firmly nonexpansive operator is averaged.

The averaged operators are sometimes defined in a different, but equivalent, way, using the following characterization.

Lemma 5.14. *An operator T is averaged if and only if, for some operator N that is nonexpansive in the Euclidean norm, and $\alpha \in (0, 1)$, we have*

$$T = (1 - \alpha)I + \alpha N.$$

Proof: We assume first that there is $\alpha \in (0, 1)$ and nonexpansive operator N such that $T = (1 - \alpha)I + \alpha N$, and so $G = I - T = \alpha(I - N)$. Since N is nonexpansive, $I - N$ is $\frac{1}{2}$-ism and $G = \alpha(I - N)$ is $\frac{1}{2\alpha}$-ism. Conversely, assume that G is ν-ism for some $\nu > \frac{1}{2}$. Let $\alpha = \frac{1}{2\nu}$ and write $T = (1 - \alpha)I + \alpha N$ for $N = I - \frac{1}{\alpha}G$. Since $I - N = \frac{1}{\alpha}G$, $I - N$ is $\alpha\nu$-ism. Consequently $I - N$ is $\frac{1}{2}$-ism and N is nonexpansive. $\qquad\square$

An averaged operator is easily constructed from a given nonexpansive operator N by taking a convex combination of N and the identity I. The beauty of the class of averaged operators is that it contains many operators, such as P_C, that are not originally defined in this way. As we shall show later, finite products of averaged operators are again averaged, so the product of finitely many orthogonal projections is averaged.

5.5.1   Gradient Operators

Another type of operator that is averaged can be derived from gradient operators.

Definition 5.15. An operator T is monotone if

$$\langle Tx - Ty, x - y \rangle \geq 0,$$

for all x and y.

Firmly nonexpansive operators on R^J are monotone operators. Let $g(x)$: $R^J \to R$ be a differentiable convex function and $f(x) = \nabla g(x)$ its gradient. The operator ∇g is also monotone. If ∇g is nonexpansive, then, as we shall see later in Theorem 17.12, ∇g is firmly nonexpansive. If, for some $L > 0$, ∇g is L-Lipschitz for the 2-norm, that is,

$$\|\nabla g(x) - \nabla g(y)\|_2 \le L\|x - y\|_2$$

for all x and y, then $\frac{1}{L}\nabla g$ is nonexpansive, therefore firmly nonexpansive, and the operator $T = I - \gamma \nabla g$ is averaged, for $0 < \gamma < \frac{2}{L}$.

5.5.2 The Krasnoselskii-Mann Theorem

For any averaged operator T, convergence of the sequence $\{T^k x^0\}$ to a fixed point of T, whenever fixed points of T exist, is guaranteed by the Krasnoselskii-Mann (KM) Theorem [124]:

Theorem 5.16. *Let T be averaged. Then the sequence $\{T^k x^0\}$ converges to a fixed point of T, whenever Fix(T) is nonempty.*

Proof: We know that $T = (1 - \alpha)I + \alpha N$ for some nonexpansive operator N and scalar $\alpha \in (0, 1)$. Let z be a fixed point of N. The iterative step becomes

$$x^{k+1} = Tx^k = (1 - \alpha)x^k + \alpha N x^k. \tag{5.8}$$

The identity in Equation 5.1 is the key to proving Theorem 5.16.

Using $Tz = z$ and $(I - T)z = 0$ and setting $G = I - T$ we have

$$\|z - x^k\|_2^2 - \|Tz - x^{k+1}\|_2^2 = 2\mathrm{Re}(\langle Gz - Gx^k, z - x^k \rangle) - \|Gz - Gx^k\|_2^2.$$

Since, by Lemma 5.14, G is $\frac{1}{2\alpha}$-ism, we have

$$\|z - x^k\|_2^2 - \|z - x^{k+1}\|_2^2 \ge (\frac{1}{\alpha} - 1)\|x^k - x^{k+1}\|_2^2. \tag{5.9}$$

Consequently the sequence $\{x^k\}$ is bounded, the sequence $\{\|z - x^k\|_2\}$ is decreasing and the sequence $\{\|x^k - x^{k+1}\|_2\}$ converges to zero. Let x^* be a cluster point of $\{x^k\}$. Then we have $Tx^* = x^*$, so we may use x^* in place of the arbitrary fixed point z. It follows then that the sequence $\{\|x^* - x^k\|_2\}$ is decreasing; since a subsequence converges to zero, the entire sequence converges to zero. The proof is complete. $\square$

A version of the KM Theorem 5.16, with variable coefficients, appears in Reich's paper [136].

5.6   Affine Linear Operators

It may not always be easy to decide if a given operator is averaged. The class of affine linear operators provides an interesting illustration of the problem.

The affine operator $Tx = Bx + d$ will be nonexpansive, a strict contraction, firmly nonexpansive, or averaged precisely when the linear operator given by multiplication by the matrix B is the same.

5.6.1   The Hermitian Case

As we shall see later, in Theorem 7.10, when B is Hermitian, we can determine if B belongs to these classes by examining its eigenvalues λ:

- B is nonexpansive if and only if $-1 \leq \lambda \leq 1$, for all λ;

- B is a strict contraction if and only if $-1 < \lambda < 1$, for all λ;

- B is averaged if and only if $-1 < \lambda \leq 1$, for all λ;

- B is firmly nonexpansive if and only if $0 \leq \lambda \leq 1$, for all λ.

Affine linear operators T that arise, for instance, in splitting methods for solving systems of linear equations, generally have non-Hermitian linear part B. Deciding if such operators belong to these classes is more difficult. Instead, we can ask if the operator is *paracontractive*, with respect to some norm.

5.7   Paracontractive Operators

By examining the properties of the orthogonal projection operators P_C, we were led to the useful class of averaged operators. The orthogonal projections also belong to another useful class, the paracontractions.

Definition 5.17. An operator T is called paracontractive, with respect to a given norm, if, for every fixed point y of T, we have

$$||Tx - y|| < ||x - y||, \tag{5.10}$$

unless $Tx = x$.

Paracontractive operators are studied by Censor and Reich in [57].

Proposition 5.18. *The operators $T = P_C$ are paracontractive, with respect to the Euclidean norm.*

Proof: It follows from Cauchy's Inequality that

$$||P_C x - P_C y||_2 \leq ||x - y||_2,$$

with equality if and only if

$$P_C x - P_C y = \alpha(x - y),$$

for some scalar α with $|\alpha| = 1$. But, because

$$0 \leq \text{Re}(\langle P_C x - P_C y, x - y \rangle) = \alpha ||x - y||_2^2,$$

it follows that $\alpha = 1$, and so

$$P_C x - x = P_C y - y.$$

When we ask if a given operator T is paracontractive, we must specify the norm. We often construct the norm specifically for the operator involved, as we did earlier in our discussion of strict contractions, in Equation (5.2). To illustrate, we consider the case of affine operators.

5.7.1 Linear and Affine Paracontractions

Let the matrix B be diagonalizable and let the columns of V be an eigenvector basis. Then we have $V^{-1}BV = D$, where D is the diagonal matrix having the eigenvalues of B along its diagonal.

Lemma 5.19. *A square matrix B is diagonalizable if all its eigenvalues are distinct.*

Proof: Let B be J by J. Let λ_j be the eigenvalues of B, $Bx^j = \lambda_j x^j$, and $x^j \neq 0$, for $j = 1, ..., J$. Let x^m be the first eigenvector that is in the span of $\{x_j | j = 1, ..., m - 1\}$. Then

$$x^m = a_1 x^1 + ... a_{m-1} x^{m-1},$$

for some constants a_j that are not all zero. Multiply both sides by λ_m to get

$$\lambda_m x^m = a_1 \lambda_m x^1 + ... a_{m-1} \lambda_m x^{m-1}.$$

From

$$\lambda_m x^m = A x^m = a_1 \lambda_1 x^1 + ... a_{m-1} \lambda_{m-1} x^{m-1},$$

it follows that

$$a_1(\lambda_m - \lambda_1) x^1 + ... + a_{m-1}(\lambda_m - \lambda_{m-1}) x^{m-1} = 0,$$

from which we can conclude that some x^n in $\{x^1, ..., x^{m-1}\}$ is in the span of the others. This is a contradiction. $\square$

We see from this lemma that almost all square matrices B are diagonalizable. Indeed, all Hermitian B are diagonalizable. If B has real entries, but is not symmetric, then the eigenvalues of B need not be real, and the eigenvectors of B can have nonreal entries. Consequently, we must consider B as a linear operator on C^J, if we are to talk about diagonalizability. For example, consider the real matrix

$$B = \begin{bmatrix} 0 & 1 \\ -1 & 0 \end{bmatrix}.$$

Its eigenvalues are $\lambda = i$ and $\lambda = -i$. The corresponding eigenvectors are $(1, i)^T$ and $(1, -i)^T$. The matrix B is then diagonalizable as an operator on C^2, but not as an operator on R^2.

Proposition 5.20. *Let T be an affine linear operator whose linear part B is diagonalizable, and $|\lambda| < 1$ for all eigenvalues λ of B that are not equal to one. Then the operator T is paracontractive, with respect to the norm given by Equation (5.2).*

Proof: This is Exercise 5.5.

We see from Proposition 5.20 that, for the case of affine operators T whose linear part is not Hermitian, instead of asking if T is averaged, we can ask if T is paracontractive; since B will almost certainly be diagonalizable, we can answer this question by examining the eigenvalues of B.

Unlike the class of averaged operators, the class of paracontractive operators is not necessarily closed to finite products, unless those factor operators have a common fixed point.

5.7.2   The Elsner-Koltracht-Neumann Theorem

Our interest in paracontractions is due to the Elsner-Koltracht-Neumann (EKN) Theorem [85]:

Theorem 5.21. *Let T be paracontractive with respect to some vector norm. If T has fixed points, then the sequence $\{T^k x^0\}$ converges to a fixed point of T, for all starting vectors x^0.*

We follow the development in [85].

Theorem 5.22. *Suppose that there is a vector norm on $\mathcal{X}$, with respect to which each T_i is a paracontractive operator, for $i = 1, ..., I$, and that $F = \cap_{i=1}^{I} \text{Fix}(T_i)$ is not empty. For $k = 0, 1, ...$, let $i(k) = k(\text{mod } I) + 1$, and $x^{k+1} = T_{i(k)} x^k$. The sequence $\{x^k\}$ converges to a member of F, for every starting vector x^0.*

Proof: Let $y \in F$. Then, for $k = 0, 1, ...,$

$$||x^{k+1} - y|| = ||T_{i(k)}x^k - y|| \leq ||x^k - y||,$$

so that the sequence $\{||x^k - y||\}$ is decreasing; let $d \geq 0$ be its limit. Since the sequence $\{x^k\}$ is bounded, we select an arbitrary cluster point, x^*. Then $d = ||x^* - y||$, from which we can conclude that

$$||T_i x^* - y|| = ||x^* - y||,$$

and $T_i x^* = x^*$, for $i = 1, ..., I$; therefore, $x^* \in F$. Replacing y, an arbitrary member of F, with x^*, we have that $||x^k - x^*||$ is decreasing. But, a subsequence converges to zero, so the whole sequence must converge to zero. This completes the proof. $\square$

Corollary 5.23. *If T is paracontractive with respect to some vector norm, and T has fixed points, then the iterative sequence $\{T^k x^0\}$ converges to a fixed point of T, for every starting vector x^0.*

Corollary 5.24. *If $T = T_I T_{I-1} \cdots T_2 T_1$, and $F = \cap_{i=1}^I \text{Fix}(T_i)$ is not empty, then $F = \text{Fix}(T)$.*

Proof: The sequence $x^{k+1} = T_{i(k)} x^k$ converges to a member of $\text{Fix}(T)$, for every x^0. Select x^0 in F. $\square$

Corollary 5.25. *The product T of two or more paracontractive operators T_i, $i = 1, ..., I$ is again a paracontractive operator, if $F = \cap_{i=1}^I \text{Fix}(T_i)$ is not empty.*

Proof: Suppose that for $T = T_I T_{I-1} \cdots T_2 T_1$, and $y \in F = \text{Fix}(T)$, we have

$$||Tx - y|| = ||x - y||.$$

Then, since

$$||T_I(T_{I-1} \cdots T_1)x - y|| \leq ||T_{I-1} \cdots T_1 x - y||$$

$$\leq ... \leq ||T_1 x - y|| \leq ||x - y||,$$

it follows that

$$||T_i x - y|| = ||x - y||,$$

and $T_i x = x$, for each i. Therefore, $Tx = x$. $\square$

5.8   Exercises

5.1. Show that a strict contraction can have at most one fixed point.

5.2. Let T be a strict contraction. Show that the sequence $\{T^k x_0\}$ is a Cauchy sequence. *Hint*: Consider

$$||x^k - x^{k+n}|| \leq ||x^k - x^{k+1}|| + ... + ||x^{k+n-1} - x^{k+n}||,$$

and use

$$||x^{k+m} - x^{k+m+1}|| \leq r^m ||x^k - x^{k+1}||.$$

Since $\{x^k\}$ is a Cauchy sequence, it has a limit, say $\hat{x}$. Let $e^k = \hat{x} - x^k$. Show that $\{e^k\} \to 0$, as $k \to +\infty$, so that $\{x^k\} \to \hat{x}$. Finally, show that $T\hat{x} = \hat{x}$.

5.3. Suppose that we want to solve the equation

$$x = \frac{1}{2} e^{-x}.$$

Let $Tx = \frac{1}{2} e^{-x}$ for x in R. Show that T is a strict contraction, when restricted to nonnegative values of x, so that, provided we begin with $x^0 > 0$, the sequence $\{x^k = Tx^{k-1}\}$ converges to the unique solution of the equation. *Hint*: Use the mean value theorem from calculus.

5.4. Show that, if the operator T is α-averaged and $1 > \beta > \alpha$, then T is β-averaged.

5.5. Prove Proposition 5.20.

5.6. Show that, if B is a linear averaged operator, then $|\lambda| < 1$ for all eigenvalues λ of B that are not equal to one.

6 | Problems and Algorithms

In almost all the applications we shall consider, the basic problem is to find a vector x satisfying certain constraints. These constraints usually include exact or approximate consistency with measured data, as well as additional requirements, such as having nonnegative entries.

6.1   Systems of Linear Equations

In remote-sensing problems, including magnetic-resonance imaging, transmission and emission tomography, acoustic and radar array processing, and elsewhere, the data we have measured is related to the object we wish to recover by linear transformation, often involving the Fourier transform. In the vector case, in which the object of interest is discretized, the vector b of measured data is related to the vector x we seek by linear equations that we write as $Ax = b$. The matrix A need not be square, there can be infinitely many solutions, or no solutions at all. We may want to calculate a minimum-norm solution, in the under-determined case, or a least-squares solution, in the over-determined case. The vector x may be the vectorization of a two-dimensional image, in which case I, the number of rows, and J, the number of columns of A, can be in the thousands, precluding the use of noniterative solution techniques. We may have additional prior knowledge about x, such as its entries are nonnegative, which we want to impose as constraints. There is usually noise in measured data, so we may not want an exact solution of $Ax = b$, even if such solutions exist, but prefer a regularized approximate solution. What we need then are iterative algorithms to solve these problems involving linear constraints.

6.1.1   Exact Solutions

When $J \geq I$, the system $Ax = b$ typically has exact solutions. To calculate one of these, we can choose among many iterative algorithms.

The algebraic reconstruction technique (ART). The ART associates the ith equation in the system with the hyperplane

$$H_i = \{x | (Ax)_i = b_i\}.$$

With P_i the orthogonal projection onto H_i, and $i = k(\mathrm{mod}\, I) + 1$, the ART is as follows:

Algorithm 6.1 (ART). *With x^0 arbitrary and having calculated x^k, let*

$$x^{k+1} = P_i x^k.$$

The operators P_i are averaged, so the product

$$T = P_I P_{I-1} \cdots P_2 P_1$$

is also averaged and convergence of the ART follows from Theorem 5.16. The ART is also an optimization method, in the sense that it minimizes $||x - x^0||_2$ over all x with $Ax = b$.

Cimmino's Algorithm. We can also use the operators P_i in a simultaneous manner; this algorithm is *Cimmino's Algorithm* [64]:

Algorithm 6.2 (Cimmino). *With x^0 arbitrary and having calculated x^k, let*

$$x^{k+1} = \frac{1}{I} \sum_{i=1}^{I} P_i x^k.$$

Once again, convergence follows from Theorem 5.16, since the operator

$$T = \frac{1}{I} \sum_{i=1}^{I} P_i$$

is averaged. Cimmino's Algorithm also minimizes $||x - x^0||_2$ over all x with $Ax = b$, but tends to converge more slowly than ART, especially if ART is implemented using a random ordering of the equations or relaxation. One advantage that Cimmino's Algorithm has over the ART is that, in the inconsistent case, in which $Ax = b$ has no solutions, Cimmino's Algorithm converges to a least-squares solution of $Ax = b$, while the ART produces a limit cycle of multiple vectors.

Note that $Ax = b$ has solutions precisely when the square system $AA^\dagger z = b$ has a solution; for $J \geq I$, if A has full rank I (which is most of the time) the matrix $AA^\dagger$ will be invertible and the latter system will have a unique solution $z = (AA^\dagger)^{-1}b$. Then $x = A^\dagger z$ is the *minimum-norm solution* of the system $Ax = b$.

Projected ART. If we require a solution of $Ax = b$ that lies in the closed, convex set C, we can modify both the ART and Cimmino's Algorithm to achieve this end; all we need to do is to replace x^{k+1} with $P_C x^{k+1}$, the orthogonal projection of x^{k+1} onto C. These modified algorithms are the *projected ART* and *projected Cimmino's Algorithm*, respectively. Convergence is again the result of Theorem 5.16.

6.1.2 Optimization and Approximate Solutions

When $I > J$ and the system $Ax = b$ has no exact solutions, we can calculate the least-squares solution closest to x^0 using Cimmino's Algorithm. When all the rows of A are normalized to have Euclidean length one, the iterative step of Cimmino's Algorithm can be written as

$$x^{k+1} = x^k + \frac{1}{I} A^\dagger (b - Ax^k).$$

Cimmino's Algorithm is a special case of Landweber's Algorithm.

Algorithm 6.3 (Landweber). *For arbitrary x^0, and any scalar γ in the interval $(0, 2/L)$, where L is the largest eigenvalue of the matrix $A^\dagger A$, let*

$$x^{k+1} = x^k + \gamma A^\dagger (b - Ax^k).$$

The sequence $\{x^k\}$ converges to the least-squares solution closest to x^0. Landweber's Algorithm can be written as $x^{k+1} = Tx^k$, for the operator T defined by

$$Tx = (I - \gamma A^\dagger A)x + \gamma A^\dagger b.$$

This operator is affine linear and is an averaged operator, since its linear part, the matrix $B = I - \gamma A^\dagger A$, is averaged for any γ in $(0, 2/L)$. Convergence then follows from Theorem 5.16. When the rows of A have Euclidean length one, the trace of $AA^\dagger$ is I, the number of rows in A, so $L \leq I$. Therefore, the choice of $\gamma = \frac{1}{I}$ used in Cimmino's Algorithm is permissible, but usually much smaller than the optimal choice.

To minimize $\|Ax - b\|_2$ over x in the closed, convex set C we can use the *Projected Landweber's Algorithm.*

Algorithm 6.4 (Projected Landweber).

$$x^{k+1} = P_C(x^k + \gamma A^\dagger (b - Ax^k)).$$

Since P_C is an averaged operator, the operator

$$Tx = P_C(x + \gamma A^\dagger (b - Ax))$$

is averaged for all γ in $(0, 2/L)$. Convergence again follows from Theorem 5.16, whenever minimizers exist. Note that when $Ax = b$ has solutions in C, the Projected Landweber's Algorithm converges to such a solution.

6.1.3   Approximate Solutions and the Nonnegativity Constraint

For the real system $Ax = b$, consider the *nonnegatively constrained least-squares* problem of minimizing the function $||Ax - b||_2$, subject to the constraints $x_j \geq 0$ for all j; this is a nonnegatively constrained least-squares approximate solution. As noted previously, we can solve this problem using a slight modification of the ART. Although there may be multiple solutions $\hat{x}$, we know, at least, that $A\hat{x}$ is the same for all solutions.

According to the Karush-Kuhn-Tucker Theorem [134], the vector $A\hat{x}$ must satisfy the condition

$$\sum_{i=1}^{I} A_{ij}((A\hat{x})_i - b_i) = 0 \qquad (6.1)$$

for all j for which $\hat{x}_j > 0$ for some nonnegative solution $\hat{x}$. Let S be the set of all indices j for which there exists a nonnegative solution $\hat{x}$ with $\hat{x}_j > 0$. Then Equation (6.1) must hold for all j in S. Let Q be the matrix obtained from A by deleting those columns whose index j is not in S. Then $Q^T(A\hat{x} - b) = 0$. If Q has full rank and the cardinality of S is greater than or equal to I, then Q^T is one-to-one and $A\hat{x} = b$. We have proven the following result.

Theorem 6.5. *Suppose that A has the full-rank property, that is, A and every matrix Q obtained from A by deleting columns have full rank. Suppose there is no nonnegative solution of the system of equations $Ax = b$. Then there is a subset S of the set $\{j = 1, 2, ..., J\}$ with cardinality at most $I - 1$ such that, if $\hat{x}$ is any minimizer of $||Ax - b||_2$ subject to $x \geq 0$, then $\hat{x}_j = 0$ for j not in S. Therefore, $\hat{x}$ is unique.*

When $\hat{x}$ is a vectorized two-dimensional image and $J > I$, the presence of at most $I - 1$ positive pixels makes the resulting image resemble stars in the sky; for that reason this theorem and the related result ([34]) for the EMML algorithm, to be discussed in the next section, are sometimes called *night sky* theorems. The zero-valued pixels typically appear scattered throughout the image. This behavior occurs with all the algorithms discussed so far that impose nonnegativity, whenever the real system $Ax = b$ has no nonnegative solutions.

6.1.4   Splitting Methods

As we noted previously, the system $Ax = b$ has solutions if and only if the square system $AA^{\dagger}z = b$ has solutions. The *splitting methods* apply to square systems $Sz = h$. The idea is to decompose S into $S = M - K$, where M is easily inverted. Then

$$Sz = Mz - Kz = h.$$

The operator T given by

$$Tz = M^{-1}Kz + M^{-1}h$$

is affine linear and is averaged whenever the operator $M^{-1}K$ is averaged. When $M^{-1}K$ is not Hermitian, if $M^{-1}K$ is a paracontraction, with respect to some norm, we can use Theorem 5.21.

Particular choices of M and K lead to Jacobi's method, the Gauss-Seidel method (GS), and the more general Jacobi and Gauss-Seidel over-relaxation methods (JOR and SOR). For the case of S nonnegative-definite, the JOR algorithm is equivalent to Landweber's Algorithm and the SOR is closely related to the relaxed ART method. Convergence of both JOR and SOR in this case follows from Theorem 5.16.

6.2 Positive Solutions of Linear Equations

Suppose now that the entries of the matrix A are nonnegative, those of b are positive, and we seek a solution x with nonnegative entries. We can, of course, use the projected algorithms discussed in the previous section. Alternatively, we can use algorithms designed specifically for nonnegative problems and based on cross-entropy, rather than on the Euclidean distance between vectors.

6.2.1 Cross-Entropy

For $a > 0$ and $b > 0$, let the cross-entropy or Kullback-Leibler distance from a to b be

$$KL(a,b) = a\log\frac{a}{b} + b - a,$$

$KL(a,0) = +\infty$, and $KL(0,b) = b$. Extend to nonnegative vectors co-ordinate-wise, so that

$$KL(x,z) = \sum_{j=1}^{J} KL(x_j, z_j).$$

Unlike the Euclidean distance, the KL distance is not symmetric; $KL(Ax,b)$ and $KL(b,Ax)$ are distinct, and we can obtain different approximate solutions of $Ax = b$ by minimizing these two distances with respect to nonnegative x.

6.2.2 The EMML and SMART algorithms

The *expectation maximization maximum likelihood* (EMML) algorithm minimizes $KL(b,Ax)$, while the *simultaneous multiplicative* ART (SMART) minimizes $KL(Ax, b)$. These methods were developed for application to tomographic image reconstruction, although they have much more general uses. Whenever there are nonnegative solutions of $Ax = b$, SMART converges to the nonnegative solution that minimizes $KL(x, x^0)$; the EMML also converges to a nonnegative solution, but no explicit description of that solution is known.

6.2.3   Acceleration

Both the EMML and SMART algorithms are simultaneous, like Cimmino's Algorithm, and use all the equations in each step of the iteration. Like Cimmino's Algorithm, they are slow to converge. In the consistent case, the ART can usually be made to converge much faster than Cimmino's Algorithm, and analogous successive- and block-projection methods for accelerating the EMML and SMART methods have been developed, including the *multiplicative* ART (MART), the *rescaled block-iterative* SMART (RBI-SMART), and the *rescaled block-iterative* EMML (RBI-EMML). These methods can be viewed as involving projections onto hyperplanes, but the projections are entropic, not orthogonal, projections.

6.2.4   Entropic Projections onto Hyperplanes

Let H_i be the hyperplane

$$H_i = \{x | (Ax)_i = b_i\}.$$

For any nonnegative z, denote by $x = P_i^e z$ the nonnegative vector in H_i that minimizes the entropic distance $KL(x, z)$. Generally, we cannot express $P_i^e z$ in closed form. On the other hand, if we ask for the nonnegative vector $x = Q_i^e z$ in H_i for which the weighted entropic distance

$$\sum_{j=1}^{J} A_{ij} KL(x_j, z_j)$$

is minimized, we find that $x = Q_i^e z$ can be written explicitly:

$$x_j = z_j \frac{b_i}{(Az)_i}.$$

We can use these weighted entropic projection operators Q_i^e to derive the MART, the SMART, the EMML, the RBI-SMART, and the RBI-EMML methods.

6.3   Sensitivity to Noise

In many applications of these iterative methods, the vector b consists of measurements, and therefore, is noisy. Even though exact solutions of $Ax = b$ may exist, they may not be useful, because they are the result of over-fitting the answer to noisy data. It is important to know where sensitivity to noise can come from and how to modify the algorithms to lessen the sensitivity. Ill-conditioning in the matrix A can lead to sensitivity to noise and *regularization* can help to make the solution less sensitive to noise and other errors.

6.3.1   Norm Constraints

For example, in the inconsistent case, when we seek a least-squares solution of $Ax = b$, we minimize $||Ax - b||_2$. To avoid over-fitting to noisy data we can minimize

$$||Ax - b||_2^2 + \epsilon^2||x||_2^2,$$

for some small ϵ. In the consistent case, instead of calculating the exact solution that minimizes $||x - x^0||_2$, we can calculate the minimizer of

$$||Ax - b||_2^2 + \epsilon^2||x - x^0||_2^2.$$

These approaches to regularization involve the addition of a penalty term to the function being minimized. Such regularization can often be obtained through a Bayesian *maximum a posteriori probability* (MAP) approach.

Noise in the data can manifest itself in a variety of ways; we have seen what can happen when we impose positivity on the calculated least-squares solution, that is, when we minimize $||Ax - b||_2$ over all nonnegative vectors x. Theorem 6.5 tells us that when $J > I$, but $Ax = b$ has no nonnegative solutions, the nonnegatively constrained least-squares solution can have at most $I - 1$ nonzero entries, regardless of how large J is. This phenomenon also occurs with several other approximate methods, such as those that minimize the cross-entropy distance.

6.4   Convex Sets as Constraints

Constraints on x often take the form of inclusion in certain convex sets. These sets may be related to the measured data, or incorporate other aspects of x known a priori. There are several related problems that then arise.

6.4.1   The Convex Feasibility Problem

Such constraints can often be formulated as requiring that the desired x lie within the intersection C of a finite collection $\{C_1, ..., C_I\}$ of convex sets.

When the number of convex sets is large and the intersection C small, any member of C may be sufficient for our purposes. Finding such x is the *convex feasibility problem* (CFP).

6.4.2 Constrained Optimization

When the intersection C is large, simply obtaining an arbitrary member of C may not be enough; we may require, in addition, that the chosen x optimize some cost function. For example, we may seek the x in C that minimizes $||x - x^0||_2^2$. This is *constrained optimization*.

6.4.3 Proximity Function Minimization

When the collection of convex sets has empty intersection, we may minimize a *proximity function*, such as

$$f(x) = \sum_{i=1}^{I} ||P_{C_i} x - x||_2^2. \tag{6.2}$$

When the set C is nonempty, the smallest value of $f(x)$ is zero and is attained at any member of C. When C is empty, the minimizers of $f(x)$ provide a reasonable approximate solution to the CFP.

6.4.4 The Moreau Envelope and Proximity Operators

Following Combettes and Wajs [69], we say that the *Moreau envelope* of index $\gamma > 0$ of the closed, proper convex function $f(x)$ is the continuous convex function

$$g(x) = \inf\{f(y) + \frac{1}{2\gamma}||x - y||_2^2\}, \tag{6.3}$$

with the infimum taken over all y in R^N. In Rockafellar's book [139], and elsewhere, it is shown that the infimum is attained at a unique y, usually denoted $\text{prox}_{\gamma f}(x)$. The proximity operators $\text{prox}_{\gamma f}(\cdot)$ are firmly nonexpansive [69] and generalize the orthogonal projections onto closed, convex sets, as we now show.

Consider the function $f(x) = \iota_C(x)$, the *indicator function* of the closed, convex set C, taking the value zero for x in C, and $+\infty$ otherwise. Then $\text{prox}_{\gamma f}(x) = P_C(x)$, the orthogonal projection of x onto C.

6.4.5 The Split Feasibility Problem

An interesting variant of the CFP is the *split feasibility problem* (SFP) [54]. Let A be an I by J (possibly complex) matrix. The SFP is to find a member of a closed, convex set C in C^J for which Ax is a member of a second closed, convex set Q in C^I. When there is no such x, we can obtain an approximate solution by minimizing the proximity function

$$g(x) = ||P_Q Ax - Ax||_2^2,$$

over all x in C, whenever such minimizers exist.

6.5 Algorithms Based on Orthogonal Projection

The CFP can be solved using the *successive orthogonal projection* (SOP) method.

Algorithm 6.6 (SOP). *For arbitrary x^0, let*

$$x^{k+1} = P_I P_{I-1} \cdots P_2 P_1 x^k,$$

where $P_i = P_{C_i}$ is the orthogonal projection onto C_i.

For nonempty C, convergence of the SOP to a solution of the CFP will follow, once we have established that, for any x^0, the iterative sequence $\{T^k x^0\}$ converges to a fixed point of T, where

$$T = P_I P_{I-1} \cdots P_2 P_1.$$

Since T is an averaged operator, the convergence of the SOP to a member of C follows from the KM Theorem 5.16, provided C is nonempty.

The SOP is useful when the sets C_i are easily described and the P_i are easily calculated, but P_C is not. The SOP converges to the member of C closest to x^0 when the C_i are hyperplanes, but not in general.

When $C = \cap_{i=1}^I C_i$ is empty and we seek to minimize the proximity function $f(x)$ in Equation (6.2), we can use the simultaneous orthogonal projection (SIMOP) method:

Algorithm 6.7 (SIMOP). *For arbitrary x^0, let*

$$x^{k+1} = \frac{1}{I} \sum_{i=1}^I P_i x^k.$$

The operator

$$T = \frac{1}{I} \sum_{i=1}^I P_i$$

is also averaged, so this iteration converges, by Theorem 5.16, whenever $f(x)$ has a minimizer.

The CQ algorithm is an iterative method for solving the SFP [45, 46].

Algorithm 6.8 (CQ). *For arbitrary x^0, let*

$$x^{k+1} = P_C(x^k - \gamma A^\dagger(I - P_Q)Ax^k).\tag{6.4}$$

The operator

$$T = P_C(I - \gamma A^\dagger(I - P_Q)A)$$

is averaged whenever γ is in the interval $(0, 2/L)$, where L is the largest eigenvalue of $A^\dagger A$, and so the CQ algorithm converges to a fixed point of T, whenever such fixed points exist. When the SFP has a solution, the CQ algorithm converges to a solution; when it does not, the CQ algorithm converges to a minimizer, over C, of the proximity function $g(x) = \|P_Q Ax - Ax\|_2$, whenever such minimizers exist. The function $g(x)$ is convex and, according to [3], its gradient is

$$\nabla g(x) = A^\dagger(I - P_Q)Ax.$$

The convergence of the CQ algorithm then follows from Theorem 5.16. In [69] Combettes and Wajs use proximity operators to generalize the CQ algorithm.

6.5.1 Projecting onto the Intersection of Convex Sets

When the intersection $C = \cap_{i=1}^I C_i$ is large, and just finding any member of C is not sufficient for our purposes, we may want to calculate the orthogonal projection of x^0 onto C using the operators P_{C_i}. We cannot use the SOP unless the C_i are hyperplanes; instead we can use Dykstra's Algorithm or the Halpern-Lions-Wittmann-Bauschke (HLWB) Algorithm. Dykstra's Algorithm employs the projections P_{C_i}, but not directly on x^k, but on translations of x^k. It is motivated by the following lemma:

Lemma 6.9. *If $x = c + \sum_{i=1}^I p_i$, where, for each i, $c = P_{C_i}(c + p_i)$, then $c = P_C x$.*

Bregman discovered an iterative algorithm for minimizing a more general convex function $f(x)$ over x with $Ax = b$ and also x with $Ax \geq b$ [17]. These algorithms are based on his extension of the SOP to include projections with respect to generalized distances, such as entropic distances.

6.6 Steepest Descent Minimization

Suppose that we want to minimize a real-valued function $g : R^J \to R$. At each x the direction of greatest decrease of g is the negative of the gradient,

$-\nabla g(x)$. The steepest descent method has the iterative step

$$x^{k+1} = x^k - \alpha_k \nabla g(x^k), \tag{6.5}$$

where, ideally, the *step-length parameter* α_k would be chosen so as to minimize $g(x)$ in the chosen direction, that is, the choice of $\alpha = \alpha_k$ would minimize

$$g(x^k - \alpha \nabla g(x^k)).$$

In practice, it is difficult, if not impossible, to determine the optimal value of α_k at each step. Therefore, a line search is usually performed to find a suitable α_k, meaning that values of $g(x^k - \alpha \nabla g(x^k))$ are calculated, for some finite number of α values, to determine a suitable choice for α_k.

6.6.1 Fixed Step-Length Methods

For practical reasons, we are often interested in iterative algorithms that avoid line searches. Some of the minimization algorithms we shall study take the form

$$x^{k+1} = x^k - \alpha \nabla g(x^k),$$

where the α is a constant, selected at the beginning of the iteration. Such iterative algorithms have the form $x^{k+1} = Tx^k$, for T the operator defined by

$$Tx = x - \alpha \nabla g(x).$$

When properly chosen, the α will not be the optimal step-length parameter for every step of the iteration, but will be sufficient to guarantee convergence. In addition, the resulting iterative sequence is often *monotonically decreasing*, which means that

$$g(x^{k+1}) < g(x^k),$$

for each k. As we have seen, if g is convex and its gradient is L-Lipschitz, then α can be chosen so that the operator T is averaged.

6.6.2 Employing Positivity

Suppose that we want to minimize the function $g : R^J \to R$, but only over nonnegative vectors z. While $z_j > 0$, let $x_j = \log z_j$ and consider $g(z)$ as a function $f(x)$ of the real vector x. Then

$$\frac{\partial f}{\partial x_j}(x) = \frac{\partial g}{\partial z_j}(z) z_j,$$

and the steepest descent iteration for f, given by Equation (6.5), becomes

$$z_j^{k+1} = z_j^k \exp\left(-\alpha_k z_j^k \frac{\partial g}{\partial z_j}(z^k)\right),$$

which we can write as

$$z_j^{k+1} = z_j^k \exp\left(-\alpha_{k,j} \nabla g(z^k)_j\right), \tag{6.6}$$

using

$$\alpha_{k,j} = \alpha_k z_j^k.$$

We shall discuss other iterative monotone methods, such as the EMML and SMART algorithms, that can be viewed as generalized steepest descent methods, either having the form of Equation (6.6), or one closely related to that form. In these cases, the step-length parameter α_k is replaced by ones that also vary with the entry index j. While this may seem even more complicated to implement, for the algorithms mentioned, these $\alpha_{k,j}$ are automatically calculated as part of the algorithm, with no line searches involved.

6.6.3 Constrained Optimization

If our goal is to minimize $g(x)$ over only those x that are in the closed, convex set C, then we may consider a *projected gradient descent* method.

Algorithm 6.10 (Projected Steepest Descent). *For arbitrary x^0, let*

$$x^{k+1} = P_C(x^k - \gamma \nabla g(x^k)).$$

When the operator $Tx = x - \gamma \nabla g(x)$ is averaged, so is $P_C T$, so the KM Theorem 5.16 will apply once again.

6.7 Bregman Projections and the SGP

If $f : R^J \to R$ is convex and differentiable, then, for all x and y, we have

$$D_f(x, y) = f(x) - f(y) - \langle \nabla f(y), x - y \rangle \geq 0.$$

If $\hat{x}$ minimizes $f(x)$ over x with $Ax = b$, then

$$\nabla f(\hat{x}) + A^\dagger c = 0,$$

for some vector c. Bregman's idea is to use $D_f(x, y)$ to define generalized projections, and then to mimic the SOP to solve for $\hat{x}$. Simply requiring that $f(x)$ be convex and differentiable is not sufficient for a complete theory and additional requirements are necessary; see the appendix on Bregman-Legendre functions and Bregman projections.

Definition 6.11. For each i, let $P_i^f z$ be the point in the hyperplane

$$H_i = \{x | (Ax)_i = b_i\}$$

that minimizes $D_f(x, z)$. Then $P_i^f z$ is the Bregman projection of z onto H_i.

Then

$$\nabla f(P_i^f z) - \nabla f(z) = \lambda_i a^i,$$

for some λ_i, where a_i is the ith column of $A^\dagger$.

Bregman's *successive generalized projection* (SGP) method is the following:

Algorithm 6.12 (SGP). *For x^0 in the interior of the domain of f, let*

$$x^{k+1} = \nabla f^{-1}(\nabla f(x^k) + \lambda_k a^i),$$

for some scalar λ_k and $\imath = k(\bmod I) + 1$.

The sequence $\{x^k\}$ will converge to x with $Ax = b$, provided solutions exist, and when x^0 is chosen so that $x^0 = A^\dagger d$, for some d, the sequence will converge to the solution that minimizes $f(x)$. Bregman also uses Bregman distances to obtain a primal-dual algorithm for minimizing $f(x)$ over all x with $Ax \geq b$. Dykstra's Algorithm can be extended to include Bregman projections; this extended algorithm is then equivalent to the generalization of Bregman's primal-dual algorithm to minimize $f(x)$ over the intersection of closed, convex sets.

6.7.1   Bregman's Approach to Linear Programming

Bregman's primal-dual algorithm suggests a method for approximating the solution of the basic problem in linear programming, to minimize a linear function $c^T x$, over all x with $Ax \geq b$. Other solution methods exist for this problem, as well. Associated with the basic *primary* problem is a *dual* problem. Both the primary and dual problems can be stated in their *canonical forms* or their *standard forms*. The primary and dual problems are connected by the Weak Duality and Strong Duality theorems. The simplex method is the best-known solution procedure.

6.7.2   The Multiple-Distance SGP (MSGP)

As we noted earlier, both the EMML and SMART algorithms can be viewed in terms of weighted entropic projections onto hyperplanes. Unlike the

SGP, the weighted entropic distances used vary with the hyperplane, suggesting that it may be possible to extend the SGP algorithm to include
Bregman projections in which the function f is replaced by f_i that depends on the set C_i. It is known, however, that merely replacing the single
Bregman function f with f_i that varies with the i is not enough to guarantee convergence. The *multiple-distance* SGP (MSGP) algorithm to be
discussed later achieves convergence by using a dominating Bregman distance $D_h(x, y)$ with

$$D_h(x, y) \geq D_{f_i}(x, y),$$

for each i, and a generalized notion of relaxation. The MSGP leads to
an interior-point method, the IPA, for minimizing certain convex functions
over convex sets.

6.8   Applications

Iterative algorithms are necessary in many areas of applications. The editorial [118] provides a brief introduction to the many uses of iterative
methods in medical imaging. Transmission and emission tomography involve the solving of large-scale systems of linear equations, or optimizing
convex functions of thousands of variables. Magnetic-resonance imaging
produces data that is related to the object of interest by means of the
Fourier transform or the Radon transform. Hyperspectral imaging leads to
several problems involving limited Fourier-transform data. Iterative data-
extrapolation algorithms can be used to incorporate prior knowledge about
the object being reconstructed, as well as to improve resolution. Entropy-
based iterative methods are used to solve the mixture problems common
to remote-sensing, as illustrated by sonar and radar array processing, as
well as hyperspectral imaging.

III | Operators

7 | Averaged and Paracontractive Operators

Many well-known algorithms in optimization, signal processing, and image reconstruction are iterative in nature. The Jacobi, Gauss-Seidel, and successive over-relaxation (SOR) procedures for solving large systems of linear equations, *projection onto convex sets* (POCS) methods and iterative optimization procedures, such as entropy and likelihood maximization, are the primary examples. It is a pleasant fact that convergence of many of these algorithms is a consequence of the Krasnoselskii-Mann (KM) Theorem 5.16 for averaged operators or the Elsner-Koltracht-Neumann (EKN) Theorem 5.21 for paracontractions. In this chapter we take a closer look at averaged nonexpansive operators and paracontractive nonexpansive operators. Later, we examine the more general class of operators that are paracontractions, with respect to Bregman distances.

7.1 Solving Linear Systems of Equations

An important class of operators are the *affine linear* ones, having the form

$$Tx = Bx + h,$$

where B is linear, so that Bx is the multiplication of the vector x by the matrix B, and h is a fixed vector. Affine linear operators occur in iterative methods for solving linear systems of equations.

7.1.1 Landweber's Algorithm

The iterative step in Landweber's Algorithm for solving the system $Ax = b$ is

$$x^{k+1} = x^k + \gamma A^\dagger (b - Ax^k),$$

where γ is a selected parameter. We can write the Landweber iteration as

$$x^{k+1} = Tx^k,$$

for

$$Tx = (I - \gamma A^\dagger A)x + A^\dagger b = Bx + h.$$

Landweber's Algorithm actually solves the square linear system $A^\dagger A = A^\dagger b$ for a least-squares solution of $Ax = b$. When there is a unique solution or unique least-squares solution of $Ax = b$, say $\hat{x}$, then the error at the kth step is $e^k = \hat{x} - x^k$ and we see that

$$Be^k = e^{k+1}.$$

We want $e^k \to 0$, and so we want $||B||_2 < 1$; this means that both T and B are Euclidean strict contractions. Since B is Hermitian, B will be a strict contraction if and only $||B||_2 < 1$, where $||B||_2 = \rho(B)$ is the matrix norm induced by the Euclidean vector norm.

On the other hand, when there are multiple solutions of $Ax = b$, the solution found by Landweber's Algorithm will be the one closest to the starting vector. In this case, we cannot define e^k and we do not want $||B||_2 < 1$; that is, we do not need that B be a strict contraction, but something weaker. As we shall see, since B is Hermitian, B will be averaged whenever γ lies in the interval $(0, 2/\rho(B))$.

7.1.2   Splitting Algorithms

Affine linear operators also occur in splitting algorithms for solving a square system of linear equations, $Sx = b$. We write $S = M - K$, with M invertible.

Algorithm 7.1 (Splitting). *For x^0 arbitrary, let*

$$x^{k+1} = M^{-1}Kx^k + M^{-1}b.$$

This iterative step can be written as

$$x^{k+1} = Tx^k,$$

for the affine linear operator

$$Tx = M^{-1}Kx + M^{-1}b = Bx + h.$$

When S is invertible, there is a unique solution of $Sx = b$, say $\hat{x}$, and we can define the error $e^k = \hat{x} - x^k$. Then $e^{k+1} = Be^k$, and again we want

$||B||_2 < 1$, that is, B is a strict contraction. However, if S is not invertible and there are multiple solutions, then we do not want B to be a strict contraction. Since B is usually not Hermitian, deciding if B is averaged may be difficult. Therefore, we may instead ask if there is a vector norm with respect to which B is paracontractive.

We begin, in the next section, a detailed discussion of averaged operators.

7.2 Averaged Operators

As we have seen, the fact that a nonexpansive operator N has fixed points is not sufficient to guarantee convergence of the orbit sequence $\{N^k x^0\}$; additional conditions are needed. Requiring the operator to be a strict contraction is quite restrictive; most of the operators we are interested in here have multiple fixed points, so are not strict contractions, in any norm. For example, if $T = P_C$, then $C = \text{Fix}(T)$. Motivated by the KM Theorem 5.16, we concentrate on averaged operators, by which we shall always mean with respect to the Euclidean norm.

7.2.1 General Properties of Averaged Operators

We present now the fundamental properties of averaged operators, in preparation for the proof that the class of averaged operators is closed to finite products.

Note that we can establish that a given operator is averaged by showing that there is an α in the interval $(0, 1)$ such that the operator

$$\frac{1}{\alpha}(A - (1 - \alpha)I)$$

is nonexpansive. Using this approach, we can easily show that if T is a strict contraction, then T is averaged.

Lemma 7.2. *Let $T = (1 - \alpha)A + \alpha N$ for some $\alpha \in (0,1)$. If A is averaged and N is nonexpansive then T is averaged.*

Proof: Let $A = (1 - \beta)I + \beta M$ for some $\beta \in (0,1)$ and nonexpansive operator M. Let $1 - \gamma = (1 - \alpha)(1 - \beta)$. Then we have

$$T = (1 - \gamma)I + \gamma[(1 - \alpha)\beta\gamma^{-1}M + \alpha\gamma^{-1}N].$$

Since the operator $K = (1 - \alpha)\beta\gamma^{-1}M + \alpha\gamma^{-1}N$ is easily shown to be nonexpansive and the convex combination of two nonexpansive operators is again nonexpansive, T is averaged. $\qquad\square$

Corollary 7.3. *If A and B are averaged and α is in the interval $[0, 1]$, then the operator $T = (1 - \alpha)A + \alpha B$ formed by taking the convex combination of A and B is averaged.*

Corollary 7.4. *Let $T = (1 - \alpha)F + \alpha N$ for some $\alpha \in (0, 1)$. If F is firmly nonexpansive and N is Euclidean nonexpansive, then T is averaged.*

The orthogonal projection operators P_H onto hyperplanes $H = H(a, \gamma)$ are sometimes used with *relaxation*, which means that P_H is replaced by the operator

$$T = (1 - \omega)I + \omega P_H,$$

for some ω in the interval $(0, 2)$. Clearly, if ω is in the interval $(0, 1)$, then T is averaged, by definition, since P_H is nonexpansive. We want to show that, even for ω in the interval $[1, 2)$, T is averaged. To do this, we consider the operator $R_H = 2P_H - I$, which is reflection through H; that is,

$$P_H x = \frac{1}{2}(x + R_H x),$$

for each x.

Lemma 7.5. *The operator $R_H = 2P_H - I$ is an isometry; that is,*

$$\|R_H x - R_H y\|_2 = \|x - y\|_2,$$

for all x and y, so that R_H is nonexpansive.

Lemma 7.6. *For $\omega = 1 + \gamma$ in the interval $[1, 2)$, we have*

$$(1 - \omega)I + \omega P_H = \alpha I + (1 - \alpha)R_H,$$

for $\alpha = \frac{1-\gamma}{2}$; therefore, $T = (1 - \omega)I + \omega P_H$ is averaged.

The product of finitely many nonexpansive operators is again nonexpansive, while the product of finitely many firmly nonexpansive operators, even orthogonal projections, need not be firmly nonexpansive. It is a helpful fact that the product of finitely many averaged operators is again averaged.

If $A = (1-\alpha)I + \alpha N$ is averaged and B is averaged, then $T = AB$ has the form $T = (1 - \alpha)B + \alpha NB$. Since B is averaged and NB is nonexpansive, it follows from Lemma 7.2 that T is averaged. Summarizing, we have

Proposition 7.7. *If A and B are averaged, then $T = AB$ is averaged.*

It is possible for $\text{Fix}(AB)$ to be nonempty while $\text{Fix}(A) \cap \text{Fix}(B)$ is empty; however, if the latter is nonempty, it must coincide with $\text{Fix}(AB)$ [9, 21]:

Proposition 7.8. *Let A and B be averaged operators and suppose that $Fix(A) \cap Fix(B)$ is nonempty. Then $Fix(A) \cap Fix(B) = Fix(AB) = Fix(BA)$.*

Proof: Let $I - A$ be ν_A-ism and $I - B$ be ν_B-ism, where both ν_A and ν_B are taken greater than $\frac{1}{2}$. Let z be in $Fix(A) \cap Fix(B)$ and x in $Fix(BA)$. Then

$$||z - x||_2^2 \geq ||z - Ax||_2^2 + (2\nu_A - 1)||Ax - x||_2^2$$

$$\geq ||z - BAx||_2^2 + (2\nu_B - 1)||BAx - Ax||_2^2 + (2\nu_A - 1)||Ax - x||_2^2$$

$$= ||z - x||_2^2 + (2\nu_B - 1)||BAx - Ax||_2^2 + (2\nu_A - 1)||Ax - x||_2^2.$$

Therefore $||Ax - x||_2 = 0$ and $||BAx - Ax||_2 = ||Bx - x||_2 = 0$. $\square$

7.2.2 Averaged Linear Operators

Affine linear operators have the form $Tx = Bx + d$, where B is a matrix. The operator T is averaged if and only if B is averaged. It is useful, then, to consider conditions under which B is averaged.

When B is averaged, there is a positive α in $(0, 1)$ and a Euclidean nonexpansive operator N, with

$$B = (1 - \alpha)I + \alpha N.$$

Therefore

$$N = \frac{1}{\alpha}B + (1 - \frac{1}{\alpha})I \tag{7.1}$$

is nonexpansive. Clearly, N is a linear operator; that is, N is multiplication by a matrix, which we also denote N. When is such a linear operator N nonexpansive?

Lemma 7.9. *A linear operator N is nonexpansive, in the Euclidean norm, if and only if $||N||_2 = \sqrt{\rho(N^\dagger N)}$, the matrix norm induced by the Euclidean vector norm, does not exceed one.*

We know that B is averaged if and only if its complement, $I - B$, is ν-ism for some $\nu > \frac{1}{2}$. Therefore,

$$Re(\langle (I - B)x, x \rangle) \geq \nu ||(I - B)x||_2^2,$$

for all x. This implies that $x^\dagger(I - B)x \geq 0$, for all x. Since this quadratic form can be written as

$$x^\dagger(I - B)x = x^\dagger(I - Q)x,$$

for $Q = \frac{1}{2}(B + B^\dagger)$, it follows that $I - Q$ must be nonnegative definite. Moreover, if B is averaged, then B is nonexpansive, so that $||B||_2 \leq 1$. Since $||B||_2 = ||B^\dagger||_2$, and $||Q||_2 \leq \frac{1}{2}(||B||_2 + ||B^\dagger||_2)$, it follows that Q must be Euclidean nonexpansive. In fact, since N is Euclidean nonexpansive if and only if $N^\dagger$ is, B is averaged if and only if $B^\dagger$ is averaged. Consequently, if the linear operator B is averaged, then so is the Hermitian operator Q, and so the eigenvalues of Q must lie in the interval $(-1, 1]$. We also know from Exercise 5.6 that, if B is averaged, then $|\lambda| < 1$, unless $\lambda = 1$, for every eigenvalue λ of B.

7.2.3   Hermitian Linear Operators

We are particularly interested in linear operators B that are Hermitian, in which case N will also be Hermitian. Therefore, we shall assume, throughout this subsection, that B is Hermitian, so that all of its eigenvalues are real. It follows from our discussion relating matrix norms to spectral radii that a Hermitian N is nonexpansive if and only if $\rho(N) \leq 1$. We now derive conditions on the eigenvalues of B that are equivalent to B being an averaged linear operator.

For any (necessarily real) eigenvalue λ of B, the corresponding eigenvalue of N is

$$\nu = \frac{1}{\alpha}\lambda + (1 - \frac{1}{\alpha}).$$

It follows that $|\nu| \leq 1$ if and only if

$$1 - 2\alpha \leq \lambda \leq 1.$$

Therefore, the Hermitian linear operator B is averaged if and only if there is α in $(0, 1)$ such that

$$-1 < 1 - 2\alpha \leq \lambda \leq 1,$$

for all eigenvalues λ of B. This is equivalent to saying that

$$-1 < \lambda \leq 1,$$

for all eigenvalues λ of B. The choice

$$\alpha_0 = \frac{1 - \lambda_{\min}}{2}$$

is the smallest α for which

$$N = \frac{1}{\alpha}B + (1 - \frac{1}{\alpha})I$$

will be nonexpansive; here $\lambda_{\min}$ denotes the smallest eigenvalue of B. So, α_0 is the smallest α for which B is α-averaged.

The linear operator B will be firmly nonexpansive if and only if it is $\frac{1}{2}$-averaged. Therefore, B will be firmly nonexpansive if and only if $0 \leq \lambda \leq 1$, for all eigenvalues λ of B. Since B is Hermitian, we can say that B is firmly nonexpansive if and only if B and $I - B$ are nonnegative definite. We summarize the situation for Hermitian B as follows.

Theorem 7.10. *Let B be Hermitian. Then B is nonexpansive if and only if $-1 \leq \lambda \leq 1$, for all eigenvalues λ; B is averaged if and only if $-1 < \lambda \leq 1$, for all eigenvalues λ; B is a strict contraction if and only if $-1 < \lambda < 1$, for all eigenvalues λ; and B is firmly nonexpansive if and only if $0 \leq \lambda \leq 1$, for all eigenvalues λ.*

7.3 Paracontractive Operators

An affine linear operator $Tx = Bx + d$ is an averaged nonexpansive operator if and only if its linear part, B, is also averaged. A Hermitian B is averaged if and only if $-1 < \lambda \leq 1$, for each eigenvalue λ of B. When B is not Hermitian, deciding if B is averaged is harder. In such cases, we can ask if there is some vector norm, with respect to which B is paracontractive. As we shall see, if B is diagonalizable, then B is paracontractive if $|\lambda| < 1$, for every eigenvalue λ of B that is not equal to one. Then we can use Theorem 5.21 to establish convergence of the iterative sequence $\{T^k x^0\}$.

7.3.1   Paracontractions and Convex Feasibility

Recall that an operator T on $\mathcal{X}$ is paracontractive, with respect to some vector norm $\| \cdot \|$, if, for every fixed point y of T and for every x, we have

$$\|Tx - y\| < \|x - y\|,$$

unless $Tx = x$. Note that T can be paracontractive without being continuous, hence without being nonexpansive. We shall restrict our attention here to those paracontractive operators that are continuous.

Let C_i, $i = 1, ..., I$, be nonempty, closed convex sets in $\mathcal{X}$, with nonempty intersection C. The orthogonal projection $P_i = P_{C_i}$ onto C_i is paracontractive, with respect to the Euclidean norm, for each i. The product $T = P_I P_{I-1} \cdots P_1$ is also paracontractive, since C is nonempty. The SOP algorithm converges to a member of C, for any starting vector x^0, as a consequence of Theorem 5.21. For the SOP to be a practical procedure, we need to be able to calculate easily the orthogonal projection onto each C_i.

The *cyclic subgradient projection* method (CSP) (see [61]) provides a practical alternative to the SOP, for sets C_i of the form

$$C_i = \{x | g_i(x) \le b_i\},$$

where g_i is a convex function on $\mathcal{X}$. In the case in which g is differentiable, for each i, let

$$T_i x = x - \omega \alpha_i(x) \nabla g_i(x),$$

for

$$\alpha_i(x) = \max(g_i(x) - b_i, 0) / ||\nabla g_i(x)||^2.$$

From [85] we have

Theorem 7.11. *For $0 < \omega < 2$, the operators T_i are paracontractive, with respect to the Euclidean norm.*

Proof: A vector y is a fixed point of T_i if and only if $g_i(y) \le 0$, so if and only if $y \in C_i$. Let x be a vector outside of C_i, and let $\alpha = \alpha_i(x)$. Since g_i has no relative minimum outside of C_i, $T_i x$ is well defined. We want to show that $||T_i x - y|| < ||x - y||$. This is equivalent to showing that

$$\omega^2 \alpha^2 ||\nabla g_i(x)||^2 \le 2\omega \alpha \langle \nabla g_i(x), x - y \rangle,$$

which, in turn, is equivalent to showing that

$$\omega(g_i(x) - b_i) \le \langle \nabla g_i(x), x - y \rangle. \tag{7.2}$$

Since $g_i(y) \le b_i$ and g_i is convex, we have

$$(g_i(x) - \beta) \le (g_i(x) - g_i(y)) \le \langle \nabla g_i(x), x - y \rangle.$$

Inequality (7.2) follows immediately. $\square$

The CSP algorithm has the iterative step

$$x^{k+1} = T_{i(k)} x^k,$$

where $i(k) = k(\mod I) + 1$. Since each of the operators T_i is paracontractive, the sequence converges to a member of C, whenever C is nonempty, as a consequence of Theorem 5.21.

Let A be an I by J real matrix, and for each i let $g_i(x) = (Ax)_i$. Then the gradient of g_i is $\nabla g_i(x) = a^i$, the ith column of A^T. The set C_i is the half-space $C = \{x | (Ax)_i \le b_i\}$, and the operator T_i is the orthogonal projection onto C_i. The CSP algorithm in this case becomes the Agmon-Motzkin-Schoenberg (AMS) Algorithm for finding x with $Ax \le b$.

7.4   Linear and Affine Paracontractions

Recall that the linear operator B is diagonalizable if $\mathcal{X}$ has a basis of eigenvectors of B. In that case let the columns of V be such an eigenvector basis. Then we have $V^{-1}BV = D$, where D is the diagonal matrix having the eigenvalues of B along its diagonal.

7.4.1   Back-Propagation-of-Error Methods

Suppose that A is I by J, with $J > I$ and that $Ax = b$ has infinitely many solutions. A *back-propagation-of-error* approach leads to an algorithm with the iterative step

$$x^{k+1} = x^k + \gamma C^\dagger (b - Ax^k),$$

where C is some I by J matrix. The algorithm can then be written in the form $x^{k+1} = T^k x^0$, for T the affine operator given by

$$Tx = (I - \gamma C^\dagger A)x + \gamma C^\dagger b.$$

Since $Ax = b$ has multiple solutions, A has a nontrivial null space, so that some of the eigenvalues of $B = (I - \gamma C^\dagger A)$ are equal to one. As we shall see, if γ is chosen so that $|\lambda| < 1$, for all the remaining eigenvalues of B, and B is diagonalizable, then T will be paracontractive, with respect to some vector norm, and the iterative sequence $\{x^k\}$ will converge to a solution. For such a γ to exist, it is necessary that, for all nonzero eigenvalues $\mu = a + bi$ of the matrix $C^\dagger A$, the real parts a be nonzero and have the same sign, which we may, without loss of generality, assume to be positive. Then we need to select γ in the intersection of the intervals $(0, 2a/(a^2+b^2))$, taken over every eigenvalue μ. When $C = A$, all the nonzero eigenvalues of $C^\dagger A = A^\dagger A$ are positive, so such a γ exists. As C deviates from A, the eigenvalues of $C^\dagger A$ begin to change. We are asking that the C not deviate from A enough to cause the real part of an eigenvalue to become negative.

7.4.2   Defining the Norm

Suppose that $Tx = Bx + d$ is an affine linear operator whose linear part B is diagonalizable, and $|\lambda| < 1$ for all eigenvalues λ of B that are not equal to one. Let $\{u^1, ..., u^J\}$ be linearly independent eigenvectors of B. For each x, we have

$$x = \sum_{j=1}^{J} a_j u^j,$$

for some coefficients a_j. Define

$$||x|| = \sum_{j=1}^{J} |a_j|.$$

We know from Proposition 5.20 that T is paracontractive with respect to this norm. It follows from Theorem 5.21 that the iterative sequence $\{T^k x^0\}$ will converge to a fixed point of T, whenever T has fixed points.

7.4.3   Proof of Convergence

It is not difficult to prove convergence directly, as we now show.

Proof (of Convergence): Let the eigenvalues of B be λ_j, for $j = 1, ..., J$, with associated linearly independent eigenvectors u^j. Define a norm on vectors x by

$$||x|| = \sum_{j=1}^{J} |a_j|,$$

for

$$x = \sum_{j=1}^{J} a_j u^j.$$

Assume that $\lambda_j = 1$, for $j = K+1, ..., J$, and that $|\lambda_j| < 1$, for $j = 1, ..., K$. Let

$$d = \sum_{j=1}^{J} d_j u^j.$$

Let $\hat{x}$ be an arbitrary fixed point of T, with

$$\hat{x} = \sum_{j=1}^{J} \hat{a}_j u^j.$$

From $T\hat{x} = \hat{x}$ we have

$$\sum_{j=1}^{J} \hat{a}_j u^j = \sum_{j=1}^{J} (\lambda_j \hat{a}_j + d_j) u^j.$$

Then with

$$x^k = \sum_{j=1}^{J} a_{jk} u^j,$$

and

$$x^{k+1} = Bx^k + h = \sum_{j=1}^{J} (\lambda_j a_{jk} + d_j) u^j,$$

we have

$$x^k - \hat{x} = \sum_{j=1}^{J} (a_{jk} - \hat{a}_j) u^j,$$

and

$$x^{k+1} - \hat{x} = \sum_{j=1}^{K} \lambda_j (a_{jk} - \hat{a}_j) u^j + \sum_{j=K+1}^{J} (a_{jk} - \hat{a}_j) u^j.$$

Therefore,

$$||x^k - \hat{x}|| = \sum_{j=1}^{K} |a_{jk} - \hat{a}| + \sum_{j=K+1}^{J} |a_{jk} - \hat{a}_j|,$$

while

$$||x^{k+1} - \hat{x}|| = \sum_{j=1}^{K} |\lambda_j| |a_{jk} - \hat{a}| + \sum_{j=K+1}^{J} |a_{jk} - \hat{a}_j|.$$

Consequently,

$$||x^k - \hat{x}|| - ||x^{k+1} - \hat{x}|| = \sum_{j=1}^{K} (1 - |\lambda_j|) |a_{jk} - \hat{a}_j|.$$

It follows that the sequence $\{||x^k - \hat{x}||\}$ is decreasing, and that the sequences $\{|a_{jk} - \hat{a}_j|\}$ converge to zero, for each $j = 1, ..., K$.

Since the sequence $\{x^k\}$ is then bounded, select a cluster point, x^*, with

$$x^* = \sum_{j=1}^{J} a_j^* u^j.$$

Then we must have

$$\{|a_{jk} - a_j^*|\} \to 0,$$

for $j = 1, ..., K$. It follows that $\hat{a}_j = a_j^*$, for $j = 1, ..., K$. Therefore,

$$\hat{x} - x^* = \sum_{j=K+1}^{J} c_j u^j,$$

for $c_j = \hat{a}_j - a_j^*$. We can conclude, therefore, that

$$\hat{x} - B\hat{x} = x^* - Bx^*,$$

so that x^* is another solution of the system $(I - B)x = d$. Therefore, the sequence $\{||x^k - x^*||\}$ is decreasing; but a subsequence converges to zero, so the entire sequence must converge to zero. We conclude that $\{x^k\}$ converges to the solution x^*. $\square$

It is worth noting that the condition that B be diagonalizable cannot be omitted. Consider the nondiagonalizable matrix

$$B = \begin{bmatrix} 1 & 1 \\ 0 & 1 \end{bmatrix},$$

and the affine operator

$$Tx = Bx + (1,0)^T.$$

The fixed points of T are the solutions of $(I - B)x = (1,0)^T$, which are the vectors of the form $x = (a, -1)^T$. With starting vector $x^0 = (1,0)^T$, we find that $x^k = (k-1)x^0$, so that the sequence $\{x^k\}$ does not converge to a fixed point of T. There is no vector norm with respect to which T is paracontractive.

If T is an affine linear operator with diagonalizable linear part, then T is paracontractive whenever T is averaged, as we know from Exercise 5.6. We see from that exercise that, for the case of affine operators T whose linear part is not Hermitian, instead of asking if T is averaged, we can ask if T is paracontractive; since B will almost certainly be diagonalizable, we can answer this question by examining the eigenvalues of B.

7.5   Other Classes of Operators

As we have seen, the class of nonexpansive operators is too broad, and the class of strict contractions too narrow, for our purposes. The KM Theorem 5.16 encourages us to focus on the intermediate class of averaged operators, and the EKN Theorem 5.21 makes the paracontractions also worth consideration. While this is certainly a fruitful approach, it is not the only possible one. In [77] De Pierro and Iusem take a somewhat different approach, basing their class of operators on properties of orthogonal projections onto convex sets. We can use the Cauchy-Schwarz Inequality and the fact that $T = P_C$ is firmly nonexpansive to show that

$$||Tx - Ty||_2 = ||x - y||_2 \tag{7.3}$$

implies that

$$Tx - Ty = x - y, \tag{7.4}$$

and

$$\langle Tx - x, x - y \rangle = 0. \tag{7.5}$$

De Pierro and Iusem consider operators $Q : R^J \to R^J$ that are nonexpansive and for which the property in Equation (7.3) implies both Equations (7.4) and (7.5). They then show that this class is closed to finite products and convex combinations.

IV | Algorithms

8 | The Algebraic Reconstruction Technique

We begin our detailed discussion of algorithms with a simple problem, solving a system of linear equations, and a simple method, the *algebraic reconstruction technique* (ART).

The ART was introduced by Gordon, Bender, and Herman [94] as a method for image reconstruction in transmission tomography. It was noticed somewhat later that the ART is a special case of Kaczmarz's Algorithm [110]. For $i = 1, ..., I$, let L_i be the set of pixel indices j for which the jth pixel intersects the ith line segment, and let $|L_i|$ be the cardinality of the set L_i. Let $A_{ij} = 1$ for j in L_i, and $A_{ij} = 0$ otherwise. With $i = k(\mathrm{mod}\, I) + 1$, the iterative step of the ART algorithm is

$$x_j^{k+1} = x_j^k + \frac{1}{|L_i|}(b_i - (Ax^k)_i),$$

for j in L_i, and

$$x_j^{k+1} = x_j^k,$$

if j is not in L_i. In each step of ART, we take the error, $b_i - (Ax^k)_i$, associated with the current x^k and the ith equation, and distribute it equally over each of the pixels that intersects L_i.

A somewhat more sophisticated version of ART allows A_{ij} to include the length of the ith line segment that lies within the jth pixel; A_{ij} is taken to be the ratio of this length to the length of the diagonal of the jth pixel.

More generally, ART can be viewed as an iterative method for solving an arbitrary system of linear equations, $Ax = b$.

8.1 The ART

Let A be a complex matrix with I rows and J columns, and let b be a member of C^I. We want to solve the system $Ax = b$.

For each index value i, let H_i be the hyperplane of J-dimensional vectors given by

$$H_i = \{x | (Ax)_i = b_i\},$$

and P_i the orthogonal projection operator onto H_i. Let x^0 be arbitrary and, for each nonnegative integer k, let $i(k) = k(\mathrm{mod}\, I) + 1$. The iterative step of the ART is

$$x^{k+1} = P_{i(k)} x^k.$$

Because the ART uses only a single equation at each step, it has been called a *row-action* method.

8.1.1   Calculating the ART

Given any vector z, the vector in H_i closest to z, in the sense of the Euclidean distance, has the entries

$$x_j = z_j + \overline{A_{ij}}(b_i - (Az)_i)/ \sum_{m=1}^{J} |A_{im}|^2. \tag{8.1}$$

To simplify our calculations, we shall assume, throughout this chapter, that the rows of A have been rescaled to have Euclidean length one; that is

$$\sum_{j=1}^{J} |A_{ij}|^2 = 1,$$

for each $i = 1, ..., I$, and that the entries of b have been rescaled accordingly, to preserve the equations $Ax = b$. The ART is then the following: begin with an arbitrary vector x^0; for each nonnegative integer k, having found x^k, the next iterate x^{k+1} has entries

$$x_j^{k+1} = x_j^k + \overline{A_{ij}}(b_i - (Ax^k)_i). \tag{8.2}$$

When the system $Ax = b$ has exact solutions, the ART converges to the solution closest to x^0, in the 2-norm. How fast the algorithm converges will depend on the ordering of the equations and on whether or not we use relaxation. In selecting the equation ordering, the important thing is to avoid particularly bad orderings, in which the hyperplanes H_i and H_{i+1} are nearly parallel.

8.1.2   Full-Cycle ART

We also consider the *full-cycle* ART, with iterative step $z^{k+1} = Tz^k$, for

$$T = P_I P_{I-1} \cdots P_2 P_1.$$

When the system $Ax = b$ has solutions, the fixed points of T are solutions. When there are no solutions of $Ax = b$, the operator T will still have fixed points, but they will no longer be exact solutions.

8.1.3 Relaxed ART

The ART employs orthogonal projections onto the individual hyperplanes. If we permit the next iterate to fall short of the hyperplane, or somewhat beyond it, we get a relaxed version of ART. The relaxed ART algorithm is as follows:

Algorithm 8.1 (Relaxed ART). With $\omega \in (0, 2)$, x^0 arbitrary, and $i = k(\mathrm{mod}\, I)$ +1, let

$$x_j^{k+1} = x_j^k + \omega \overline{A_{ij}}(b_i - (Ax^k)_i). \tag{8.3}$$

The relaxed ART converges to the solution closest to x^0, in the consistent case. In the inconsistent case, it does not converge, but subsequences associated with the same i converge to distinct vectors, forming a limit cycle.

8.1.4 Constrained ART

Let C be a closed, nonempty convex subset of C^J and $P_C x$ the orthogonal projection of x onto C. If there are solutions of $Ax = b$ that lie within C, we can find them using the constrained ART algorithm:

Algorithm 8.2 (Constrained ART). With x^0 arbitrary and $i = k(\mathrm{mod}\, I) + 1$, let

$$x_j^{k+1} = P_C(x_j^k + \overline{A_{ij}}(b_i - (Ax^k)_i)). \tag{8.4}$$

For example, if A and b are real and we seek a nonnegative solution to $Ax = b$, we can use

Algorithm 8.3 (Nonnegative ART). With x^0 arbitrary and $i = k(\mathrm{mod}\, I) + 1$, let

$$x_j^{k+1} = (x_j^k + A_{ij}(b_i - (Ax^k)_i))_+, \tag{8.5}$$

where, for any real number a, $a_+ = \max\{a, 0\}$.

The constrained ART converges to a solution of $Ax = b$ within C, whenever such solutions exist.

Noise in the data can manifest itself in a variety of ways; we have seen what can happen when we impose positivity on the calculated least-squares

solution, that is, when we minimize $||Ax - b||_2$ over all nonnegative vectors
x. Theorem 6.5 tells us that when $J > I$, but $Ax = b$ has no nonnegative
solutions, the nonnegatively constrained least-squares solution can have at
most $I - 1$ nonzero entries, regardless of how large J is. This phenomenon
also occurs with several other approximate methods, such as those that
minimize the cross-entropy distance.

8.2   When $Ax = b$ Has Solutions

For the consistent case, in which the system $Ax = b$ has exact solutions,
we have the following result.

Theorem 8.4. *Let $A\hat{x} = b$ and let x^0 be arbitrary. Let $\{x^k\}$ be generated
by Equation (8.2). Then the sequence $\{||\hat{x} - x^k||_2\}$ is decreasing and $\{x^k\}$
converges to the solution of $Ax = b$ closest to x^0.*

The proof of the following lemma follows immediately from the definition of the ART iteration.

Lemma 8.5. *Let x^0 and y^0 be arbitrary and $\{x^k\}$ and $\{y^k\}$ be the sequences
generated by applying the ART algorithm, beginning with x^0 and y^0, respectively; that is, $y^{k+1} = P_{i(k)}y^k$. Then*

$$||x^0 - y^0||_2^2 - ||x^I - y^I||_2^2 = \sum_{i=1}^{I} |(Ax^{i-1})_i - (Ay^{i-1})_i|^2. \qquad (8.6)$$

Proof (of Theorem 8.4): Let $A\hat{x} = b$. Let $v_i^r = (Ax^{rI+i-1})_i$ and $v^r =
(v_1^r, ..., v_I^r)^T$, for $r = 0, 1, ...$. It follows from Equation (8.6) that the se-
quence $\{||\hat{x} - x^{rI}||_2\}$ is decreasing and the sequence $\{v^r - b\} \to 0$. So
$\{x^{rI}\}$ is bounded; let $x^{*,0}$ be a cluster point. Then, for $i = 1, 2, ..., I$, let
$x^{*,i}$ be the successor of $x^{*,i-1}$ using the ART algorithm. It follows that
$(Ax^{*,i-1})_i = b_i$ for each i, from which we conclude that $x^{*,0} = x^{*,i}$ for all
i and that $Ax^{*,0} = b$. Using $x^{*,0}$ in place of the arbitrary solution $\hat{x}$, we
have that the sequence $\{||x^{*,0} - x^k||_2\}$ is decreasing. But a subsequence
converges to zero, so $\{x^k\}$ converges to $x^{*,0}$. By Equation (8.6), the dif-
ference $||\hat{x} - x^k||_2^2 - ||\hat{x} - x^{k+1}||_2^2$ is independent of which solution $\hat{x}$ we
pick; consequently, so is $||\hat{x} - x^0||_2^2 - ||\hat{x} - x^{*,0}||_2^2$. It follows that $x^{*,0}$ is the
solution closest to x^0. This completes the proof. $\square$

8.3   When $Ax = b$ Has No Solutions

When there are no exact solutions, the ART does not converge to a single
vector, but, for each fixed i, the subsequence $\{x^{nI+i}, n = 0, 1, ...\}$ converges

to a vector z^i and the collection $\{z^i \,|i = 1, ..., I\}$ is called the *limit cycle*. This was shown by Tanabe [149] and also follows from the results of De Pierro and Iusem [77]. For simplicity, we assume that $I > J$, and that the matrix A has full rank, which implies that $Ax = 0$ if and only if $x = 0$. Because the operator $T = P_I P_{i-1} \cdots P_2 P_1$ is averaged, this subsequential convergence to a limit cycle will follow from the KM Theorem 5.16, once we have established that T has fixed points. A different proof of subsequential convergence is given in the discussion of the AMS Algorithm in Chapter 23 (see also [48]).

8.3.1 Subsequential Convergence of ART

We know from Lemma 3.24 that the operator T is affine linear and has the form

$$Tx = Bx + d, \tag{8.7}$$

where B is the matrix

$$B = (I - a^I (a^I)^\dagger) \cdots (I - a^1 (a^1)^\dagger),$$

and d a vector.

The matrix $I - B$ is invertible, since if $(I - B)x = 0$, then $Bx = x$. It follows that x is in H_{i0} for each i, which means that $\langle a^i, x \rangle = 0$ for each i. Therefore $Ax = 0$, and so $x = 0$.

Lemma 8.6. *The operator T in Equation (8.7) is strictly nonexpansive, meaning that*

$$||x - y||_2 \geq ||Tx - Ty||_2,$$

with equality if and only if $x = Tx$ and $y = Ty$.

Proof: Write $Tx - Ty = Bx - By = B(x - y)$. Since B is the product of orthogonal projections, B is averaged. Therefore, there is $\alpha > 0$ with

$$||x - y||_2^2 - ||Bx - By||_2^2 \geq (\frac{1}{\alpha} - 1)||(I - B)x - (I - B)y||_2^2. \qquad \square$$

The function $||x - Tx||_2$ has minimizers, since $||x - Tx||_2^2 = ||x - Bx - d||_2^2$ is quadratic in x. For any such minimizer z we will have

$$||z - Tz||_2 = ||Tz - T^2 z||_2.$$

Since T is strictly nonexpansive, it follows that $z = Tz$.

Lemma 8.7. *Let* $AA^\dagger = L + D + L^\dagger$, *for diagonal matrix* D *and lower triangular matrix* L. *The operator* T *in Equation (8.7) can be written as*

$$Tx = (I - A^\dagger(L+D)^{-1})x + A^\dagger(L+D)^{-1}b.$$

As we shall see, this formulation of the operator T provides a connection between the full-cycle ART for $Ax = b$ and the Gauss-Seidel method, as applied to the system $AA^\dagger z = b$, as Dax has pointed out [74].

The ART limit cycle will vary with the ordering of the equations, and contains more than one vector unless an exact solution exists. There are several open questions about the limit cycle.

Open Question 1. For a fixed ordering, does the limit cycle depend on the initial vector x^0? If so, how?

8.3.2   The Geometric Least-Squares Solution

When the system $Ax = b$ has no solutions, it is reasonable to seek an approximate solution, such as the *least-squares* solution, $x_{LS} = (A^\dagger A)^{-1}A^\dagger b$, which minimizes $||Ax - b||_2$. It is important to note that the system $Ax = b$ has solutions if and only if the related system $WAx = Wb$ has solutions, where W denotes an invertible matrix; when solutions of $Ax = b$ exist, they are identical to those of $WAx = Wb$. But, when $Ax = b$ does not have solutions, the least-squares solutions of $Ax = b$ need not be unique, but usually are. The least-squares solutions of $WAx = Wb$ need not be identical to those of $Ax = b$. In the typical case in which $A^\dagger A$ is invertible, the unique least-squares solution of $Ax = b$ is

$$(A^\dagger A)^{-1}A^\dagger b,$$

while the unique least-squares solution of $WAx = Wb$ is

$$(A^\dagger W^\dagger W A)^{-1}A^\dagger W^\dagger b,$$

and these need not be the same.

A simple example is the following. Consider the system

$$x = 1$$

$$x = 2,$$

which has the unique least-squares solution $x = 1.5$, and the system

$$2x = 2$$

$$x = 2,$$

which has the least-squares solution $x = 1.2$.

Definition 8.8. The geometric least-squares solution of $Ax = b$ is the least-squares solution of $WAx = Wb$, for W the diagonal matrix whose entries are the reciprocals of the Euclidean lengths of the rows of A.

In our example above, the geometric least-squares solution for the first system is found by using $W_{11} = 1 = W_{22}$, so is again $x = 1.5$, while the geometric least-squares solution of the second system is found by using $W_{11} = 0.5$ and $W_{22} = 1$, so that the geometric least-squares solution is $x = 1.5$, not $x = 1.2$.

Open Question 2. If there is a unique geometric least-squares solution, where is it, in relation to the vectors of the limit cycle? Can it be calculated easily, from the vectors of the limit cycle?

There is a partial answer to the second question. In [38] (see also [48]) it was shown that if the system $Ax = b$ has no exact solution, and if $I = J+1$, then the vectors of the limit cycle lie on a sphere in J-dimensional space having the least-squares solution at its center. This is not true more generally, however.

8.4   Regularized ART

If the entries of b are noisy but the system $Ax = b$ remains consistent (which can easily happen in the underdetermined case, with $J > I$), the ART begun at $x^0 = 0$ converges to the solution having minimum Euclidean norm, but this norm can be quite large. The resulting solution is probably useless. Instead of solving $Ax = b$, we *regularize* by minimizing, for example, the function

$$F_\epsilon(x) = ||Ax - b||_2^2 + \epsilon^2 ||x||_2^2.$$

The solution to this problem is the vector

$$\hat{x}_\epsilon = (A^\dagger A + \epsilon^2 I)^{-1} A^\dagger b.$$

However, we do not want to calculate $A^\dagger A + \epsilon^2 I$ when the matrix A is large. Fortunately, there are ways to find $\hat{x}_\epsilon$, using only the matrix A and the ART algorithm.

We discuss two methods for using ART to obtain regularized solutions of $Ax = b$. The first one is presented in [48], while the second one is due to Eggermont, Herman, and Lent [84].

In our first method we use ART to solve the system of equations given in matrix form by

$$[A^\dagger \quad \gamma I] \begin{bmatrix} u \\ v \end{bmatrix} = 0.$$

We begin with $u^0 = b$ and $v^0 = 0$. Then, the lower component of the limit vector is $v^\infty = -\gamma \hat{x}_\epsilon$.

The method of Eggermont et al. is similar. In their method we use ART to solve the system of equations given in matrix form by

$$[A \quad \gamma I] \begin{bmatrix} x \\ v \end{bmatrix} = b.$$

We begin at $x^0 = 0$ and $v^0 = 0$. Then, the limit vector has for its upper component $x^\infty = \hat{x}_\epsilon$ as before, and that $\gamma v^\infty = b - A\hat{x}_\epsilon$.

Open Question 3. In both the consistent and inconsistent cases, the sequence $\{x^k\}$ of ART iterates is bounded, as Tanabe [149], and De Pierro and Iusem [77] have shown. The proof is easy in the consistent case. Is there an easy proof for the inconsistent case?

8.5   Avoiding the Limit Cycle

Generally, the greater the minimum value of $||Ax - b||_2^2$ the more the vectors of the LC are distinct from one another. There are several ways to avoid the LC in ART and to obtain a least-squares solution. One way is the *double ART* (DART) [42].

8.5.1   Double ART (DART)

We know that any b can be written as $b = A\hat{x} + \hat{w}$, where $A^T \hat{w} = 0$ and $\hat{x}$ is a minimizer of $||Ax - b||_2^2$. The vector $\hat{w}$ is the orthogonal projection of b onto the null space of the matrix transformation $A^\dagger$. Therefore, in Step 1 of DART we apply the ART algorithm to the consistent system of linear equations $A^\dagger w = 0$, beginning with $w^0 = b$. The limit is $w^\infty = \hat{w}$, the member of the null space of $A^\dagger$ closest to b. In Step 2, apply ART to the consistent system of linear equations $Ax = b - w^\infty = A\hat{x}$. The limit is then the minimizer of $||Ax - b||_2$ closest to x^0. Notice that we could also obtain the least-squares solution by applying ART to the system $A^\dagger y = A^\dagger b$, starting with $y^0 = 0$, to obtain the minimum-norm solution, which is $y = A\hat{x}$, and then applying ART to the system $Ax = y$.

8.5.2   Strongly Under-Relaxed ART

Another method for avoiding the LC is *strong under-relaxation*, due to Censor, Eggermont, and Gordon [53]. Let $t > 0$. Replace the iterative step in ART with

$$x_j^{k+1} = x_j^k + t\overline{A_{ij}}(b_i - (Ax^k)_i).$$

In [53] it is shown that, as $t \to 0$, the vectors of the LC approach the geometric least squares solution closest to x^0; a short proof is in [38]. Bertsekas [14] uses strong under-relaxation to obtain convergence of more general incremental methods.

9 | Simultaneous and Block-Iterative ART

The ART is a sequential algorithm, using only a single equation from the system $Ax = b$ at each step of the iteration. In this chapter we consider iterative procedures for solving $Ax = b$ in which several or all of the equations are used at each step. Such methods are called *block-iterative* and *simultaneous* algorithms, respectively. As before, we shall assume that the equations have been normalized so that the rows of A have Euclidean length one.

9.1 Cimmino's Algorithm

The ART seeks a solution of $Ax = b$ by projecting the current vector x^k orthogonally onto the next hyperplane $H(a^{i(k)}, b_{i(k)})$ to get x^{k+1}. In Cimmino's Algorithm, we project the current vector x^k onto each of the hyperplanes and then average the result to get x^{k+1}. The algorithm begins with an arbitrary x^0; the iterative step is then

$$x^{k+1} = \frac{1}{I} \sum_{i=1}^{I} P_i x^k, \qquad (9.1)$$

where P_i is the orthogonal projection onto $H(a^i, b_i)$. The iterative step can then be written as

$$x^{k+1} = x^k + \frac{1}{I} A^\dagger (b - Ax^k). \qquad (9.2)$$

As we saw in our discussion of the ART, when the system $Ax = b$ has no solutions, the ART does not converge to a single vector, but to a limit cycle. One advantage of many simultaneous algorithms, such as Cimmino's, is that they do converge to a least-squares solution in the inconsistent case.

Cimmino's Algorithm has the form $x^{k+1} = Tx^k$, for the operator T given by

$$Tx = (I - \frac{1}{I}A^\dagger A)x + \frac{1}{I}A^\dagger b.$$

Experience with Cimmino's Algorithm shows that it is slow to converge. In the next section we consider how we might accelerate the algorithm.

9.2   The Landweber Algorithms

Landweber's Algorithm [13, 115], with the iterative step

$$x^{k+1} = x^k + \gamma A^\dagger (b - Ax^k), \tag{9.3}$$

converges to the least-squares solution closest to the starting vector x^0, provided that $0 < \gamma < 2/\lambda_{\max}$, where $\lambda_{\max}$ is the largest eigenvalue of the nonnegative-definite matrix $A^\dagger A$. Loosely speaking, the larger γ is, the faster the convergence. However, precisely because A is large, calculating the matrix $A^\dagger A$, not to mention finding its largest eigenvalue, can be prohibitively expensive. The matrix A is said to be sparse if most of its entries are zero. Useful upper bounds for $\lambda_{\max}$ are then given by Theorem 4.15.

9.2.1   Finding the Optimum γ

The operator

$$Tx = x + \gamma A^\dagger (b - Ax) = (I - \gamma A^\dagger A)x + \gamma A^\dagger b$$

is affine linear and is averaged if and only if its linear part, the Hermitian matrix

$$B = I - \gamma A^\dagger A,$$

is averaged. To guarantee this we need $0 \le \gamma < 2/\lambda_{\max}$. Should we always try to take γ near its upper bound, or is there an optimum value of γ? To answer this question we consider the eigenvalues of B for various values of γ.

Lemma 9.1. *If $\gamma < 0$, then none of the eigenvalues of B is less than one.*

Lemma 9.2. *For*

$$0 \le \gamma \le \frac{2}{\lambda_{\max} + \lambda_{\min}},$$

we have

$$\rho(B) = 1 - \gamma \lambda_{\min};$$

the smallest value of $\rho(B)$ occurs when

$$\gamma = \frac{2}{\lambda_{\max} + \lambda_{\min}},$$

and equals

$$\frac{\lambda_{\max} - \lambda_{\min}}{\lambda_{\max} + \lambda_{\min}}.$$

Similarly, for

$$\gamma \geq \frac{2}{\lambda_{\max} + \lambda_{\min}},$$

we have

$$\rho(B) = \gamma\lambda_{\max} - 1;$$

the smallest value of $\rho(B)$ occurs when

$$\gamma = \frac{2}{\lambda_{\max} + \lambda_{\min}},$$

and equals

$$\frac{\lambda_{\max} - \lambda_{\min}}{\lambda_{\max} + \lambda_{\min}}.$$

We see from this lemma that, if $0 \leq \gamma < 2/\lambda_{\max}$, and $\lambda_{\min} > 0$, then $||B||_2 = \rho(B) < 1$, so that B is a strict contraction. We minimize $||B||_2$ by taking

$$\gamma = \frac{2}{\lambda_{\max} + \lambda_{\min}},$$

in which case we have

$$||B||_2 = \frac{\lambda_{\max} - \lambda_{\min}}{\lambda_{\max} + \lambda_{\min}} = \frac{c - 1}{c + 1},$$

for $c = \lambda_{\max}/\lambda_{\min}$, the *condition number* of the positive-definite matrix $A^\dagger A$. The closer c is to one, the smaller the norm $||B||_2$, and the faster the convergence.

On the other hand, if $\lambda_{\min} = 0$, then $\rho(B) = 1$ for all γ in the interval $(0, 2/\lambda_{\max})$. The matrix B is still averaged, but it is no longer a strict contraction. For example, consider the orthogonal projection P_0 onto the hyperplane $H_0 = H(a, 0)$, where $||a||_2 = 1$. This operator can be written

$$P_0 = I - aa^\dagger.$$

The largest eigenvalue of $aa^\dagger$ is $\lambda_{\max} = 1$; the remaining ones are zero. The relaxed projection operator

$$B = I - \gamma aa^\dagger$$

has $\rho(B) = 1 - \gamma > 1$, if $\gamma < 0$, and for $\gamma \geq 0$, we have $\rho(B) = 1$. The operator B is averaged, in fact, it is firmly nonexpansive, but it is not a strict contraction.

9.2.2   The Projected Landweber's Algorithm

When we require a nonnegative approximate solution x for the real system $Ax = b$ we can use a modified version of Landweber's Algorithm, called the Projected Landweber's Algorithm [13], in this case having the iterative step

$$x^{k+1} = (x^k + \gamma A^\dagger(b - Ax^k))_+, \tag{9.4}$$

where, for any real vector a, we denote by $(a)_+$ the nonnegative vector whose entries are those of a, for those that are nonnegative, and are zero otherwise. The Projected Landweber's Algorithm converges to a vector that minimizes $||Ax - b||_2$ over all nonnegative vectors x, for the same values of γ.

The Projected Landweber's Algorithm is actually more general. For any closed, nonempty convex set C in X, define the iterative sequence

$$x^{k+1} = P_C(x^k + \gamma A^\dagger(b - Ax^k)).$$

This sequence converges to a minimizer of the function $||Ax - b||_2$ over all x in C, whenever such minimizers exist.

Both Landweber's and Projected Landweber's Algorithms are special cases of the CQ algorithm [45], which, in turn, is a special case of the more general iterative fixed point algorithm, the Krasnoselskii-Mann (KM) method, with convergence governed by the KM Theorem 5.16.

9.3   The Block-Iterative ART

The ART is generally faster than the simultaneous versions, particularly when relaxation or random ordering of the equations is included. On the other hand, the simultaneous methods, such as Landweber's Algorithm, converge to an approximate solution in the inconsistent case, and lend themselves to parallel processing. We turn now to block-iterative versions of ART, which use several equations at each step of the iteration. The ART can also be inefficient, in that it fails to make use of the way in which the equations are actually stored and retrieved within the computer. *Block-iterative* ART (BI-ART) can be made more efficient than the ART, without much loss of speed of convergence.

Let the index set $\{i = 1, ..., I\}$ be partitioned into N subsets, or blocks, $B_1,...,B_N$, for some positive integer N, with $1 \leq N \leq I$. Let I_n be the cardinality of B_n. Let A_n be the I_n by J matrix obtained from A by discarding all rows except those whose index is in B_n. Similarly, let b^n be the I_n by 1 vector obtained from b. For $k = 0, 1, ...$, let $n = k(\mathrm{mod}\,N) + 1$.

Algorithm 9.3 (Block-Iterative ART). The block-iterative ART (BI-ART) has the iterative step

$$x^{k+1} = x^k + \frac{1}{I_n} A_n^\dagger (b^n - A_n x^k). \tag{9.5}$$

9.4   The Rescaled Block-Iterative ART

The use of the weighting $1/I_n$ in the block-iterative ART is not necessary; we do have some choice in the selection of the weighting factor. The *rescaled* BI-ART (RBI-ART) algorithm is the following:

Algorithm 9.4. Let x^0 be arbitrary, and $n = k(\mathrm{mod}\,N) + 1$. Then let

$$x^{k+1} = x^k + \gamma_n A_n^\dagger (b^n - A_n x^k), \tag{9.6}$$

for $0 < \gamma_n < 2/L_n$, where L_n is the largest eigenvalue of the matrix $A_n^\dagger A_n$.

How we select the blocks and the parameters γ_n will determine the speed of convergence of the RBI-ART.

9.5   Convergence of the RBI-ART

Suppose now that the system is consistent and that $A\hat{x} = b$. Then

$$\begin{aligned}
||\hat{x} - x^k||_2^2 - ||\hat{x} - x^{k+1}||_2^2 &= 2\gamma_n \mathrm{Re}\langle \hat{x} - x^k, A_n^\dagger (b^n - A_n x^k)\rangle \\
&\quad - \gamma_n^2 ||A_n^\dagger (b^n - A_n x^k)||_2^2 \\
&= 2\gamma_n ||b^n - A_n x^k||_2^2 - \gamma_n^2 ||A_n^\dagger (b^n - A_n x^k)||_2^2.
\end{aligned}$$

Therefore, we have

$$||\hat{x} - x^k||_2^2 - ||\hat{x} - x^{k+1}||_2^2 \geq (2\gamma_n - \gamma_n^2 L_n)||b^n - A_n x^k||_2^2. \tag{9.7}$$

It follows that the sequence $\{||\hat{x} - x^k||_2^2\}$ is decreasing and that the sequence $\{||b^n - A_n x^k||_2^2\}$ converges to 0. The sequence $\{x^k\}$ is then bounded; let x^* be any cluster point of the subsequence $\{x^{mN}\}$. Then let

$$x^{*,n} = x^{*,n-1} + \gamma_n A_n^\dagger (b^n - A_n x^{*,n-1}),$$

for $n = 1, 2, ..., N$. It follows that $x^{*,n} = x^*$ for all n and that $Ax^* = b$. Replacing the arbitrary solution $\hat{x}$ with x^*, we find that the sequence $\{||x^* - x^k||_2^2\}$ is decreasing; but a subsequence converges to zero. Consequently, the sequence $\{||x^* - x^k||_2^2\}$ converges to zero. We can therefore conclude that the RBI-ART converges to a solution, whenever the system is consistent. In fact, since we have shown that the difference $||\hat{x} - x^k||_2^2 - ||\hat{x} - x^{k+1}||_2^2$ is nonnegative and independent of the solution $\hat{x}$ that we choose, we know that the difference $||\hat{x} - x^0||_2^2 - ||\hat{x} - x^*||_2^2$ is also nonnegative and independent of $\hat{x}$. It follows that x^* is the solution closest to x^0.

From Inequality (9.7) we see that we make progress toward a solution to the extent that the right side of the inequality,

$$(2\gamma_n - \gamma_n^2 L_n)||b^n - A_n x^k||_2^2,$$

is large. One conclusion we draw from this is that we want to avoid ordering the blocks so that the quantity $||b^n - A_n x^k||_2^2$ is small. We also want to select γ_n reasonably large, subject to the bound $\gamma_n < 2/L_n$; the maximum of $2\gamma_n - \gamma_n^2 L_n$ is at $\gamma_n = L_n^{-1}$. Because the rows of A_n have length one, the trace of $A_n^\dagger A_n$ is I_n, the number of rows in A_n. Since L_n is not greater than this trace, we have $L_n \leq I_n$, so the choice of $\gamma_n = 1/I_n$ used in BI-ART is acceptable, but possibly far from optimal, particularly if A_n is sparse.

Inequality (9.7) can be used to give a rough measure of the speed of convergence of the RBI-ART. The term $||b^n - A_n x^k||_2^2$ is on the order of I_n, while the term $2\gamma_n - \gamma_n^2 L_n$ has $1/L_n$ for its maximum, so, very roughly, is on the order of $1/I_n$. Consequently, the improvement made in one step of the BI-ART is on the order of one. One complete cycle of the BI-ART, that is, one complete pass through all the blocks, then corresponds to an improvement on the order of N, the number of blocks. It is a "rule of thumb" that block-iterative methods are capable of improving the speed of convergence by a factor of the number of blocks, if unfortunate ordering of the blocks and selection of the equations within the blocks are avoided, and the parameters are well chosen.

To obtain good choices for the γ_n, we need to have a good estimate of L_n. As we have seen, such estimates are available for sparse matrices.

9.6   Using Sparseness

Let s_{nj} be the number of nonzero elements in the jth column of A_n, and let s_n be the maximum of the s_{nj}. We know then that $L_n \leq s_n$. Therefore, we can choose $\gamma_n < 2/s_n$.

Suppose, for the sake of illustration, that each column of A has s nonzero elements, for some $s < I$, and we let $r = s/I$. Suppose also that $I_n = I/N$ and that N is not too large. Then s_n is approximately equal to $rI_n = s/N$. On the other hand, unless A_n has only zero entries, we know that $s_n \geq 1$. Therefore, it is no help to select N for which $s/N < 1$. For a given degree of sparseness s we need not select N greater than s. The more sparse the matrix A, the fewer blocks we need to gain the maximum advantage from the rescaling, and the more we can benefit from parallelizability in the calculations at each step of the RBI-ART.

10 | Jacobi and Gauss-Seidel Methods

Linear systems $Ax = b$ need not be square but can be associated with two square systems, $A^\dagger Ax = A^\dagger b$, the so-called *normal equations*, and $AA^\dagger z = b$, sometimes called the *Björck-Elfving equations* [74]. In this chapter we consider two well-known iterative algorithms for solving square systems of linear equations, the Jacobi method and the Gauss-Seidel method. Both these algorithms are easy to describe and to motivate. They both require not only that the system be square, that is, have the same number of unknowns as equations, but that it satisfy additional constraints needed for convergence.

Both the Jacobi and the Gauss-Seidel algorithms can be modified to apply to any square system of linear equations, $Sz = h$. The resulting algorithms, the *Jacobi over-relaxation* (JOR) and *successive over-relaxation* (SOR) methods, involve the choice of a parameter. The JOR and SOR will converge for more general classes of matrices, provided that the parameter is appropriately chosen.

When we say that an iterative method is convergent, or converges, under certain conditions, we mean that it converges for any consistent system of the appropriate type, and for any starting vector; any iterative method will converge if we begin at the right answer.

10.1 The Jacobi and Gauss-Seidel Methods: An Example

Suppose we wish to solve the 3 by 3 system

$$S_{11}z_1 + S_{12}z_2 + S_{13}z_3 = h_1,$$
$$S_{21}z_1 + S_{22}z_2 + S_{23}z_3 = h_2,$$
$$S_{31}z_1 + S_{32}z_2 + S_{33}z_3 = h_3,$$

which we can rewrite as

$$z_1 = S_{11}^{-1}[h_1 - S_{12}z_2 - S_{13}z_3],$$
$$z_2 = S_{22}^{-1}[h_2 - S_{21}z_1 - S_{23}z_3],$$
$$z_3 = S_{33}^{-1}[h_3 - S_{31}z_1 - S_{32}z_2],$$

assuming that the diagonal terms S_{mm} are not zero. Let $z^0 = (z_1^0, z_2^0, z_3^0)^T$ be an initial guess for the solution. We then insert the entries of z^0 on the right sides and use the left sides to define the entries of the next guess z^1. This is one full cycle of the Jacobi method.

The Gauss-Seidel method is similar. Let $z^0 = (z_1^0, z_2^0, z_3^0)^T$ be an initial guess for the solution. We then insert z_2^0 and z_3^0 on the right side of the first equation, obtaining a new value z_1^1 on the left side. We then insert z_3^0 and z_1^1 on the right side of the second equation, obtaining a new value z_2^1 on the left. Finally, we insert z_1^1 and z_2^1 into the right side of the third equation, obtaining a new z_3^1 on the left side. This is one full cycle of the *Gauss-Seidel* (GS) method.

10.2   Splitting Methods

The Jacobi and Gauss-Seidel methods are particular cases of a more general approach, known as splitting methods. Splitting methods apply to square systems of linear equations. Let S be an arbitrary N by N square matrix, written as $S = M - K$. Then the linear system of equations $Sz = h$ is equivalent to $Mz = Kz + h$. If M is invertible, then we can also write $z = M^{-1}Kz + M^{-1}h$. This last equation suggests a class of iterative methods for solving $Sz = h$ known as *splitting methods*. The idea is to select a matrix M so that the equation

$$Mz^{k+1} = Kz^k + h$$

can be easily solved to get z^{k+1}; in the Jacobi method, M is diagonal, and in the Gauss-Seidel method, M is triangular. Then we write

$$z^{k+1} = M^{-1}Kz^k + M^{-1}h. \tag{10.1}$$

From $K = M - S$, we can write Equation (10.1) as

$$z^{k+1} = z^k + M^{-1}(h - Sz^k). \tag{10.2}$$

Suppose that S is invertible and $\hat{z}$ is the unique solution of $Sz = h$. The error we make at the kth step is $e^k = \hat{z} - z^k$, so that $e^{k+1} = M^{-1}Ke^k$. We want the error to decrease with each step, which means that we should seek M and K so that $\|M^{-1}K\| < 1$. If S is not invertible and there are multiple solutions of $Sz = h$, then we do not want $M^{-1}K$ to be a strict contraction, but only averaged or paracontractive. The operator T defined by

$$Tz = M^{-1}Kz + M^{-1}h = Bz + d$$

is an affine linear operator and will be a strict contraction or an averaged operator whenever $B = M^{-1}K$ is.

It follows from our previous discussion concerning linear averaged operators that, if $B = B^\dagger$ is Hermitian, then B is averaged if and only if

$$-1 < \lambda \leq 1,$$

for all (necessarily real) eigenvalues λ of B.

In general, though, the matrix $B = M^{-1}K$ will not be Hermitian, and deciding if such a non-Hermitian matrix is averaged is not a simple matter. We do know that, if B is averaged, so is $B^\dagger$; consequently, the Hermitian matrix $Q = \frac{1}{2}(B + B^\dagger)$ is also averaged. Therefore, $I - Q = \frac{1}{2}(M^{-1}S + (M^{-1}S)^\dagger)$ is ism, and so is nonnegative definite. We have $-1 < \lambda \leq 1$, for any eigenvalue λ of Q.

Alternatively, we can use Theorem 5.21. According to that theorem, if B has a basis of eigenvectors, and $|\lambda| < 1$ for all eigenvalues λ of B that are not equal to one, then $\{z^k\}$ will converge to a solution of $Sz = h$, whenever solutions exist.

In what follows we shall write an arbitrary square matrix S as

$$S = L + D + U,$$

where L is the strictly lower triangular part of S, D the diagonal part, and U the strictly upper triangular part. When S is Hermitian, we have

$$S = L + D + L^\dagger.$$

We list now several examples of iterative algorithms obtained by the splitting method. In the remainder of the chapter we discuss these methods in more detail.

10.3 Some Examples of Splitting Methods

As we shall now see, the Jacobi and Gauss-Seidel methods, as well as their over-relaxed versions, JOR and SOR, are splitting methods.

The Jacobi method. The Jacobi method uses $M = D$ and $K = -L - U$, under the assumption that D is invertible. The matrix B is

$$B = M^{-1}K = -D^{-1}(L + U). \tag{10.3}$$

The Gauss-Seidel method. The Gauss-Seidel (GS) method uses the splitting $M = D + L$, so that the matrix B is

$$B = I - (D + L)^{-1}S. \tag{10.4}$$

The Jacobi over-relaxation method (JOR). The JOR uses the splitting

$$M = \frac{1}{\omega}D$$

and

$$K = M - S = (\frac{1}{\omega} - 1)D - L - U.$$

The matrix B is

$$B = M^{-1}K = (I - \omega D^{-1}S). \tag{10.5}$$

The successive over-relaxation method (SOR). The SOR uses the splitting $M = (\frac{1}{\omega}D + L)$, so that

$$B = M^{-1}K = (D + \omega L)^{-1}[(1 - \omega)D - \omega U]$$

or

$$B = I - \omega(D + \omega L)^{-1}S,$$

or

$$B = (I + \omega D^{-1}L)^{-1}[(1 - \omega)I - \omega D^{-1}U]. \tag{10.6}$$

10.4   The Jacobi Algorithm and JOR

The matrix B in Equation (10.3) is not generally averaged and the Jacobi iterative scheme will not converge, in general. Additional conditions need to be imposed on S in order to guarantee convergence. One such condition is that S be strictly diagonally dominant. In that case, all the eigenvalues of $B = M^{-1}K$ can be shown to lie inside the unit circle of the complex plane, so that $\rho(B) < 1$. It follows from Lemma 4.22 that B is a strict contraction with respect to some vector norm, and the Jacobi iteration converges. If, in addition, S is Hermitian, the eigenvalues of B are in the interval $(-1, 1)$, and so B is a strict contraction with respect to the Euclidean norm.

Alternatively, one has the Jacobi over-relaxation method, which is essentially a special case of Landweber's Algorithm and involves an arbitrary parameter.

For S an N by N matrix, the Jacobi method can be written as

$$z_m^{\text{new}} = S_{mm}^{-1}[h_m - \sum_{j \neq m} S_{mj} z_j^{\text{old}}],$$

for $m = 1, ..., N$. With D the invertible diagonal matrix with entries $D_{mm} = S_{mm}$ we can write one cycle of the Jacobi method as

$$z^{\text{new}} = z^{\text{old}} + D^{-1}(h - S z^{\text{old}}).$$

The Jacobi over-relaxation method has the following full-cycle iterative step:

$$z^{\text{new}} = z^{\text{old}} + \omega D^{-1}(h - S z^{\text{old}});$$

choosing $\omega = 1$ we get the Jacobi method. Convergence of the JOR iteration will depend, of course, on properties of S and on the choice of ω. When S is Hermitian, nonnegative-definite, for example, $S = A^\dagger A$ or $S = AA^\dagger$, we can say more.

10.4.1   The JOR in the Nonnegative-Definite Case

When S is nonnegative-definite and the system $Sz = h$ is consistent, the JOR converges to a solution for any $\omega \in (0, 2/\rho(D^{-1/2}SD^{-1/2}))$, where $\rho(Q)$ denotes the largest eigenvalue of the nonnegative-definite matrix Q. For nonnegative-definite S, the convergence of the JOR method is implied by the KM Theorem 5.16, since the JOR is equivalent to Landweber's Algorithm in these cases.

The JOR method, as applied to $Sz = AA^\dagger z = b$, is equivalent to Landweber's Algorithm, applied to the system $Ax = b$.

Lemma 10.1. *If $\{z^k\}$ is the sequence obtained from the JOR, then the sequence $\{A^\dagger z^k\}$ is the sequence obtained by applying Landweber's Algorithm to the system $D^{-1/2}Ax = D^{-1/2}b$, where D is the diagonal part of the matrix $S = AA^\dagger$.*

If we select $\omega = 1/I$ we obtain Cimmino's Algorithm. Since the trace of the matrix $D^{-1/2}SD^{-1/2}$ equals I, we know that $\omega = 1/I$ is not greater than the largest eigenvalue of the matrix $D^{-1/2}SD^{-1/2}$ and so this choice of ω is acceptable and Cimmino's Algorithm converges whenever there are solutions of $Ax = b$. In fact, it can be shown that Cimmino's Algorithm converges to a least-squares approximate solution generally.

Similarly, the JOR method applied to the system $A^{\dagger}Ax = A^{\dagger}b$ is equivalent to Landweber's Algorithm, applied to the system $Ax = b$.

Lemma 10.2. *Show that, if $\{z^k\}$ is the sequence obtained from the JOR, then the sequence $\{D^{1/2}z^k\}$ is the sequence obtained by applying Landweber's Algorithm to the system $AD^{-1/2}x = b$, where D is the diagonal part of the matrix $S = A^{\dagger}A$.*

10.5   The Gauss-Seidel Method and SOR

In general, the full-cycle iterative step of the Gauss-Seidel method is the following:

$$z^{\text{new}} = z^{\text{old}} + (D + L)^{-1}(h - Sz^{\text{old}}),$$

where $S = D + L + U$ is the decomposition of the square matrix S into its diagonal, lower triangular, and upper triangular parts. The GS method does not converge without restrictions on the matrix S. As with the Jacobi method, strict diagonal dominance is a sufficient condition.

10.5.1   The Nonnegative-Definite Case

Now we consider the square system $Sz = h$, assuming that $S = L+D+L^{\dagger}$ is Hermitian and nonnegative-definite, so that $x^{\dagger}Sx \geq 0$, for all x. It is easily shown that all the entries of D are nonnegative. We assume that all the diagonal entries of D are positive, so that $D + L$ is invertible. The Gauss-Seidel iterative step is $z^{k+1} = Tz^k$, where T is the affine linear operator given by $Tz = Bz + d$, for $B = -(D + L)^{-1}L^{\dagger}$ and $d = (D + L)^{-1}h$.

Proposition 10.3. *Let λ be an eigenvalue of B that is not equal to one. Then $|\lambda| < 1$.*

If B is diagonalizable, then there is a norm with respect to which T is paracontractive, so, by the EKN Theorem 5.21, the GS iteration converges to a solution of $Sz = h$, whenever solutions exist.

Proof (of Proposition 10.3): Let $Bv = \lambda v$, for v nonzero. Then $-Bv = (D + L)^{-1}L^{\dagger}v = -\lambda v$, so that

$$L^{\dagger}v = -\lambda(D + L)v,$$

and

$$Lv = -\bar{\lambda}(D + L)^{\dagger}v.$$

Therefore,

$$v^\dagger L^\dagger v = -\lambda v^\dagger (D + L)v.$$

Adding $v^\dagger (D + L)v$ to both sides, we get

$$v^\dagger S v = (1 - \lambda)v^\dagger (D + L)v.$$

Since the left side of the equation is real, so is the right side. Therefore

$$(1 - \overline{\lambda})(D + L)^\dagger v = (1 - \lambda)v^\dagger (D + L)v$$

$$= (1 - \lambda)v^\dagger D v + (1 - \lambda)v^\dagger L v$$

$$= (1 - \lambda)v^\dagger D v - (1 - \lambda)\overline{\lambda}v^\dagger (D + L)^\dagger v.$$

So we have

$$[(1 - \overline{\lambda}) + (1 - \lambda)\overline{\lambda}]v^\dagger (D + L)^\dagger v = (1 - \lambda)v^\dagger D v,$$

or

$$(1 - |\lambda|^2)v^\dagger (D + L)^\dagger v = (1 - \lambda)v^\dagger D v.$$

Multiplying by $(1 - \overline{\lambda})$ on both sides, we get, on the left side,

$$(1 - |\lambda|^2)v^\dagger (D + L)^\dagger v - (1 - |\lambda|^2)\overline{\lambda}v^\dagger (D + L)^\dagger v,$$

which is equal to

$$(1 - |\lambda|^2)v^\dagger (D + L)^\dagger v + (1 - |\lambda|^2)v^\dagger L v,$$

and, on the right side, we get

$$|1 - \lambda|^2 v^\dagger D v.$$

Consequently, we have

$$(1 - |\lambda|^2)v^\dagger S v = |1 - \lambda|^2 v^\dagger D v.$$

Since $v^\dagger S v \geq 0$ and $v^\dagger D v > 0$, it follows that $1 - |\lambda|^2 \geq 0$. If $|\lambda| = 1$, then $|1 - \lambda|^2 = 0$, so that $\lambda = 1$. This completes the proof. $\square$

Note that $\lambda = 1$ if and only if $Sv = 0$. Therefore, if S is invertible, the affine linear operator T is a strict contraction, and the GS iteration converges to the unique solution of $Sz = h$.

10.5.2   Successive Over-Relaxation

The successive over-relaxation method has the following full-cycle iterative step:

$$z^{\text{new}} = z^{\text{old}} + (\omega^{-1}D + L)^{-1}(h - Sz^{\text{old}});$$

the choice of $\omega = 1$ gives the GS method. Convergence of the SOR iteration will depend, of course, on properties of S and on the choice of ω.

Using the form

$$B = (D + \omega L)^{-1}[(1 - \omega)D - \omega U]$$

we can show that

$$|\det(B)| = |1 - \omega|^{N}.$$

From this and the fact that the determinant of B is the product of its eigenvalues, we conclude that $\rho(B) > 1$ if $\omega < 0$ or $\omega > 2$.

When S is Hermitian, nonnegative-definite, as, for example, when we take $S = A^{\dagger}A$ or $S = AA^{\dagger}$, we can say more.

10.5.3   The SOR for Nonnegative-Definite S

When S is nonnegative-definite and the system $Sz = h$ is consistent, the SOR converges to a solution for any $\omega \in (0, 2)$. This follows from the convergence of the ART algorithm, since, for such S, the SOR is equivalent to the ART.

Now we consider the SOR method applied to the Björck-Elfving equations $AA^{\dagger}z = b$. Rather than count a full cycle as one iteration, we now count as a single step the calculation of a single new entry. Therefore, for $k = 0, 1, \ldots$ the $k + 1$-st step replaces the value z_{i}^{k} only, where $i = k(\bmod I) + 1$. We have

$$z_{i}^{k+1} = (1 - \omega)z_{i}^{k} + \omega D_{ii}^{-1}\left(b_{i} - \sum_{n=1}^{i-1} S_{in}z_{n}^{k} - \sum_{n=i+1}^{I} S_{in}z_{n}^{k}\right)$$

and $z_{n}^{k+1} = z_{n}^{k}$ for $n \neq i$. Now we calculate $x^{k+1} = A^{\dagger}z^{k+1}$:

$$x_{j}^{k+1} = x_{j}^{k} + \omega D_{ii}^{-1}\overline{A_{ij}}(b_{i} - (Ax^{k})_{i}).$$

This is one step of the relaxed algebraic reconstruction technique applied to the original system of equations $Ax = b$. The relaxed ART converges to a solution, when solutions exist, for any $\omega \in (0, 2)$.

When $Ax = b$ is consistent, so is $AA^{\dagger}z = b$. We consider now the case in which $S = AA^{\dagger}$ is invertible. Since the relaxed ART sequence

$\{x^k = A^\dagger z^k\}$ converges to a solution x^∞, for any $\omega \in (0,2)$, the sequence $\{AA^\dagger z^k\}$ converges to b. Since $S = AA^\dagger$ is invertible, the SOR sequence $\{z^k\}$ then converges to $S^{-1}b$.

11 | Conjugate-Direction Methods in Optimization

Finding the least-squares solution of a possibly inconsistent system of linear equations $Ax = b$ is equivalent to minimizing the quadratic function $f(x) = \frac{1}{2}||Ax - b||_2^2$ and so can be viewed within the framework of optimization. Iterative optimization methods can then be used to provide, or at least suggest, algorithms for obtaining the least-squares solution. The *conjugate gradient method* is one such method.

11.1 Iterative Minimization

Iterative methods for minimizing a real-valued function $f(x)$ over the vector variable x usually take the following form: having obtained x^{k-1}, a new direction vector d^k is selected, an appropriate scalar $\alpha_k > 0$ is determined, and the next member of the iterative sequence is given by

$$x^k = x^{k-1} + \alpha_k d^k. \tag{11.1}$$

Ideally, one would choose the α_k to be the value of α for which the function $f(x^{k-1} + \alpha d^k)$ is minimized. It is assumed that the direction d^k is a *descent direction*; that is, for small positive α the function $f(x^{k-1} + \alpha d^k)$ is strictly decreasing. Finding the optimal value of α at each step of the iteration is difficult, if not impossible, in most cases, and approximate methods, using line searches, are commonly used.

Lemma 11.1. *For each k we have*

$$\nabla f(x^k) \cdot d^k = 0. \tag{11.2}$$

Proof: Differentiate the function $f(x^{k-1} + \alpha d^k)$ with respect to the variable α. $\qquad\square$

Since the gradient $\nabla f(x^k)$ is orthogonal to the previous direction vector d^k and also because $-\nabla f(x)$ is the direction of greatest decrease of $f(x)$, the choice of $d^{k+1} = -\nabla f(x^k)$ as the next direction vector is a reasonable one. With this choice we obtain Cauchy's *steepest descent method* [123]:

Algorithm 11.2 (Steepest Descent). Let x^0 be arbitrary. Then let

$$x^{k+1} = x^k - \alpha_{k+1} \nabla f(x^k).$$

The steepest descent method need not converge in general and even when it does, it can do so slowly, suggesting that there may be better choices for the direction vectors. For example, the Newton-Raphson method [130] employs the following iteration:

$$x^{k+1} = x^k - \nabla^2 f(x^k)^{-1} \nabla f(x^k),$$

where $\nabla^2 f(x)$ is the Hessian matrix for $f(x)$ at x. To investigate further the issues associated with the selection of the direction vectors, we consider the more tractable special case of quadratic optimization.

11.2   Quadratic Optimization

Let A be an arbitrary real I by J matrix. The linear system of equations $Ax = b$ need not have any solutions, and we may wish to find a least-squares solution $x = \hat{x}$ that minimizes

$$f(x) = \frac{1}{2} \|b - Ax\|_2^2. \tag{11.3}$$

The vector b can be written

$$b = A\hat{x} + \hat{w},$$

where $A^T \hat{w} = 0$ and a least-squares solution is an exact solution of the linear system $Qx = c$, with $Q = A^T A$ and $c = A^T b$. We shall assume that Q is invertible and there is a unique least-squares solution; this is the typical case.

We consider now the iterative scheme described by Equation (11.1) for $f(x)$ as in Equation (11.3). For this $f(x)$ the gradient becomes

$$\nabla f(x) = Qx - c.$$

The optimal α_k for the iteration can be obtained in closed form.

Lemma 11.3. *The optimal* α_k *is*

$$\alpha_k = \frac{r^k \cdot d^k}{d^k \cdot Q d^k}, \tag{11.4}$$

where $r^k = c - Q x^{k-1}$.

Lemma 11.4. *Let* $||x||_Q^2 = x \cdot Qx$ *denote the square of the Q-norm of* x. *Then*

$$||\hat{x} - x^{k-1}||_Q^2 - ||\hat{x} - x^k||_Q^2 = (r^k \cdot d^k)^2/d^k \cdot Qd^k \geq 0$$

for any direction vectors d^k.

If the sequence of direction vectors $\{d^k\}$ is completely general, the iterative sequence need not converge. However, if the set of direction vectors is finite and spans R^J and we employ them cyclically, convergence follows.

Theorem 11.5. *Let* $\{d^1, ..., d^J\}$ *be any finite set whose span is all of* R^J. *Let* α_k *be chosen according to Equation (11.4). Then, for* $k = 0, 1, ...,$ $j = k(\text{mod } J) + 1$, *and any* x^0, *the sequence defined by*

$$x^k = x^{k-1} + \alpha_k d^j$$

converges to the least-squares solution.

Proof: The sequence $\{||\hat{x} - x^k||_Q^2\}$ is decreasing and, therefore, the sequence $\{(r^k \cdot d^k)^2/d^k \cdot Qd^k$ must converge to zero. Therefore, the vectors x^k are bounded, and for each $j = 1, ..., J$, the subsequences $\{x^{mJ+j}, m = 0, 1, ...\}$ have cluster points, say $x^{*,j}$ with

$$x^{*,j} = x^{*,j-1} + \frac{(c - Qx^{*,j-1}) \cdot d^j}{d^j \cdot Qd^j} d^j.$$

Since

$$r^{mJ+j} \cdot d^j \to 0,$$

it follows that, for each $j = 1, ..., J$,

$$(c - Qx^{*,j}) \cdot d^j = 0.$$

Therefore,

$$x^{*,1} = ... = x^{*,J} = x^*$$

with $Qx^* = c$. Consequently, x^* is the least-squares solution and the sequence $\{||x^* - x^k||_Q\}$ is decreasing. But a subsequence converges to zero; therefore, $\{||x^* - x^k||_Q\} \to 0$. This completes the proof. $\square$

There is an interesting corollary to this theorem that pertains to a modified version of the ART algorithm. For $k = 0, 1, ...$ and $i = k(\text{mod } M) + 1$

and with the rows of A normalized to have length one, the ART iterative step is

$$x^{k+1} = x^k + (b_i - (Ax^k)_i)a^i,$$

where a^i is the ith column of A^T. When $Ax = b$ has no solutions, the ART algorithm does not converge to the least-squares solution; rather, it exhibits subsequential convergence to a limit cycle. However, using the previous theorem, we can show that the following modification of the ART, which we shall call the *least-squares* ART (LS-ART), converges to the least-squares solution for every x^0:

$$x^{k+1} = x^k + \frac{r^{k+1} \cdot a^i}{a^i \cdot Qa^i} a^i.$$

In the quadratic case the steepest descent iteration has the form

$$x^k = x^{k-1} + \frac{r^k \cdot r^k}{r^k \cdot Qr^k} r^k.$$

We have the following result.

Theorem 11.6. *The steepest descent method converges to the least-squares solution.*

Proof: As in the proof of the previous theorem, we have

$$\|\hat{x} - x^{k-1}\|_Q^2 - \|\hat{x} - x^k\|_Q^2 = (r^k \cdot d^k)^2/d^k \cdot Qd^k \geq 0,$$

where now the direction vectors are $d^k = r^k$. So, the sequence $\{\|\hat{x} - x^k\|_Q^2\}$ is decreasing, and therefore the sequence $\{(r^k \cdot r^k)^2/r^k \cdot Qr^k\}$ must converge to zero. The sequence $\{x^k\}$ is bounded; let x^* be a cluster point. It follows that $c - Qx^* = 0$, so that x^* is the least-squares solution $\hat{x}$. The rest of the proof follows as in the proof of the previous theorem.   □

11.3   Conjugate Bases for R^J

If the set $\{v^1, ..., v^J\}$ is a basis for R^J, then any vector x in R^J can be expressed as a linear combination of the basis vectors; that is, there are real numbers $a_1, ..., a_J$ for which

$$x = a_1 v^1 + a_2 v^2 + ... + a_J v^J.$$

For each x the coefficients a_j are unique. To determine the a_j we write

$$x \cdot v^m = a_1 v^1 \cdot v^m + a_2 v^2 \cdot v^m + ... + a_J v^J \cdot v^m,$$

for $m = 1, ..., M$. Having calculated the quantities $x \cdot v^m$ and $v^j \cdot v^m$, we solve the resulting system of linear equations for the a_j.

If the set $\{u^1, ..., u^M\}$ is an orthogonal basis, that is, then $u^j \cdot u^m = 0$, unless $j = m$, then the system of linear equations is now trivial to solve. The solution is $a_j = x \cdot u^j / u^j \cdot u^j$, for each j. Of course, we still need to compute the quantities $x \cdot u^j$.

The least-squares solution of the linear system of equations $Ax = b$ is

$$\hat{x} = (A^T A)^{-1} A^T b = Q^{-1} c.$$

To express $\hat{x}$ as a linear combination of the members of an orthogonal basis $\{u^1, ..., u^J\}$ we need the quantities $\hat{x} \cdot u^j$, which usually means that we need to know $\hat{x}$ first. For a special kind of basis, a Q-conjugate basis, knowing $\hat{x}$ ahead of time is not necessary; we need only know Q and c. Therefore, we can use such a basis to find $\hat{x}$. This is the essence of the conjugate gradient method (CGM), in which we calculate a conjugate basis and, in the process, determine $\hat{x}$.

11.3.1 Conjugate Directions

From Equation (11.2) we have

$$(c - Qx^{k+1}) \cdot d^k = 0,$$

which can be expressed as

$$(\hat{x} - x^{k+1}) \cdot Qd^k = (\hat{x} - x^{k+1})^T Qd^k = 0.$$

Definition 11.7. Two vectors x and y are said to be *Q-orthogonal* (or *Q-conjugate*, or just *conjugate*), if $x \cdot Qy = 0$.

So, the least-squares solution that we seek lies in a direction from x^{k+1} that is Q-orthogonal to d^k. This suggests that we can do better than steepest descent if we take the next direction to be Q-orthogonal to the previous one, rather than just orthogonal. This leads us to *conjugate direction methods*.

Lemma 11.8. *Say that the set $\{p^1, ..., p^n\}$ is a conjugate set for R^J if $p^i \cdot Qp^j = 0$ for $i \neq j$. Any conjugate set that does not contain zero is linearly independent. If $p^n \neq 0$ for $n = 1, ..., J$, then the least-squares vector $\hat{x}$ can be written as*

$$\hat{x} = a_1 p^1 + ... + a_J p^J,$$

with $a_j = c \cdot p^j / p^j \cdot Qp^j$ for each j.

Proof: Use the Q-inner product $\langle x, y \rangle_Q = x \cdot Qy$.

Therefore, once we have a conjugate basis, computing the least-squares solution is trivial. Generating a conjugate basis can obviously be done using the standard Gram-Schmidt approach.

11.3.2   The Gram-Schmidt Method

Let $\{v^1, ..., v^J\}$ be a linearly independent set of vectors in the space R^M, where $J \le M$. The Gram-Schmidt method uses the v^j to create an orthogonal basis $\{u^1, ..., u^J\}$ for the span of the v^j. Begin by taking $u^1 = v^1$. For $j = 2, ..., J$, let

$$u^j = v^j - \frac{u^1 \cdot v^j}{u^1 \cdot u^1} u^1 - \cdots - \frac{u^{j-1} \cdot v^j}{u^{j-1} \cdot u^{j-1}} u^{j-1}.$$

To apply this approach to obtain a conjugate basis, we would simply replace the dot products $u^k \cdot v^j$ and $u^k \cdot u^k$ with the Q-inner products, that is,

$$p^j = v^j - \frac{p^1 \cdot Qv^j}{p^1 \cdot Qp^1} p^1 - \cdots - \frac{p^{j-1} \cdot Qv^j}{p^{j-1} \cdot Qp^{j-1}} p^{j-1}. \tag{11.5}$$

Even though the Q-inner products can always be written as $x \cdot Qy = Ax \cdot Ay$, so that we need not compute the matrix Q, calculating a conjugate basis using Gram-Schmidt is not practical for large J. There is a way out, fortunately.

If we take $p^1 = v^1$ and $v^j = Qp^{j-1}$, we have a much more efficient mechanism for generating a conjugate basis, namely a three-term recursion formula [123]. The set $\{p^1, Qp^1, ..., Qp^{J-1}\}$ need not be a linearly independent set, in general, but, if our goal is to find $\hat{x}$, and not really to calculate a full conjugate basis, this does not matter, as we shall see.

Theorem 11.9. *Let $p^1 \ne 0$ be arbitrary. Let p^2 be given by*

$$p^2 = Qp^1 - \frac{Qp^1 \cdot Qp^1}{p^1 \cdot Qp^1} p^1,$$

so that $p^2 \cdot Qp^1 = 0$. Then, for $n \ge 2$, let p^{n+1} be given by

$$p^{n+1} = Qp^n - \frac{Qp^n \cdot Qp^n}{p^n \cdot Qp^n} p^n - \frac{Qp^{n-1} \cdot Qp^n}{p^{n-1} \cdot Qp^{n-1}} p^{n-1}. \tag{11.6}$$

Then, the set $\{p^1, ..., p^J\}$ is a conjugate set for R^J. If $p^n \ne 0$ for each n, then the set is a conjugate basis for R^J.

Proof: We consider the induction step of the proof. Assume that $\{p^1, ..., p^n\}$ is a Q-orthogonal set of vectors; we then show that $\{p^1, ..., p^{n+1}\}$ is also, provided that $n \le J - 1$. It is clear from Equation (11.6) that

$$p^{n+1} \cdot Qp^n = p^{n+1} \cdot Qp^{n-1} = 0.$$

For $j \le n - 2$, we have

$$p^{n+1} \cdot Qp^j = p^j \cdot Qp^{n+1} = p^j \cdot Q^2 p^n - ap^j \cdot Qp^n - bp^j \cdot Qp^{n-1},$$

for constants a and b. The second and third terms on the right side are then zero because of the induction hypothesis. The first term is also zero; we have

$$p^j \cdot Q^2 p^n = (Qp^j) \cdot Qp^n = 0,$$

because Qp^j is in the span of the vectors $\{p^1, ..., p^{j+1}\}$, and so is Q-orthogonal to p^n.   □

The calculations in the three-term recursion formula Equation (11.6) also occur in the Gram-Schmidt approach in Equation (11.5); the point is that Equation (11.6) uses only the first three terms, in every case.

11.4   The Conjugate Gradient Method

The main idea in the *conjugate gradient method* (CGM) is to build the conjugate set as we calculate the least-squares solution using the iterative algorithm

$$x^n = x^{n-1} + \alpha_n p^n. \tag{11.7}$$

The α_n is chosen so as to minimize the function of α defined by $f(x^{n-1} + \alpha p^n)$, and so we have

$$\alpha_n = \frac{r^n \cdot p^n}{p^n \cdot Qp^n},$$

where $r^n = c - Qx^{n-1}$. Since the function $f(x) = \frac{1}{2}\|Ax - b\|_2^2$ has for its gradient $\nabla f(x) = A^T(Ax - b) = Qx - c$, the residual vector $r^n = c - Qx^{n-1}$ is the direction of steepest descent from the point $x = x^{n-1}$. The CGM combines the use of the negative gradient directions from the steepest descent method with the use of a conjugate basis of directions, by using the r^{n+1} to construct the next direction p^{n+1} in such a way as to form a conjugate set $\{p_1, ..., p^J\}$.

As before, there is an efficient recursive formula that provides the next direction: let $p^1 = r^1 = (c - Qx^0)$ and

$$p^{n+1} = r^{n+1} - \frac{r^{n+1} \cdot Qp^n}{p^n \cdot Qp^n} p^n. \tag{11.8}$$

Since the α_n is the optimal choice and

$$r^{n+1} = -\nabla f(x^n),$$

we have, according to Equation (11.2),

$$r^{n+1} \cdot p^n = 0.$$

Lemma 11.10. *For all n, $r^{n+1} = 0$ whenever $p^{n+1} = 0$, in which case we have $c = Qx^n$, so that x^n is the least-squares solution.*

In theory, the CGM converges to the least-squares solution in finitely many steps, since we either reach $p^{n+1} = 0$ or $n + 1 = J$. In practice, the CGM can be employed as a fully iterative method by cycling back through the previously used directions.

An induction proof similar to the one used to prove Theorem 11.9 establishes that the set $\{p^1, ..., p^J\}$ is a conjugate set [123, 130]. In fact, we can say more.

Theorem 11.11. *For $n = 1, 2, ..., J$ and $j = 1, ..., n - 1$ we have*

(1) $r^n \cdot r^j = 0$;

(2) $r^n \cdot p^j = 0$; and

(3) $p^n \cdot Qp^j = 0$.

The proof presented here through a series of lemmas is based on that given in [130]. The proof uses induction on the number n. Throughout the following lemmas assume that the statements in the theorem hold for some $n < J$. We prove that they hold also for $n + 1$.

Lemma 11.12. *The vector Qp^j is in the span of the vectors r^j and r^{j+1}.*

Proof: Use the fact that

$$r^{j+1} = r^j - \alpha_j Qp^j. \qquad \square$$

Lemma 11.13. *For each n, $r^{n+1} \cdot r^n = 0$.*

Proof: Establish that

$$\alpha_n = \frac{r^n \cdot r^n}{p^n \cdot Qp^n}. \qquad \square$$

Lemma 11.14. *For $j = 1, ..., n - 1$, $r^{n+1} \cdot r^j = 0$.*

Proof: Use the induction hypothesis.

Lemma 11.15. *For $j = 1, ..., n$, $r^{n+1} \cdot p^j = 0$.*

Proof: First, establish that

$$p^j = r^j - \beta_{j-1} p^{j-1},$$

where

$$\beta_{j-1} = \frac{r^j \cdot Q p^{j-1}}{p^{j-1} \cdot Q p^{j-1}},$$

and

$$r^{n+1} = r^n - \alpha_n Q p^n. \qquad \square$$

Lemma 11.16. *For $j = 1, ..., n-1$, $p^{n+1} \cdot Q p^j = 0$.*

Proof: Use

$$Q p^j = \alpha_j^{-1} (r^j - r^{j+1}). \qquad \square$$

The final step in the proof is contained in the following lemma.

Lemma 11.17. *For each n, we have $p^{n+1} \cdot Q p^n = 0$.*

Proof: Establish that

$$\beta_n = -\frac{r^{n+1} \cdot r^{n+1}}{r^n \cdot r^n}. \qquad \square$$

The convergence rate of the CGM depends on the condition number of the matrix Q, which is the ratio of its largest to its smallest eigenvalues. When the condition number is much greater than one, convergence can be accelerated by *preconditioning* the matrix Q; this means replacing Q with $P^{-1/2} Q P^{-1/2}$, for some positive-definite approximation P of Q (see [4]).

There are versions of the CGM for the minimization of nonquadratic functions. In the quadratic case the next conjugate direction p^{n+1} is built from the residual r^{n+1} and p^n. Since, in that case, $r^{n+1} = -\nabla f(x^n)$, this suggests that in the nonquadratic case we build p^{n+1} from $-\nabla f(x^n)$ and p^n. This leads to the Fletcher-Reeves method. Other similar algorithms, such as the Polak-Ribiere and the Hestenes-Stiefel methods, perform better on certain problems [130].

11.5   Exercises

11.1. Look into the extension of the CGM to non-quadratic optimization.

V | Positivity in Linear Systems

12 | The Multiplicative ART (MART)

The *multiplicative* ART (MART) [94] is an iterative algorithm closely related to the ART. It applies to systems of linear equations $Ax = b$ for which the b_i are positive and the A_{ij} are nonnegative; the solution x we seek will have nonnegative entries. It is not so easy to see the relation between ART and MART if we look at the most general formulation of MART. For that reason, we begin with a simpler case, in which the relation is most clearly visible.

12.1 A Special Case of MART

We begin by considering the application of MART to the transmission tomography problem. For $i = 1, ..., I$, let L_i be the set of pixel indices j for which the jth pixel intersects the ith line segment, and let $|L_i|$ be the cardinality of the set L_i. Let $A_{ij} = 1$ for j in L_i, and $A_{ij} = 0$ otherwise. With $i = k(\operatorname{mod} I) + 1$, the iterative step of the ART algorithm is

$$x_j^{k+1} = x_j^k + \frac{1}{|L_i|}(b_i - (Ax^k)_i),$$

for j in L_i, and

$$x_j^{k+1} = x_j^k,$$

if j is not in L_i. In each step of ART, we take the error, $b_i - (Ax^k)_i$, associated with the current x^k and the ith equation, and distribute it equally over each of the pixels that intersects L_i.

Suppose, now, that each b_i is positive, and we know in advance that the desired image we wish to reconstruct must be nonnegative. We can begin with $x^0 > 0$, but as we compute the ART steps, we may lose nonnegativity.

One way to avoid this loss is to correct the current x^k multiplicatively, rather than additively, as in ART. This leads to the multiplicative ART (MART).

The MART, in this case, has the iterative step

$$x_j^{k+1} = x_j^k \left(\frac{b_i}{(Ax^k)_i} \right),$$

for those j in L_i, and

$$x_j^{k+1} = x_j^k,$$

otherwise. Therefore, we can write the iterative step as

$$x_j^{k+1} = x_j^k \left(\frac{b_i}{(Ax^k)_i} \right)^{A_{ij}}.$$

12.2   MART in the General Case

Taking the entries of the matrix A to be either one or zero, depending on whether or not the jth pixel is in the set L_i, is too crude. The line L_i may just clip a corner of one pixel, but pass through the center of another. Surely, it makes more sense to let A_{ij} be the length of the intersection of line L_i with the jth pixel, or, perhaps, this length divided by the length of the diagonal of the pixel. It may also be more realistic to consider a strip, instead of a line. Other modifications to A_{ij} may be made, in order to better describe the physics of the situation. Finally, all we can be sure of is that A_{ij} will be nonnegative, for each i and j. In such cases, what is the proper form for the MART?

The MART, which can be applied only to nonnegative systems, is a sequential, or row-action, method that uses one equation only at each step of the iteration.

Algorithm 12.1 (MART). Let x^0 be any positive vector, and $i = k(\bmod I) + 1$. Having found x^k for positive integer k, define x^{k+1} by

$$x_j^{k+1} = x_j^k \left(\frac{b_i}{(Ax^k)_i} \right)^{m_i^{-1} A_{ij}}, \qquad (12.1)$$

where $m_i = \max \{A_{ij} \,|\, j = 1, 2, ..., J\}$.

Some treatments of MART leave out the m_i, but require only that the entries of A have been rescaled so that $A_{ij} \leq 1$ for all i and j. The m_i is important, however, in accelerating the convergence of MART.

The MART can be accelerated by relaxation, as well.

Algorithm 12.2 (Relaxed MART). Let x^0 be any positive vector, and $i = k(\bmod I) + 1$. Having found x^k for positive integer k, define x^{k+1} by

$$x_j^{k+1} = x_j^k \left(\frac{b_i}{(Ax^k)_i} \right)^{\gamma_i m_i^{-1} A_{ij}}, \tag{12.2}$$

where γ_i is in the interval $(0, 1)$.

As with ART, finding the best relaxation parameters is a bit of an art.

In the consistent case, by which we mean that $Ax = b$ has nonnegative solutions, we have the following convergence theorem for MART.

Theorem 12.3. *In the consistent case, the MART converges to the unique nonnegative solution of $b = Ax$ for which $\sum_{j=1}^J s_j KL(x_j, x_j^0)$ is minimized.*

Here $s_j = \sum_{i=1}^I A_{ij} > 0$. If $s_j = 1$, for all j, and the starting vector x^0 is the vector whose entries are all one, then the MART converges to the solution that maximizes the Shannon entropy,

$$SE(x) = \sum_{j=1}^J x_j \log x_j - x_j.$$

As with ART, the speed of convergence is greatly affected by the ordering of the equations, converging most slowly when consecutive equations correspond to nearly parallel hyperplanes.

Open Question 4. When there are no nonnegative solutions, MART does not converge to a single vector, but, like ART, it is always observed to produce a limit cycle of vectors. Unlike ART, there is no proof of the existence of a limit cycle for MART.

12.3 ART and MART as Sequential Projection Methods

We know from our discussion of the ART that the iterative ART step can be viewed as the orthogonal projection of the current vector, x^k, onto H_i, the hyperplane associated with the ith equation. Can we view MART in a similar way? Yes, but we need to consider a different measure of closeness between nonnegative vectors.

12.3.1 Cross-Entropy or the Kullback-Leibler Distance

For positive numbers u and v, the Kullback-Leibler distance [114] from u to v is

$$KL(u, v) = u \log \frac{u}{v} + v - u. \tag{12.3}$$

We also define $KL(0,0) = 0$, $KL(0,v) = v$, and $KL(u,0) = +\infty$. The KL distance is extended to nonnegative vectors component-wise, so that, for nonnegative vectors x and z, we have

$$KL(x,z) = \sum_{j=1}^{J} KL(x_j, z_j). \tag{12.4}$$

One of the most useful facts about the KL distance is contained in the following lemma.

Lemma 12.4. *For nonnegative vectors x and z, with $z_+ = \sum_{j=1}^{J} z_j > 0$, we have*

$$KL(x,z) = KL(x_+, z_+) + KL(x, \frac{x_+}{z_+}z). \tag{12.5}$$

Given the vector x^k, we find the vector z in H_i for which the KL distance $f(z) = KL(x^k, z)$ is minimized; this z will be the KL projection of x^k onto H_i. Using a Lagrange multiplier, we find that

$$0 = \frac{\partial f}{\partial z_j}(z) - \lambda_i A_{ij},$$

for some constant λ_i, so that

$$0 = -\frac{x_j^k}{z_j} + 1 - \lambda_i A_{ij},$$

for each j. Multiplying by z_j, we get

$$z_j - x_j^k = z_j A_{ij} \lambda_i. \tag{12.6}$$

For the special case in which the entries of A_{ij} are zero or one, we can solve Equation (12.6) for z_j. We have

$$z_j - x_j^k = z_j \lambda_i,$$

for each $j \in L_i$, and $z_j = x_j^k$, otherwise. Multiply both sides by A_{ij} and sum on j to get

$$b_i(1 - \lambda_i) = (Ax^k)_i.$$

Therefore,

$$z_j = x_j^k \frac{b_i}{(Ax^k)_i},$$

which is clearly x_j^{k+1}. So, at least in the special case we have been discussing, MART consists of projecting, in the KL sense, onto each of the hyperplanes in succession.

12.3.2   Weighted KL Projections

For the more general case in which the entries A_{ij} are arbitrary nonnegative numbers, we cannot directly solve for z_j in Equation (12.6). There is an alternative, though. Instead of minimizing $KL(x, z)$, subject to $(Az)_i = b_i$, we minimize the weighted KL distance

$$\sum_{j=1}^{J} A_{ij} KL(x_j, z_j),$$

subject to the same constraint on z. The optimal z is $Q_i^e x$, which we shall denote here by $Q_i x$, the weighted KL projection of x onto the ith hyperplane. Again using a Lagrange multiplier approach, we find that

$$0 = -A_{ij}(\frac{x_j}{z_j} + 1) - A_{ij}\lambda_i,$$

for some constant λ_i. Multiplying by z_j, we have

$$A_{ij}z_j - A_{ij}x_j = A_{ij}z_j\lambda_i. \tag{12.7}$$

Summing over the index j, we get

$$b_i - (Ax)_i = b_i\lambda_i,$$

from which it follows that

$$1 - \lambda_i = (Ax)_i/b_i.$$

Substituting for λ_i in Equation (12.7), we obtain

$$z_j = (Q_i x)_j = x_j\frac{b_i}{(Ax)_i}, \tag{12.8}$$

for all j for which $A_{ij} \neq 0$.

Note that the MART step does not define x^{k+1} to be this weighted KL projection of x^k onto the hyperplane H_i; that is,

$$x_j^{k+1} \neq (Q_i x^k)_j,$$

except for those j for which $\frac{A_{ij}}{m_i} = 1$. What is true is that the MART step involves relaxation. Writing

$$x_j^{k+1} = (x_j^k)^{1-m_i^{-1}A_{ij}} \left(x_j^k \frac{b_i}{(Ax^k)_i}\right)^{m_i^{-1}A_{ij}},$$

we see that x_j^{k+1} is a weighted geometric mean of x_j^k and $(Q_i x^k)_j$.

12.4   Proof of Convergence for MART

We assume throughout this proof that $\hat{x}$ is a nonnegative solution of $Ax = b$. For $i = 1, 2, ..., I$, let

$$G_i(x, z) = KL(x, z) + m_i^{-1}KL((Ax)_i, b_i) - m_i^{-1}KL((Ax)_i, (Az)_i).$$

Lemma 12.5. *For all i, we have $G_i(x, z) \geq 0$ for all x and z.*

Proof: Use Equation (12.5).   □

Then $G_i(x, z)$, viewed as a function of z, is minimized by $z = x$, as we see from the equation

$$G_i(x, z) = G_i(x, x) + KL(x, z) - m_i^{-1}KL((Ax)_i, (Az)_i). \qquad (12.9)$$

Viewed as a function of x, $G_i(x, z)$ is minimized by $x = z'$, where

$$z'_j = z_j \left(\frac{b_i}{(Az)_i} \right)^{m_i^{-1}A_{ij}},$$

as we see from the equation

$$G_i(x, z) = G_i(z', z) + KL(x, z'). \qquad (12.10)$$

We note that $x^{k+1} = (x^k)'$.

Now we calculate $G_i(\hat{x}, x^k)$ in two ways, using, first, the definition, and, second, Equation (12.10). From the definition, we have

$$G_i(\hat{x}, x^k) = KL(\hat{x}, x^k) - m_i^{-1}KL(b_i, (Ax^k)_i).$$

From Equation (12.10), we have

$$G_i(\hat{x}, x^k) = G_i(x^{k+1}, x^k) + KL(\hat{x}, x^{k+1}).$$

Therefore,

$$KL(\hat{x}, x^k) - KL(\hat{x}, x^{k+1}) = G_i(x^{k+1}, x^k) + m_i^{-1}KL(b_i, (Ax^k)_i). \ (12.11)$$

From Equation (12.11) we can conclude several things:

(1) the sequence $\{KL(\hat{x}, x^k)\}$ is decreasing;

(2) the sequence $\{x^k\}$ is bounded, and therefore has a cluster point, x^*; and

(3) the sequences $\{G_i(x^{k+1}, x^k)\}$ and $\{m_i^{-1}KL(b_i, (Ax^k)_i)\}$ converge decreasingly to zero, and so $b_i = (Ax^*)_i$ for all i.

Since $b = Ax^*$, we can use x^* in place of the arbitrary solution $\hat{x}$ to conclude that the sequence $\{KL(x^*, x^k)\}$ is decreasing. But, a subsequence converges to zero, so the entire sequence must converge to zero, and therefore $\{x^k\}$ converges to x^*. Finally, since the right side of Equation (12.11) is independent of which solution $\hat{x}$ we have used, so is the left side. Summing over k on the left side, we find that

$$KL(\hat{x}, x^0) - KL(\hat{x}, x^*)$$

is independent of which $\hat{x}$ we use. We can conclude then that minimizing $KL(\hat{x}, x^0)$ over all solutions $\hat{x}$ has the same answer as minimizing $KL(\hat{x}, x^*)$ over all such $\hat{x}$; but the solution to the latter problem is obviously $\hat{x} = x^*$. This concludes the proof of convergence.

12.5 Comments on the Rate of Convergence of MART

We can see from Equation (12.11),

$$KL(\hat{x}, x^k) - KL(\hat{x}, x^{k+1}) = G_i(x^{k+1}, x^k) + m_i^{-1} KL(b_i, (Ax^k)_i),$$

that the decrease in distance to a solution that occurs with each step of MART depends on m_i^{-1} and on $KL(b_i, (Ax^k)_i)$; the latter measures the extent to which the current vector x^k solves the current equation. We see then that it is reasonable to select m_i as we have done, namely, as the smallest positive number c_i for which $A_{ij}/c_i \leq 1$ for all j. We also see that it is helpful if the equations are ordered in such a way that $KL(b_i, (Ax^k)_i)$ is fairly large, for each k. It is not usually necessary to determine an optimal ordering of the equations; the important thing is to avoid ordering the equations so that successive hyperplanes have nearly parallel normal vectors.

13 Rescaled Block-Iterative (RBI) Methods

Image reconstruction problems in tomography are often formulated as statistical likelihood maximization problems in which the pixel values of the desired image play the role of parameters. Iterative algorithms based on cross-entropy minimization, such as the *expectation maximization maximum likelihood* (EMML) method and the *simultaneous multiplicative algebraic reconstruction technique* (SMART) can be used to solve such problems. Because the EMML and SMART are slow to converge for large amounts of data typical in imaging problems, acceleration of the algorithms using blocks of data or ordered subsets has become popular. There are a number of different ways to formulate these block-iterative versions of EMML and SMART, involving the choice of certain normalization and regularization parameters. These methods are not faster merely because they are block-iterative; the correct choice of the parameters is crucial. The purpose of this chapter is to discuss these different formulations in detail sufficient to reveal the precise roles played by the parameters and to guide the user in choosing them.

13.1 Overview

The algorithms we discuss here have interesting histories, which we sketch in this section.

13.1.1 The SMART and Its Variants

Like the ART, the MART has a simultaneous version, called the SMART. Like MART, SMART applies only to nonnegative systems of equations. Unlike MART, SMART is a simultaneous algorithm that uses all equations in each step of the iteration. The SMART was discovered in 1972, independently, by Darroch and Ratcliff, working in statistics, [73] and by Schmidlin [141] in medical imaging; neither work makes reference to MART. Darroch and Ratcliff do consider block-iterative versions of their algorithm, in which only some of the equations are used at each step, but their convergence proof involves unnecessary restrictions on the system matrix. Censor and Segman [59] seem to be the first to present the SMART and its block-iterative variants explicitly as generalizations of MART.

13.1.2   The EMML and Its Variants

The expectation maximization maximum likelihood (EMML) method turns out to be closely related to the SMART, although it has quite a different history. The EMML algorithm we discuss here is actually a special case of a more general approach to likelihood maximization, usually called the EM algorithm [75]; the book by McLachnan and Krishnan [125] is a good source for the history of this more general algorithm.

It was noticed by Rockmore and Macovski [140] that certain image reconstruction problems posed by medical tomography could be formulated as statistical parameter estimation problems. Following up on this idea, Shepp and Vardi [143] suggested the use of the EM algorithm for solving the reconstruction problem in emission tomography. In [116], Lange and Carson presented an EM-type iterative method for transmission tomographic image reconstruction, and pointed out a gap in the convergence proof given in [143] for the emission case. In [152], Vardi, Shepp, and Kaufman repaired the earlier proof, relying on techniques due to Csiszár and Tusnády [70]. In [117] Lange, Bahn, and Little improved the transmission and emission algorithms, by including regularization to reduce the effects of noise. The question of uniqueness of the solution in the inconsistent case was resolved in [34].

The MART and SMART were initially designed to apply to consistent systems of equations. Darroch and Ratcliff did not consider what happens in the inconsistent case, in which the system of equations has no nonnegative solutions; this issue was resolved in [34], where it was shown that the SMART converges to a nonnegative minimizer of the Kullback-Leibler distance $KL(Ax, b)$. The EMML, as a statistical parameter estimation technique, was not originally thought to be connected to any system of linear equations. In [34], it was shown that the EMML leads to a nonnegative minimizer of the Kullback-Leibler distance $KL(b, Ax)$, thereby exhibiting a close connection between the SMART and the EMML methods. Consequently, when the nonnegative system of linear equations $Ax = b$ has a nonnegative solution, the EMML converges to such a solution.

13.1.3   Block-Iterative Versions of SMART and EMML

As we have seen, Darroch and Ratcliff included what are now called block-iterative versions of SMART in their original paper [73]. Censor and Segman [59] viewed SMART and its block-iterative versions as natural extension of the MART. Consequently, block-iterative variants of SMART have been around for some time. The story with the EMML is quite different.

The paper of Holte, Schmidlin, et al. [105] compares the performance of Schmidlin's method of [141] with the EMML algorithm. Almost as an aside, they notice the accelerating effect of what they call *projection interleaving*, that is, the use of blocks. This paper contains no explicit formulas, however, and presents no theory, so one can only make educated guesses as to the precise iterative methods employed. Somewhat later, Hudson, Hutton, and Larkin [106,107] observed that the EMML can be significantly accelerated if, at each step, one employs only some of the data. They referred to this approach as the *ordered subset* EM method (OSEM). They gave a proof of convergence of the OSEM, for the consistent case. The proof relied on a fairly restrictive relationship between the matrix A and the choice of blocks, called *subset balance*. In [37] a revised version of the OSEM, called the *rescaled block-iterative* EMML (RBI-EMML), was shown to converge, in the consistent case, regardless of the choice of blocks.

13.1.4 Basic Assumptions

Methods based on cross-entropy, such as the MART, SMART, EMML, and all block-iterative versions of these algorithms apply to nonnegative systems that we denote by $Ax = b$, where b is a vector of positive entries, A is a matrix with entries $A_{ij} \geq 0$ such that for each j the sum $s_j = \sum_{i=1}^{I} A_{ij}$ is positive, and we seek a solution x with nonnegative entries. If no nonnegative x satisfies $b = Ax$ we say the system is *inconsistent*.

Simultaneous iterative algorithms employ all of the equations at each step of the iteration; block-iterative methods do not. For the latter methods we assume that the index set $\{i = 1, ..., I\}$ is the (not necessarily disjoint) union of the N sets or *blocks* B_n, $n = 1, ..., N$. We shall require that $s_{nj} = \sum_{i \in B_n} A_{ij} > 0$ for each n and each j. Block-iterative methods like ART and MART for which each block consists of precisely one element are called *row-action* or *sequential* methods. We begin our discussion with the SMART and the EMML method.

13.2   The SMART and the EMML Algorithm

Both the SMART and the EMML algorithm provide a solution of $b = Ax$ when such exist and (distinct) approximate solutions in the inconsistent case.

Algorithm 13.1 (SMART). *Let x^0 be an arbitrary positive vector. For $k = 0, 1, \ldots$ let*

$$x_j^{k+1} = x_j^k \exp \left(s_j^{-1} \sum_{i=1}^{I} A_{ij} \log \frac{b_i}{(Ax^k)_i} \right). \tag{13.1}$$

The exponential and logarithm in the SMART iterative step are computationally expensive. The EMML algorithm is similar to the SMART, but somewhat less costly to compute.

Algorithm 13.2 (EMML). *Let x^0 be an arbitrary positive vector. For $k = 0, 1, \ldots$ let*

$$x_j^{k+1} = x_j^k s_j^{-1} \sum_{i=1}^{I} A_{ij} \frac{b_i}{(Ax^k)_i}. \tag{13.2}$$

The main results concerning the SMART are given by the following theorem.

Theorem 13.3. *In the consistent case the SMART converges to the unique nonnegative solution of $b = Ax$ for which $\sum_{j=1}^{J} s_j KL(x_j, x_j^0)$ is minimized. In the inconsistent case it converges to the unique nonnegative minimizer of the distance $KL(Ax, y)$ for which $\sum_{j=1}^{J} s_j KL(x_j, x_j^0)$ is minimized; if A and every matrix derived from A by deleting columns has full rank, then there is a unique nonnegative minimizer of $KL(Ax, y)$ and at most $I - 1$ of its entries are nonzero.*

For the EMML method the main results are the following.

Theorem 13.4. *In the consistent case the EMML algorithm converges to a nonnegative solution of $b = Ax$. In the inconsistent case it converges to a nonnegative minimizer of the distance $KL(y, Ax)$; if A and every matrix derived from A by deleting columns has full rank, then there is a unique nonnegative minimizer of $KL(y, Ax)$ and at most $I - 1$ of its entries are nonzero.*

In the consistent case there may be multiple nonnegative solutions and the one obtained by the EMML algorithm will depend on the starting vector x^0; how it depends on x^0 is an open question.

These theorems are special cases of more general results on block-iterative methods that we shall prove later in this chapter.

Both the EMML and SMART are related to likelihood maximization. Minimizing the function $KL(y, Ax)$ is equivalent to maximizing the likelihood when the b_i are taken to be measurements of independent Poisson random variables having means $(Ax)_i$. The entries of x are the parameters to be determined. This situation arises in emission tomography. So the EMML is a likelihood maximizer, as its name suggests.

The connection between SMART and likelihood maximization is a bit more convoluted. Suppose that $s_j = 1$ for each j. The solution of $b = Ax$ for which $KL(x, x^0)$ is minimized necessarily has the form

$$x_j = x_j^0 \exp \left(\sum_{i=1}^{I} A_{ij} \lambda_i \right) \tag{13.3}$$

for some vector λ with entries λ_i. This *log linear* form also arises in transmission tomography, where it is natural to assume that $s_j = 1$ for each j and $\lambda_i \le 0$ for each i. We have the following lemma that helps to connect the SMART algorithm with the transmission tomography problem:

Lemma 13.5. *Minimizing $KL(d, x)$ over x as in Equation (13.3) is equivalent to minimizing $KL(x, x^0)$, subject to $Ax = Ad$.*

The solution to the latter problem can be obtained using the SMART.

With $x_+ = \sum_{j=1}^{J} x_j$ the vector A with entries $p_j = x_j / x_+$ is a probability vector. Let $d = (d_1, ..., d_J)^T$ be a vector whose entries are nonnegative integers, with $K = \sum_{j=1}^{J} d_j$. Suppose that, for each j, p_j is the probability of index j and d_j is the number of times index j was chosen in K trials. The likelihood function of the parameters λ_i is

$$L(\lambda) = \prod_{j=1}^{J} p_j^{d_j}$$

so that the log-likelihood function is

$$LL(\lambda) = \sum_{j=1}^{J} d_j \log p_j. \tag{13.4}$$

Since A is a probability vector, maximizing $L(\lambda)$ is equivalent to minimizing $KL(d, p)$ with respect to λ, which, according to the lemma above, can be solved using SMART. In fact, since all of the block-iterative versions of SMART have the same limit whenever they have the same starting vector, any of these methods can be used to solve this maximum likelihood problem. In the case of transmission tomography the λ_i must be nonpositive, so if SMART is to be used, some modification is needed to obtain such a solution.

Those who have used the SMART or the EMML on sizable problems have certainly noticed that they are both slow to converge. An important issue, therefore, is how to accelerate convergence. One popular method is through the use of *block-iterative* (or *ordered subset*) methods.

13.3   Ordered-Subset Versions

To illustrate block-iterative methods and to motivate our subsequent discussion we consider now the ordered subset EM algorithm (OSEM), which is a popular technique in some areas of medical imaging, as well as an analogous version of SMART, which we shall call here the OSSMART. The OSEM is now used quite frequently in tomographic image reconstruction, where it is acknowledged to produce usable images significantly faster then EMML. From a theoretical perspective both OSEM and OSSMART are incorrect. How to correct them is the subject of much that follows here.

The idea behind the OSEM (OSSMART) is simple: the iteration looks very much like the EMML (SMART), but at each step of the iteration the summations are taken only over the current block. The blocks are processed cyclically.

The OSEM iteration is the following: for $k = 0, 1, \ldots$ and $n = k(\bmod N)$ $+ 1$, having found x^k let

$$x_j^{k+1} = x_j^k s_{nj}^{-1} \sum_{i \in B_n} A_{ij} \frac{b_i}{(Ax^k)_i}. \tag{13.5}$$

The OSSMART has the following iterative step:

$$x_j^{k+1} = x_j^k \exp\left(s_{nj}^{-1} \sum_{i \in B_n} A_{ij} \log \frac{b_i}{(Ax^k)_i}\right). \tag{13.6}$$

In general we do not expect block-iterative algorithms to converge in the inconsistent case, but to exhibit *subsequential convergence* to a *limit cycle*, as we shall discuss later. We do, however, want them to converge to a solution in the consistent case; the OSEM and OSSMART fail to do this except when the matrix A and the set of blocks $\{B_n, n = 1, \ldots, N\}$ satisfy the condition known as *subset balance*, which means that the sums s_{nj} depend only on j and not on n. While this may be approximately valid in some special cases, it is overly restrictive, eliminating, for example, almost every set of blocks whose cardinalities are not all the same. When the OSEM does well in practice in medical imaging it is probably because the N is not large and only a few iterations are carried out.

The experience with the OSEM was encouraging, however, and strongly suggested that an equally fast, but mathematically correct, block-iterative version of EMML was to be had; this is the *rescaled block-iterative* EMML (RBI-EMML). Both RBI-EMML and an analogous corrected version of OSSMART, the RBI-SMART, provide fast convergence to a solution in the consistent case, for any choice of blocks.

13.4   The RBI-SMART

We turn next to the block-iterative versions of the SMART, which we shall denote BI-SMART. These methods were known prior to the discovery of RBI-EMML and played an important role in that discovery; the importance of rescaling for acceleration was apparently not appreciated, however.

We start by considering a formulation of BI-SMART that is general enough to include all of the variants we wish to discuss. As we shall see, this formulation is too general and will need to be restricted in certain ways to obtain convergence. Let the iterative step be

$$x_j^{k+1} = x_j^k \exp\left(\beta_{nj} \sum_{i\in B_n} \alpha_{ni} A_{ij} \log\left(\frac{b_i}{(Ax^k)_i}\right)\right), \qquad (13.7)$$

for $j = 1, 2, ..., J$, $n = k(\mathrm{mod}\,N) + 1$ and β_{nj} and α_{ni} positive. As we shall see, our convergence proof will require that β_{nj} be separable, that is, $\beta_{nj} = \gamma_j \delta_n$ for each j and n and that

$$\gamma_j \delta_n \sigma_{nj} \leq 1, \qquad (13.8)$$

for $\sigma_{nj} = \sum_{i\in B_n} \alpha_{ni} A_{ij}$. With these conditions satisfied we have the following result.

Theorem 13.6. *Let x be a nonnegative solution of $b = Ax$. For any positive vector x^0 and any collection of blocks $\{B_n,\ n = 1, ..., N\}$ the sequence $\{x^k\}$ given by Equation (13.7) converges to the unique solution of $b = Ax$ for which the weighted cross-entropy $\sum_{j=1}^{J} \gamma_j^{-1} KL(x_j, x_j^0)$ is minimized.*

The inequality in the following lemma is the basis for the convergence proof.

Lemma 13.7. *Let $b = Ax$ for some nonnegative x. Then for $\{x^k\}$ as in Equation (13.7) we have*

$$\sum_{j=1}^{J} \gamma_j^{-1} KL(x_j, x_j^k) - \sum_{j=1}^{J} \gamma_j^{-1} KL(x_j, x_j^{k+1}) \geq$$

$$\delta_n \sum_{i\in B_n} \alpha_{ni} KL(b_i, (Ax^k)_i). \quad (13.9)$$

Proof: First note that

$$x_j^{k+1} = x_j^k \exp\left(\gamma_j \delta_n \sum_{i\in B_n} \alpha_{ni} A_{ij} \log\left(\frac{b_i}{(Ax^k)_i}\right)\right), \qquad (13.10)$$

and

$$\exp\left(\gamma_j \delta_n \sum_{i\in B_n} \alpha_{ni} A_{ij} \log\left(\frac{b_i}{(Ax^k)_i}\right)\right)$$

can be written as

$$\exp\left((1 - \gamma_j \delta_n \sigma_{nj}) \log 1 + \gamma_j \delta_n \sum_{i \in B_n} \alpha_{ni} A_{ij} \log\left(\frac{b_i}{(Ax^k)_i}\right)\right),$$

which, by the convexity of the exponential function, is not greater than

$$(1 - \gamma_j \delta_n \sigma_{nj}) + \gamma_j \delta_n \sum_{i \in B_n} \alpha_{ni} A_{ij} \frac{b_i}{(Ax^k)_i}.$$

It follows that

$$\sum_{j=1}^{J} \gamma_j^{-1}(x_j^k - x_j^{k+1}) \geq \delta_n \sum_{i \in B_n} \alpha_{ni}((Ax^k)_i - b_i).$$

We also have

$$\log(x_j^{k+1}/x_j^k) = \gamma_j \delta_n \sum_{i \in B_n} \alpha_{ni} A_{ij} \log \frac{b_i}{(Ax^k)_i}.$$

Therefore

$$\sum_{j=1}^{J} \gamma_j^{-1} KL(x_j, x_j^k) - \sum_{j=1}^{J} \gamma_j^{-1} KL(x_j, x_j^{k+1})$$

$$= \sum_{j=1}^{J} \gamma_j^{-1}(x_j \log(x_j^{k+1}/x_j^k) + x_j^k - x_j^{k+1})$$

$$= \sum_{j=1}^{J} x_j \delta_n \sum_{i \in B_n} \alpha_{ni} A_{ij} \log \frac{b_i}{(Ax^k)_i} + \sum_{j=1}^{J} \gamma_j^{-1}(x_j^k - x_j^{k+1})$$

$$= \delta_n \sum_{i \in B_n} \alpha_{ni}(\sum_{j=1}^{J} x_j A_{ij}) \log \frac{b_i}{(Ax^k)_i} + \sum_{j=1}^{J} \gamma_j^{-1}(x_j^k - x_j^{k+1})$$

$$\geq \delta_n \left(\sum_{i \in B_n} \alpha_{ni}(b_i \log \frac{b_i}{(Ax^k)_i} + (Ax^k)_i - b_i) \right)$$

$$= \delta_n \sum_{i \in B_n} \alpha_{ni} KL(b_i, (Ax^k)_i).$$

This completes the proof of the lemma.   □

From the inequality (13.9) we conclude that the sequence

$$\left\{ \sum_{j=1}^{J} \gamma_j^{-1} KL(x_j, x_j^k) \right\}$$

is decreasing, that $\{x^k\}$ is therefore bounded, and the sequence

$$\left\{\sum_{i \in B_n} \alpha_{ni} KL(b_i, (Ax^k)_i)\right\}$$

is converging to zero. Let x^* be any cluster point of the sequence $\{x^k\}$. Then it is not difficult to show that $b = Ax^*$. Replacing x with x^*, we have that the sequence

$$\left\{\sum_{j=1}^{J} \gamma_j^{-1} KL(x_j^*, x_j^k)\right\}$$

is decreasing; since a subsequence converges to zero, so does the whole sequence. Therefore x^* is the limit of the sequence $\{x^k\}$. This proves that the algorithm produces a solution of $b = Ax$. To conclude further that the solution is the one for which the quantity

$$\sum_{j=1}^{J} \gamma_j^{-1} KL(x_j, x_j^0)$$

is minimized requires further work to replace the inequality (13.9) with an equation in which the right side is independent of the particular solution x chosen; see the final section of this chapter for the details.

We see from the theorem that how we select the γ_j is determined by how we wish to weight the terms in the sum $\sum_{j=1}^{J} \gamma_j^{-1} KL(x_j, x_j^0)$. In some cases we want to minimize the cross-entropy $KL(x, x^0)$ subject to $b = Ax$; in this case we would select $\gamma_j = 1$. In other cases we may have some prior knowledge as to the relative sizes of the x_j and wish to emphasize the smaller values more; then we may choose γ_j proportional to our prior estimate of the size of x_j. Having selected the γ_j, we see from the inequality (13.9) that convergence will be accelerated if we select δ_n as large as permitted by the condition $\gamma_j \delta_n \sigma_{nj} \leq 1$. This suggests that we take

$$\delta_n = 1/\min\{\sigma_{nj}\gamma_j, j = 1, ..., J\}. \tag{13.11}$$

The *rescaled* BI-SMART (RBI-SMART) as presented in [36,38,39] uses this choice, but with $\alpha_{ni} = 1$ for each n and i. For each $n = 1, ..., N$ let

$$m_n = \max\{s_{nj}s_j^{-1} | j = 1, ..., J\}.$$

The original RBI-SMART is as follows:

Algorithm 13.8 (RBI-SMART). *Let* x^0 *be an arbitrary positive vector. For* $k = 0, 1, ...,$ *let* $n = k \pmod{N} + 1$. *Then let*

$$x_j^{k+1} = x_j^k \exp\left(m_n^{-1} s_j^{-1} \sum_{i \in B_n} A_{ij} \log\left(\frac{b_i}{(Ax^k)_i}\right)\right). \tag{13.12}$$

Notice that Equation (13.12) can be written as

$$\log x_j^{k+1} = (1 - m_n^{-1} s_j^{-1} s_{nj}) \log x_j^k + m_n^{-1} s_j^{-1} \sum_{i \in B_n} A_{ij} \log\left(x_j^k \frac{b_i}{(Ax^k)_i}\right),$$

$$\tag{13.13}$$

from which we see that x_j^{k+1} is a weighted geometric mean of x_j^k and the weighted KL projections $(Q_i x^k)_j$, for $i \in B_n$. This will be helpful in deriving block-iterative versions of the EMML algorithm.

Let's look now at some of the other choices for these parameters that have been considered in the literature.

First, we notice that the OSSMART does not generally satisfy the requirements, since in Equation (13.6) the choices are $\alpha_{ni} = 1$ and $\beta_{nj} = s_{nj}^{-1}$; the only times this is acceptable is if the s_{nj} are separable; that is, $s_{nj} = r_j t_n$ for some r_j and t_n. This is slightly more general than the condition of subset balance and is sufficient for convergence of OSSMART.

In [59] Censor and Segman make the choices $\beta_{nj} = 1$ and $\alpha_{ni} > 0$ such that $\sigma_{nj} \leq 1$ for all n and j. In those cases in which σ_{nj} is much less than 1 for each n and j their iterative scheme is probably excessively relaxed; it is hard to see how one might improve the rate of convergence by altering only the weights α_{ni}, however. Limiting the choice to $\gamma_j \delta_n = 1$ reduces our ability to accelerate this algorithm.

The original SMART in Equation (13.1) uses $N = 1$, $\gamma_j = s_j^{-1}$, and $\alpha_{ni} = \alpha_i = 1$. Clearly the inequality (13.8) is satisfied; in fact it becomes an equality now.

For the row-action version of SMART, the *multiplicative* ART (MART), due to Gordon, Bender, and Herman [94], we take $N = I$ and $B_n = B_i = \{i\}$ for $i = 1, ..., I$. The MART has the iterative

$$x_j^{k+1} = x_j^k \left(\frac{b_i}{(Ax^k)_i}\right)^{m_i^{-1} A_{ij}},$$

for $j = 1, 2, ..., J$, $i = k \pmod{I} + 1$ and $m_i > 0$ chosen so that $m_i^{-1} A_{ij} \leq 1$ for all j. The smaller m_i is the faster the convergence, so a good choice is $m_i = \max\{A_{ij}|, j = 1, ..., J\}$. Although this particular choice for m_i is not explicitly mentioned in the various discussions of MART I have seen, it was used in implementations of MART from the beginning [103].

Darroch and Ratcliff included a discussion of a block-iterative version of SMART in their 1972 paper [73]. Close inspection of their version reveals that they require that $s_{nj} = \sum_{i \in B_n} A_{ij} = 1$ for all j. Since this is unlikely to be the case initially, we might try to rescale the equations or unknowns to obtain this condition. However, unless $s_{nj} = \sum_{i \in B_n} A_{ij}$ depends only on j and not on n, which is the *subset balance* property used in [107], we cannot redefine the unknowns in a way that is independent of n.

The MART fails to converge in the inconsistent case. What is always observed, but for which no proof exists, is that, for each fixed $i = 1, 2, ..., I$, as $m \to +\infty$, the MART subsequences $\{x^{mI+i}\}$ converge to separate limit vectors, say $x^{\infty,i}$. This *limit cycle* LC $= \{x^{\infty,i} | i = 1, ..., I\}$ reduces to a single vector whenever there is a nonnegative solution of $b = Ax$. The greater the minimum value of $KL(Ax, y)$ the more distinct from one another the vectors of the limit cycle are. An analogous result is observed for BI-SMART.

13.5   The RBI-EMML

As we did with SMART, we consider now a formulation of BI-EMML that is general enough to include all of the variants we wish to discuss. Once again, the formulation is too general and will need to be restricted in certain ways to obtain convergence. Let the iterative step be

$$x_j^{k+1} = x_j^k(1 - \beta_{nj}\sigma_{nj}) + x_j^k \beta_{nj} \sum_{i \in B_n} \alpha_{ni} A_{ij} \frac{b_i}{(Ax^k)_i}, \qquad (13.14)$$

for $j = 1, 2, ..., J$, $n = k(\mathrm{mod}\, N) + 1$, and β_{nj} and α_{ni} positive. As in the case of BI-SMART, our convergence proof will require that β_{nj} be separable, that is,

$$\beta_{nj} = \gamma_j \delta_n$$

for each j and n and that Inequality (13.8) holds. With these conditions satisfied we have the following result.

Theorem 13.9. *Let x be a nonnegative solution of $b = Ax$. For any positive vector x^0 and any collection of blocks $\{B_n, n = 1, ..., N\}$ the sequence $\{x^k\}$ given by Equation (13.7) converges to a nonnegative solution of $b = Ax$.*

When there are multiple nonnegative solutions of $b = Ax$ the solution obtained by BI-EMML will depend on the starting point x^0, but precisely how it depends on x^0 is an open question. Also, in contrast to the case of

BI-SMART, the solution can depend on the particular choice of the blocks. The inequality in the following lemma is the basis for the convergence proof.

Lemma 13.10. *Let* $b = Ax$ *for some nonnegative* x. *Then for* $\{x^k\}$ *as in Equation (13.14) we have*

$$\sum_{j=1}^{J} \gamma_j^{-1} KL(x_j, x_j^k) - \sum_{j=1}^{J} \gamma_j^{-1} KL(x_j, x_j^{k+1}) \geq$$

$$\delta_n \sum_{i \in B_n} \alpha_{ni} KL(b_i, (Ax^k)_i). \quad (13.15)$$

Proof: From the iterative step

$$x_j^{k+1} = x_j^k(1 - \gamma_j \delta_n \sigma_{nj}) + x_j^k \gamma_j \delta_n \sum_{i \in B_n} \alpha_{ni} A_{ij} \frac{b_i}{(Ax^k)_i}$$

we have

$$\log(x_j^{k+1}/x_j^k) = \log\left((1 - \gamma_j \delta_n \sigma_{nj}) + \gamma_j \delta_n \sum_{i \in B_n} \alpha_{ni} A_{ij} \frac{b_i}{(Ax^k)_i}\right).$$

By the concavity of the logarithm we obtain the inequality

$$\log(x_j^{k+1}/x_j^k) \geq \left((1 - \gamma_j \delta_n \sigma_{nj}) \log 1 + \gamma_j \delta_n \sum_{i \in B_n} \alpha_{ni} A_{ij} \log \frac{b_i}{(Ax^k)_i}\right),$$

or

$$\log(x_j^{k+1}/x_j^k) \geq \gamma_j \delta_n \sum_{i \in B_n} \alpha_{ni} A_{ij} \log \frac{b_i}{(Ax^k)_i}.$$

Therefore

$$\sum_{j=1}^{J} \gamma_j^{-1} x_j \log(x_j^{k+1}/x_j^k) \geq \delta_n \sum_{i \in B_n} \alpha_{ni} \left(\sum_{j=1}^{J} x_j A_{ij}\right) \log \frac{b_i}{(Ax^k)_i}.$$

Note that it is at this step that we used the separability of the β_{nj}. Also

$$\sum_{j=1}^{J} \gamma_j^{-1} (x_j^{k+1} - x_j^k) = \delta_n \sum_{i \in B_n} ((Ax^k)_i - b_i).$$

This concludes the proof of the lemma. □

From the inequality in (13.15) we conclude, as we did in the BI-SMART case, that the sequence $\{\sum_{j=1}^{J} \gamma_j^{-1} KL(x_j, x_j^k)\}$ is decreasing, that $\{x^k\}$ is therefore bounded, and the sequence $\{\sum_{i \in B_n} \alpha_{ni} KL(b_i, (Ax^k)_i)\}$ is converging to zero. Let x^* be any cluster point of the sequence $\{x^k\}$. Then it is not difficult to show that $b = Ax^*$. Replacing x with x^*, we have that the sequence

$$\left\{ \sum_{j=1}^{J} \gamma_j^{-1} KL(x_j^*, x_j^k) \right\}$$

is decreasing; since a subsequence converges to zero, so does the whole sequence. Therefore x^* is the limit of the sequence $\{x^k\}$. This proves that the algorithm produces a nonnegative solution of $b = Ax$. We are, as yet, unable to replace Inequality (13.15) with an equation in which the right side is independent of the particular solution x chosen.

Having selected the γ_j, we see from the inequality in (13.15) that convergence will be accelerated if we select δ_n as large as permitted by the condition $\gamma_j \delta_n \sigma_{nj} \leq 1$. This suggests that once again we take

$$\delta_n = 1/\min\{\sigma_{nj}\gamma_j, \ j = 1, ..., J\}. \tag{13.16}$$

The *rescaled* BI-EMML (RBI-EMML) as presented in [36, 38, 39] uses this choice, but with $\alpha_{ni} = 1$ for each n and i. The original motivation for the RBI-EMML came from consideration of Equation (13.13), replacing the geometric means with arithmetic means. This RBI-EMML is as follows:

Algorithm 13.11 (RBI-EMML). *Let x^0 be an arbitrary positive vector. For $k = 0, 1, ...,$ let $n = k(\mathrm{mod}\, N) + 1$. Then let*

$$x_j^{k+1} = (1 - m_n^{-1} s_j^{-1} s_{nj}) x_j^k + m_n^{-1} s_j^{-1} x_j^k \sum_{i \in B_n} \left(A_{ij} \frac{b_i}{(Ax^k)_i} \right). \tag{13.17}$$

Let's look now at some of the other choices for these parameters that have been considered in the literature.

First, we notice that the OSEM does not generally satisfy the requirements, since in Equation (13.5) the choices are $\alpha_{ni} = 1$ and $\beta_{nj} = s_{nj}^{-1}$; the only times this is acceptable is if the s_{nj} are separable; that is, $s_{nj} = r_j t_n$ for some r_j and t_n. This is slightly more general than the condition of subset balance and is sufficient for convergence of OSEM.

The original EMML in Equation (13.2) uses $N = 1$, $\gamma_j = s_j^{-1}$, and $\alpha_{ni} = \alpha_i = 1$. Clearly the inequality (13.8) is satisfied; in fact it becomes an equality now.

Notice that the calculations required to perform the BI-SMART are somewhat more complicated than those needed in BI-EMML. Because the

MART converges rapidly in most cases there is considerable interest in the row-action version of EMML. It was clear from the outset that using the OSEM in a row-action mode does not work. We see from the formula for BI-EMML that the proper row-action version of EMML, which we call the EM-MART, is the following:

Algorithm 13.12 (EM-MART). *Let x^0 be an arbitrary positive vector and $i = k(\mathrm{mod}\, I) + 1$. Then let*

$$x_j^{k+1} = (1 - \delta_i \gamma_j \alpha_{ii} A_{ij}) x_j^k + \delta_i \gamma_j \alpha_{ii} A_{ij} \frac{b_i}{(Ax^k)_i},$$

with

$$\gamma_j \delta_i \alpha_{ii} A_{ij} \leq 1$$

for all i and j.

The optimal choice would seem to be to take $\delta_i \alpha_{ii}$ as large as possible; that is, to select $\delta_i \alpha_{ii} = 1/\max\{\gamma_j A_{ij}, j = 1, ..., J\}$. With this choice the EM-MART is called the *rescaled* EM-MART (REM-MART).

The EM-MART fails to converge in the inconsistent case. What is always observed, but for which no proof exists, is that, for each fixed $i = 1, 2, ..., I$, as $m \to +\infty$, the EM-MART subsequences $\{x^{mI+i}\}$ converge to separate limit vectors, say $x^{\infty,i}$. This limit cycle LC $= \{x^{\infty,i} | i = 1, ..., I\}$ reduces to a single vector whenever there is a nonnegative solution of $b = Ax$. The greater the minimum value of $KL(y, Ax)$ the more distinct from one another the vectors of the limit cycle are. An analogous result is observed for BI-EMML.

We must mention a method that closely resembles the REM-MART, the *row-action maximum likelihood algorithm* (RAMLA), which was discovered independently by Browne and De Pierro [20]. The RAMLA avoids the limit cycle in the inconsistent case by using strong under-relaxation involving a decreasing sequence of relaxation parameters λ_k. The RAMLA is the following:

Algorithm 13.13 (RAMLA). *Let x^0 be an arbitrary positive vector, and $n = k(\mathrm{mod}\, N) + 1$. Let the positive relaxation parameters λ_k be chosen to converge to zero and $\sum_{k=0}^{+\infty} \lambda_k = +\infty$. Then,*

$$x_j^{k+1} = (1 - \lambda_k \sum_{i \in B_n} A_{ij}) x_j^k + \lambda_k x_j^k \sum_{i \in B_n} A_{ij} \left(\frac{b_i}{(Ax^k)_i} \right), \qquad (13.18)$$

13.6 RBI-SMART and Entropy Maximization

As we stated earlier, in the consistent case the sequence $\{x^k\}$ generated by the BI-SMART algorithm and given by Equation (13.10) converges to the unique solution of $b = Ax$ for which the distance $\sum_{j=1}^{J} \gamma_j^{-1} KL(x_j, x_j^0)$ is minimized. In this section we sketch the proof of this result as a sequence of lemmas, each of which is easily established.

Lemma 13.14. *For any nonnegative vectors a and b, with $a_+ = \sum_{m=1}^{M} a_m$ and $b_+ = \sum_{m=1}^{M} b_m > 0$, we have*

$$KL(a, b) = KL(a_+, b_+) + KL(a_+, \frac{a_+}{b_+}b). \qquad (13.19)$$

For nonnegative vectors x and z, let

$$G_n(x, z) = \sum_{j=1}^{J} \gamma_j^{-1} KL(x_j, z_j)$$

$$+ \delta_n \sum_{i \in B_n} \alpha_{ni}[KL((Ax)_i, b_i) - KL((Ax)_i, (Az)_i)]. \qquad (13.20)$$

It follows from Equation (13.19) and the inequality

$$\gamma_j^{-1} - \delta_n \sigma_{nj} \geq 1$$

that $G_n(x, z) \geq 0$ in all cases.

Lemma 13.15. *For every x we have*

$$G_n(x, x) = \delta_n \sum_{i \in B_n} \alpha_{ni} KL((Ax)_i, b_i),$$

so that

$$G_n(x, z) = G_n(x, x) + \sum_{j=1}^{J} \gamma_j^{-1} KL(x_j, z_j)$$

$$- \delta_n \sum_{i \in B_n} \alpha_{ni} KL((Ax)_i, (Az)_i). \qquad (13.21)$$

Therefore the distance $G_n(x, z)$ is minimized, as a function of z, by $z = x$. Now we minimize $G_n(x, z)$ as a function of x. The following lemma shows that the answer is

$$x_j = z_j' = z_j \exp\left(\gamma_j \delta_n \sum_{i \in B_n} \alpha_{ni} A_{ij} \log \frac{b_i}{(Az)_i}\right). \qquad (13.22)$$

Lemma 13.16. *For each x and z we have*

$$G_n(x, z) = G_n(z', z) + \sum_{j=1}^{J} \gamma_j^{-1} KL(x_j, z_j').$$
(13.23)

Proof: It is clear that $(x^k)' = x^{k+1}$ for all k. Now let $b = Pu$ for some non-negative vector u. We calculate $G_n(u, x^k)$ in two ways: using the definition we have

$$G_n(u, x^k) = \sum_{j=1}^{J} \gamma_j^{-1} KL(u_j, x_j^k) - \delta_n \sum_{i \in B_n} \alpha_{ni} KL(b_i, (Ax^k)_i),$$

while using Equation (13.23) we find that

$$G_n(u, x^k) = G_n(x^{k+1}, x^k) + \sum_{j=1}^{J} \gamma_j^{-1} KL(u_j, x_j^{k+1}).$$

Therefore

$$\sum_{j=1}^{J} \gamma_j^{-1} KL(u_j, x_j^k) - \sum_{j=1}^{J} \gamma_j^{-1} KL(u_j, x_j^{k+1})$$

$$= G_n(x^{k+1}, x^k) + \delta_n \sum_{i \in B_n} \alpha_{ni} KL(b_i, (Ax^k)_i). \quad (13.24)$$

We conclude several things from this.

First, the sequence $\{\sum_{j=1}^{J} \gamma_j^{-1} KL(u_j, x_j^k)\}$ is decreasing, so that the sequences $\{G_n(x^{k+1}, x^k)\}$ and $\{\delta_n \sum_{i \in B_n} \alpha_{ni} KL(b_i, (Ax^k)_i)\}$ converge to zero. Therefore the sequence $\{x^k\}$ is bounded and we may select an arbitrary cluster point x^*. It follows that $b = Ax^*$. We may therefore replace the generic solution u with x^* to find that $\{\sum_{j=1}^{J} \gamma_j^{-1} KL(x_j^*, x_j^k)\}$ is a decreasing sequence; but since a subsequence converges to zero, the entire sequence must converge to zero. Therefore $\{x^k\}$ converges to the solution x^*.

Finally, since the right side of Equation (13.24) does not depend on the particular choice of solution we made, neither does the left side. By *telescoping* we conclude that

$$\sum_{j=1}^{J} \gamma_j^{-1} KL(u_j, x_j^0) - \sum_{j=1}^{J} \gamma_j^{-1} KL(u_j, x_j^*)$$

is also independent of the choice of u. Consequently, minimizing the function $\sum_{j=1}^{J} \gamma_j^{-1} KL(u_j, x_j^0)$ over all solutions u is equivalent to minimizing $\sum_{j=1}^{J} \gamma_j^{-1} KL(u_j, x_j^*)$ over all solutions u; but the solution to the latter problem is obviously $u = x^*$. This completes the proof. $\square$

VI | Stability

14 | Sensitivity to Noise

When we use an iterative algorithm, we want it to solve our problem. We also want the solution in a reasonable amount of time, and we want slight errors in the measurements to cause only slight perturbations in the calculated answer. We have already discussed the use of block-iterative methods to accelerate convergence. Now we turn to regularization as a means of reducing sensitivity to noise. Because a number of regularization methods can be derived using a Bayesian *maximum a posteriori* approach, regularization is sometimes treated under the heading of MAP methods (see, for example, [48]).

14.1 Where Does Sensitivity Come From?

We illustrate the sensitivity problem that can arise when the inconsistent system $Ax = b$ has more equations than unknowns. We take A to be I by J and we calculate the least-squares solution,

$$x_{LS} = (A^\dagger A)^{-1} A^\dagger b,$$

assuming that the J by J Hermitian, nonnegative-definite matrix $Q = (A^\dagger A)$ is invertible, and therefore positive-definite.

The matrix Q has the eigenvalue/eigenvector decomposition

$$Q = \lambda_1 u_1 u_1^\dagger + \cdots + \lambda_J u_J u_J^\dagger,$$

where the (necessarily positive) eigenvalues of Q are

$$\lambda_1 \geq \lambda_2 \geq \cdots \geq \lambda_J > 0,$$

and the vectors u_j are the corresponding orthonormal eigenvectors.

14.1.1   The Singular-Value Decomposition of A

The square roots $\sqrt{\lambda_j}$ are called the *singular values* of A. The *singular-value decomposition* (SVD) of A is similar to the eigenvalue/eigenvector decomposition of Q: we have

$$A = \sqrt{\lambda_1} u_1 v_1^\dagger + \cdots + \sqrt{\lambda_J} u_J v_J^\dagger,$$

where the v_j are particular eigenvectors of $AA^\dagger$. We see from the SVD that the quantities $\sqrt{\lambda_j}$ determine the relative importance of each term $u_j v_j^\dagger$.

The SVD is commonly used for compressing transmitted or stored images. In such cases, the rectangular matrix A is a discretized image. It is not uncommon for many of the lowest singular values of A to be nearly zero, and to be essentially insignificant in the reconstruction of A. Only those terms in the SVD for which the singular values are significant need to be transmitted or stored. The resulting images may be slightly blurred, but can be restored later, as needed.

When the matrix A is a finite model of a linear imaging system, there will necessarily be model error in the selection of A. Getting the dominant terms in the SVD nearly correct is much more important (and usually much easier) than getting the smaller ones correct. The problems arise when we try to invert the system, to solve $Ax = b$ for x.

14.1.2   The Inverse of $Q = A^\dagger A$

The inverse of Q can then be written

$$Q^{-1} = \lambda_1^{-1} u_1 u_1^\dagger + \cdots + \lambda_J^{-1} u_J u_J^\dagger,$$

so that, with $A^\dagger b = c$, we have

$$x_{LS} = \lambda_1^{-1}(u_1^\dagger c) u_1 + \cdots + \lambda_J^{-1}(u_J^\dagger c) u_J.$$

Because the eigenvectors are orthonormal, we can express $\|A^\dagger b\|_2^2 = \|c\|_2^2$ as

$$\|c\|_2^2 = |u_1^\dagger c|^2 + \cdots + |u_J^\dagger c|^2,$$

and $\|x_{LS}\|_2^2$ as

$$\|x_{LS}\|_2^2 = \lambda_1^{-1}|u_1^\dagger c|^2 + \cdots + \lambda_J^{-1}|u_J^\dagger c|^2.$$

It is not uncommon for the eigenvalues of Q to be quite distinct, with some of them much larger than the others. When this is the case, we see that $\|x_{LS}\|_2$ can be much larger than $\|c\|_2$, because of the presence of the terms involving the reciprocals of the small eigenvalues. When the measurements b are essentially noise-free, we may have $|u_j^\dagger c|$ relatively small, for the indices near J, keeping the product $\lambda_j^{-1}|u_j^\dagger c|^2$ reasonable in size, but when the b becomes noisy, this may no longer be the case. The result is that those terms corresponding to the reciprocals of the smallest eigenvalues dominate the sum for x_{LS} and the norm of x_{LS} becomes quite large. The least-squares solution we have computed is essentially all noise and useless.

When we impose a nonnegativity constraint on the solution, noise in the data can manifest itself in a different way, as we saw in our discussion of the ART. When A has more columns than rows, but $Ax = b$ has no nonnegative solution, then, at least for those A having the *full-rank property*, the nonnegatively constrained least-squares solution has at most $I - 1$ nonzero entries. This happens also with the EMML and SMART solutions. As with the ART, regularization can eliminate the problem.

14.1.3 Reducing the Sensitivity to Noise

As we just saw, the presence of small eigenvalues for Q and noise in b can cause $\|x_{LS}\|_2$ to be much larger than $\|A^\dagger b\|_2$, with the result that x_{LS} is useless. In this case, even though x_{LS} minimizes $\|Ax - b\|_2$, it does so by overfitting to the noisy b. To reduce the sensitivity to noise and thereby obtain a more useful approximate solution, we can *regularize* the problem.

It often happens in applications that, even when there is an exact solution of $Ax = b$, noise in the vector b makes such as exact solution undesirable; in such cases a *regularized solution* is usually used instead. Select $\epsilon > 0$ and a vector p that is a prior estimate of the desired solution. Define

$$F_\epsilon(x) = (1 - \epsilon)\|Ax - b\|_2^2 + \epsilon\|x - p\|_2^2. \tag{14.1}$$

Lemma 14.1. *The function F_ϵ always has a unique minimizer $\hat{x}_\epsilon$, given by*

$$\hat{x}_\epsilon = ((1 - \epsilon)A^\dagger A + \epsilon I)^{-1}((1 - \epsilon)A^\dagger b + \epsilon p);$$

this is a regularized solution of $Ax = b$. Here, p is a prior estimate of the desired solution. Note that the inverse above always exists.

Note that, if $p = 0$, then

$$\hat{x}_\epsilon = (A^\dagger A + \gamma^2 I)^{-1}A^\dagger b, \tag{14.2}$$

for $\gamma^2 = \frac{\epsilon}{1 - \epsilon}$. The regularized solution has been obtained by modifying the formula for x_{LS}, replacing the inverse of the matrix $Q = A^\dagger A$ with the inverse of $Q + \gamma^2 I$. When ϵ is near zero, so is γ^2, and the matrices

Q and $Q + \gamma^2 I$ are nearly equal. What is different is that the eigenvalues of $Q + \gamma^2 I$ are $\lambda_i + \gamma^2$, so that, when the eigenvalues are inverted, the reciprocal eigenvalues are no larger than $1/\gamma^2$, which prevents the norm of x_ϵ from being too large, and decreases the sensitivity to noise.

Lemma 14.2. *Let ϵ be in $(0,1)$, and let I be the identity matrix whose dimensions are understood from the context. Then*

$$((1 - \epsilon)AA^\dagger + \epsilon I)^{-1} A = A((1 - \epsilon)A^\dagger A + \epsilon I)^{-1},$$

and, taking conjugate transposes,

$$A^\dagger((1 - \epsilon)AA^\dagger + \epsilon I)^{-1} = ((1 - \epsilon)A^\dagger A + \epsilon I)^{-1} A^\dagger.$$

Proof: Use the identity

$$A((1 - \epsilon)A^\dagger A + \epsilon I) = ((1 - \epsilon)AA^\dagger + \epsilon I)A. \qquad \square$$

Lemma 14.3. *Any vector p in R^J can be written as $p = A^\dagger q + r$, where $Ar = 0$.*

What happens to $\hat{x}_\epsilon$ as ϵ goes to zero? This will depend on which case we are in:

(1) $J \leq I$, and we assume that $A^\dagger A$ is invertible; or

(2) $J > I$, and we assume that $AA^\dagger$ is invertible.

Lemma 14.4. *In Case 1, taking limits as $\epsilon \to 0$ on both sides of the expression for $\hat{x}_\epsilon$ gives $\hat{x}_\epsilon \to (A^\dagger A)^{-1} A^\dagger b$, the least-squares solution of $Ax = b$.*

We consider Case 2 now. Write $p = A^\dagger q + r$, with $Ar = 0$. Then

$$\hat{x}_\epsilon = A^\dagger((1 - \epsilon)AA^\dagger + \epsilon I)^{-1}((1 - \epsilon)b + \epsilon q) + ((1 - \epsilon)A^\dagger A + \epsilon I)^{-1}(\epsilon r).$$

Lemma 14.5. *We have*

$$((1 - \epsilon)A^\dagger A + \epsilon I)^{-1}(\epsilon r) = r,$$

for all $\epsilon \in (0,1)$. Taking the limit of $\hat{x}_\epsilon$, as $\epsilon \to 0$, we get $\hat{x}_\epsilon \to A^\dagger(AA^\dagger)^{-1} b + r$. This is the solution of $Ax = b$ closest to p.

Proof: For the first assertion, let

$$t_\epsilon = ((1 - \epsilon)A^\dagger A + \epsilon I)^{-1}(\epsilon r).$$

Then, multiplying by A gives

$$At_\epsilon = A((1 - \epsilon)A^\dagger A + \epsilon I)^{-1}(\epsilon r).$$

Now show that $At_\epsilon = 0$. For the second assertion, draw a diagram for the case of one equation in two unknowns. $\qquad \square$

14.2   Iterative Regularization

It is often the case that the entries of the vector b in the system $Ax = b$ come from measurements, so are usually noisy. If the entries of b are noisy but the system $Ax = b$ remains consistent (which can easily happen in the under-determined case, with $J > I$), the ART begun at $x^0 = 0$ converges to the solution having minimum norm, but this norm can be quite large. The resulting solution is probably useless. Instead of solving $Ax = b$, we *regularize* by minimizing, for example, the function $F_\epsilon(x)$ given in Equation (14.1). For the case of $p = 0$, the solution to this problem is the vector $\hat{x}_\epsilon$ in Equation (14.2). However, we do not want to calculate $A^\dagger A + \gamma^2 I$, in order to solve

$$(A^\dagger A + \gamma^2 I)x = A^\dagger b,$$

when the matrix A is large. Fortunately, there are ways to find $\hat{x}_\epsilon$, using only the matrix A. We saw previously how this might be accomplished using the ART; now we show how Landweber's Algorithm can be used to calculate this regularized solution.

14.2.1   Iterative Regularization with Landweber's Algorithm

Our goal is to minimize the function in Equation (14.1), with $p = 0$. Notice that this is equivalent to minimizing the function

$$F(x) = ||Bx - c||_2^2,$$

for

$$B = \begin{bmatrix} A \\ \gamma I \end{bmatrix} \quad \text{and} \quad c = \begin{bmatrix} b \\ 0 \end{bmatrix},$$

where 0 denotes a column vector with all entries equal to zero and $\gamma = \frac{\epsilon}{1-\epsilon}$. The Landweber iteration for the problem $Bx = c$ is

$$x^{k+1} = x^k + \alpha B^T(c - Bx^k), \tag{14.3}$$

for $0 < \alpha < 2/\rho(B^T B)$, where $\rho(B^T B)$ is the spectral radius of $B^T B$. Equation (14.3) can be written as

$$x^{k+1} = (1 - \alpha\gamma^2)x^k + \alpha A^T(b - Ax^k). \tag{14.4}$$

We see from Equation (14.4) that Landweber's Algorithm for solving the regularized least-squares problem amounts to a relaxed version of Landweber's Algorithm applied to the original least-squares problem.

14.3   A Bayesian View of Reconstruction

The EMML iterative algorithm maximizes the likelihood function for the case in which the entries of the data vector $b = (b_1, ..., b_I)^T$ are assumed to be samples of independent Poisson random variables with mean values $(Ax)_i$; here, A is an I by J matrix with nonnegative entries and $x = (x_1, ..., x_J)^T$ is the vector of nonnegative parameters to be estimated. Equivalently, it minimizes the Kullback-Leibler distance $KL(b, Ax)$. This situation arises in single photon emission tomography, where the b_i are the number of photons counted at each detector i, x is the vectorized image to be reconstructed, and its entries x_j are (proportional to) the radionuclide intensity levels at each voxel j. When the signal-to-noise ratio is low, which is almost always the case in medical applications, maximizing likelihood can lead to unacceptably noisy reconstructions, particularly when J is larger than I. One way to remedy this problem is simply to halt the EMML algorithm after a few iterations, to avoid overfitting the x to the noisy data. A more mathematically sophisticated remedy is to employ a Bayesian approach and seek a maximum a posteriori (MAP) estimate of x.

In the Bayesian approach we view x as an instance of a random vector having a probability density function $f(x)$. Instead of maximizing the likelihood given the data, we now maximize the posterior likelihood, given both the data and the prior distribution for x. This is equivalent to minimizing

$$F(x) = KL(b, Ax) - \log f(x). \qquad (14.5)$$

The EMML algorithm is an example of an optimization method based on alternating minimization of a function $H(x, z) > 0$ of two vector variables. The alternating minimization works this way: let x and z be vector variables and $H(x, z) > 0$. If we fix z and minimize $H(x, z)$ with respect to x, we find that the solution is $x = z$, the vector we fixed; that is,

$$H(x, z) \geq H(z, z)$$

always. If we fix x and minimize $H(x, z)$ with respect to z, we get something new; call it Tx. The EMML algorithm has the iterative step $x^{k+1} = Tx^k$.

Obviously, we can't use an arbitrary function H; it must be related to the function $KL(b, Ax)$ that we wish to minimize, and we must be able to obtain each intermediate optimizer in closed form. The clever step is to select $H(x, z)$ so that $H(x, x) = KL(b, Ax)$, for any x. Now see what we have so far:

$$KL(b, Ax^k) = H(x^k, x^k) \geq H(x^k, x^{k+1})$$
$$\geq H(x^{k+1}, x^{k+1}) = KL(b, Ax^{k+1}).$$

That tells us that the algorithm makes $KL(b, Ax^k)$ decrease with each iteration. The proof doesn't stop here, but at least it is now plausible that the EMML iteration could minimize $KL(b, Ax)$.

The function $H(x, z)$ used in the EMML case is the KL distance

$$H(x, z) = KL(r(x), q(z)) = \sum_{i=1}^{I} \sum_{j=i}^{J} KL(r(x)_{ij}, q(z)_{ij}); \qquad (14.6)$$

we define, for each nonnegative vector x for which $(Ax)_i = \sum_{j=1}^{J} A_{ij}x_j > 0$, the arrays $r(x) = \{r(x)_{ij}\}$ and $q(x) = \{q(x)_{ij}\}$ with entries

$$r(x)_{ij} = x_j A_{ij} \frac{b_i}{(Ax)_i}$$

and

$$q(x)_{ij} = x_j A_{ij}.$$

With $x = x^k$ fixed, we minimize with respect to z to obtain the next EMML iterate x^{k+1}. Having selected the prior probability density function (pdf) $f(x)$, we want an iterative algorithm to minimize the function $F(x)$ in Equation (14.5). It would be a great help if we could mimic the alternating minimization formulation and obtain x^{k+1} by minimizing

$$KL(r(x^k), q(z)) - \log f(z) \qquad (14.7)$$

with respect to z. Unfortunately, to be able to express each new x^{k+1} in closed form, we need to choose $f(x)$ carefully.

14.4   The Gamma Prior Distribution for x

In [117] Lange et al. suggest viewing the entries x_j as samples of independent gamma-distributed random variables. A gamma-distributed random variable x takes positive values and has for its pdf the *gamma distribution* defined for positive x by

$$\gamma(x) = \frac{1}{\Gamma(\alpha)} (\frac{\alpha}{\beta})^\alpha x^{\alpha-1} e^{-\alpha x/\beta},$$

where α and β are positive parameters and Γ denotes the gamma function. The mean of such a gamma-distributed random variable is then $\mu = \beta$ and the variance is $\sigma^2 = \beta^2/\alpha$.

Lemma 14.6. *If the entries z_j of z are viewed as independent and gamma-distributed with means μ_j and variances σ_j^2, then minimizing the function in line (14.7) with respect to z is equivalent to minimizing the function*

$$KL(r(x^k), q(z)) + \sum_{j=1}^{J} \delta_j KL(\gamma_j, z_j), \tag{14.8}$$

for

$$\delta_j = \frac{\mu_j}{\sigma_j^2}, \ \gamma_j = \frac{\mu_j^2 - \sigma_j^2}{\mu_j},$$

under the assumption that the latter term is positive.

The resulting regularized EMML algorithm is the following:

Algorithm 14.7 (γ-prior Regularized EMML). Let x^0 be an arbitrary positive vector. Then let

$$x_j^{k+1} = \frac{\delta_j}{\delta_j + s_j}\gamma_j + \frac{1}{\delta_j + s_j}x_j^k \sum_{i=1}^{I} A_{ij} b_i/(Ax^k)_i, \tag{14.9}$$

where $s_j = \sum_{i=1}^{I} A_{ij}$.

We see from Equation (14.9) that the MAP iteration using the gamma priors generates a sequence of estimates, each entry of which is a convex combination or weighted arithmetic mean of the result of one EMML step and the prior estimate γ_j. Convergence of the resulting iterative sequence is established by Lange, Bahn, and Little in [117]; see also [34].

14.5   The One-Step-Late Alternative

It may well happen that we do not wish to use the gamma priors model and prefer some other $f(x)$. Because we will not be able to find a closed form expression for the z minimizing the function in Expression (14.7), we need some other way to proceed with the alternating minimization. Green [95] has offered the *one-step-late* (OSL) alternative.

When we try to minimize the function in Expression (14.7) by setting the gradient to zero we replace the variable z that occurs in the gradient of the term $-\log f(z)$ with x^k, the previously calculated iterate. Then, we can solve for z in closed form to obtain the new x^{k+1}. Unfortunately, negative entries can result and convergence is not guaranteed. There is a sizable literature on the use of MAP methods for this problem. In [43] an *interior-point algorithm* (IPA) is presented that avoids the OSL issue. In [129] the IPA is used to regularize transmission tomographic images.

14.6   Regularizing the SMART

The SMART algorithm is not derived as a maximum likelihood method, so regularized versions do not take the form of MAP algorithms. Neverthe- less, in the presence of noisy data, the SMART algorithm suffers from the same problem that afflicts the EMML, overfitting to noisy data resulting in an unacceptably noisy image. As we saw earlier, there is a close con- nection between the EMML and SMART algorithms. This suggests that a regularization method for SMART can be developed along the lines of the MAP with gamma priors used for EMML. Since the SMART is obtained by minimizing the function $KL(q(z), r(x^k))$ with respect to z to obtain x^{k+1}, it seems reasonable to attempt to derive a regularized SMART iterative scheme by minimizing

$$KL(q(z), r(x^k)) + \sum_{j=1}^{J} \delta_j KL(z_j, \gamma_j), \qquad (14.10)$$

as a function of z, for selected positive parameters δ_j and γ_j. This leads to the following algorithm:

Algorithm 14.8 (Regularized SMART). Let x^0 be an arbitrary positive vec- tor. Then let

$$\log x_j^{k+1} = \frac{\delta_j}{\delta_j + s_j} \log \gamma_j + \frac{1}{\delta_j + s_j} x_j^k \sum_{i=1}^{I} A_{ij} \log[b_i/(Ax^k)_i]. \qquad (14.11)$$

In [34] it was shown that this iterative sequence converges to a minimizer of the function

$$KL(Ax, y) + \sum_{j=1}^{J} \delta_j KL(x_j, \gamma_j).$$

It is useful to note that, although it may be possible to rederive this min- imization problem within the framework of Bayesian MAP estimation by carefully selecting a prior pdf for the vector x, we have not done so. The MAP approach is a special case of regularization through the use of penalty functions. These penalty functions need not arise through a Bayesian for- mulation of the parameter-estimation problem.

14.7   De Pierro's Surrogate-Function Method

In [76] De Pierro presents a modified EMML algorithm that includes reg- ularization in the form of a penalty function. His objective is the same as ours was in the case of regularized SMART: to embed the penalty term in the alternating minimization framework in such a way as to make it possible to obtain the next iterate in closed form. Because his *surrogate function* method has been used subsequently by others to obtain penalized likelihood algorithms [62], we consider his approach in some detail.

Let x and z be vector variables and $H(x, z) > 0$. Mimicking the behavior of the function $H(x, z)$ used in Equation (14.6), we require that if we fix z and minimize $H(x, z)$ with respect to x, the solution should be $x = z$, the vector we fixed; that is, $H(x, z) \geq H(z, z)$ always. If we fix x and minimize $H(x, z)$ with respect to z, we should get something new; call it Tx. As with the EMML, the algorithm will have the iterative step $x^{k+1} = Tx^k$.

Summarizing, we see that we need a function $H(x, z)$ with the properties (1) $H(x, z) \geq H(z, z)$ for all x and z; (2) $H(x, x)$ is the function $F(x)$ we wish to minimize; and (3) minimizing $H(x, z)$ with respect to z for fixed x is easy.

The function to be minimized is

$$F(x) = KL(b, Ax) + g(x),$$

where $g(x) \geq 0$ is some penalty function. De Pierro uses penalty functions $g(x)$ of the form

$$g(x) = \sum_{l=1}^{p} f_l(\langle s_l, x \rangle).$$

Let us define the matrix S to have for its lth row the vector s_l^T. Then $\langle s_l, x \rangle = (Sx)_l$, the lth entry of the vector Sx. Therefore,

$$g(x) = \sum_{l=1}^{p} f_l((Sx)_l).$$

Let $\lambda_{lj} > 0$ with $\sum_{j=1}^{J} \lambda_{lj} = 1$, for each l.

Assume that the functions f_l are convex. Therefore, for each l, we have

$$f_l((Sx)_l) = f_l(\sum_{j=1}^{J} S_{lj} x_j) = f_l(\sum_{j=1}^{J} \lambda_{lj}(S_{lj}/\lambda_{lj})x_j)$$

$$\leq \sum_{j=1}^{J} \lambda_{lj} f_l((S_{lj}/\lambda_{lj})x_j).$$

Therefore,

$$g(x) \leq \sum_{l=1}^{p} \sum_{j=1}^{J} \lambda_{lj} f_l((S_{lj}/\lambda_{lj})x_j).$$

So we have replaced $g(x)$ with a related function in which the x_j occur separately, rather than just in the combinations $(Sx)_l$. But we aren't quite done yet.

We would like to take for De Pierro's $H(x, z)$ the function used in the EMML algorithm, plus the function

$$\sum_{l=1}^{p} \sum_{j=1}^{J} \lambda_{lj} f_l((S_{lj}/\lambda_{lj})z_j).$$

But there is one slight problem: we need $H(z, z) = F(z)$, which we don't have yet. De Pierro's clever trick is to replace $f_l((S_{lj}/\lambda_{lj})z_j)$ with

$$f_l((S_{lj}/\lambda_{lj})z_j - (S_{lj}/\lambda_{lj})x_j + (Sx)_l).$$

So, De Pierro's function $H(x, z)$ is the sum of the $H(x, z)$ used in the EMML case and the function

$$\sum_{l=1}^{p} \sum_{j=1}^{J} \lambda_{lj} f_l((S_{lj}/\lambda_{lj})z_j - (S_{lj}/\lambda_{lj})x_j + (Sx)_l).$$

Now he has the three properties he needs. Once he has computed x^k, he minimizes $H(x^k, z)$ by taking the gradient and solving the equations for the correct $z = Tx^k = x^{k+1}$. For the choices of f_l he discusses, these intermediate calculations can either be done in closed form (the quadratic case) or with a simple Newton-Raphson iteration (the logcosh case).

14.8   Block-Iterative Regularization

We saw previously that it is possible to obtain a regularized least-squares solution $\hat{x}_\epsilon$, and thereby avoid the limit cycle, using only the matrix A and the ART algorithm. This prompts us to ask if it is possible to find regularized SMART solutions using block-iterative variants of SMART. Similarly, we wonder if it is possible to do the same for EMML.

Open Question 5. Can we use the MART to find the minimizer of the function

$$KL(Ax, b) + \epsilon KL(x, p)?$$

More generally, can we obtain the minimizer using RBI-SMART?

Open Question 6. Can we use the RBI-EMML methods to obtain the minimizer of the function

$$KL(b, Ax) + \epsilon KL(p, x)?$$

There have been various attempts to include regularization in block-iterative methods, to reduce noise sensitivity and avoid limit cycles, but all of these approaches have been ad-hoc, with little or no theoretical basis. Typically, they simply modify each iterative step by including an additional term that appears to be related to the regularizing penalty function. The case of the ART is instructive, however. In that case, we obtained the desired iterative algorithm by using an augmented set of variables, not simply by modifying each step of the original ART algorithm. How to do this for the MART and the other block-iterative algorithms is not obvious.

Recall that the RAMLA method in Equation (13.18) is similar to the RBI-EMML algorithm, but it employs a sequence of decreasing relaxation parameters, which, if properly chosen, will cause the iterates to converge to the minimizer of $KL(b, Ax)$, thereby avoiding the limit cycle. In [78] De Pierro and Yamaguchi present a regularized version of RAMLA, but without guaranteed convergence.

15 | Feedback in Block-Iterative Reconstruction

When the nonnegative system of linear equations $Ax = b$ has no nonnegative solutions we say that we are in the *inconsistent case*. In this case the SMART and EMML algorithms still converge, to a nonnegative minimizer of $KL(Ax, b)$ and $KL(b, Ax)$, respectively. On the other hand, the rescaled block-iterative versions of these algorithms, RBI-SMART and RBI-EMML, do not converge. Instead they exhibit *cyclic subsequential convergence*; for each fixed $n = 1, ..., N$, with N the number of blocks, the subsequence $\{x^{mN+n}\}$ converges to their own limits. These limit vectors then constitute the *limit cycle* (LC). The LC for RBI-SMART is not the same as for RBI-EMML, generally, and the LC varies with the choice of blocks. Our problem is to find a way to calculate the SMART and EMML limit vectors using the RBI methods. More specifically, how can we calculate the SMART and EMML limit vectors from their associated RBI limit cycles?

As is often the case with the algorithms based on the KL distance, we can turn to the ART algorithm for guidance. What happens with the ART algorithm in the inconsistent case is often closely related to what happens with RBI-SMART and RBI-EMML, although proofs for the latter methods are more difficult to obtain. For example, when the system $Ax = b$ has no solution we can prove that ART exhibits cyclic subsequential convergence to a limit cycle. The same behavior is seen with the RBI methods, but no one knows how to prove this. When the system $Ax = b$ has no solution we usually want to calculate the least-squares (LS) approximate solution. The problem then is to use the ART to find the LS solution. There are several ways to do this, as discussed in [38, 48]. We would like to be able to borrow some of these methods and apply them to the RBI problem. In this section we focus on one specific method that works for ART and we try to make it work for RBI; it is the *feedback* approach.

15.1   Feedback in ART

Suppose that the system $Ax = b$ has no solution. We apply the ART and get the limit cycle $\{z^1, z^2, ..., z^I\}$, where I is the number of equations and $z^0 = z^I$. We assume that the rows of A have been normalized so that their lengths are equal to one. Then the ART iterative step gives

$$z_j^i = z_j^{i-1} + \overline{A_{ij}}(b_i - (Az^{i-1})_j)$$

or

$$z_j^i - z_j^{i-1} = \overline{A_{ij}}(b_i - (Az^{i-1})_j).$$

Summing over the index i and using $z^0 = z^I$ we obtain zero on the left side, for each j. Consequently $A^\dagger b = A^\dagger c$, where c is the vector with entries $c_i = (Az^{i-1})_i$. It follows that the systems $Ax = b$ and $Ax = c$ have the same LS solutions and that it may help to use both b and c to find the LS solution from the limit cycle. The article [38] contains several results along these lines. One approach is to apply the ART again to the system $Ax = c$, obtaining a new LC and a new candidate for the right side of the system of equations. If we repeat this feedback procedure, each time using the LC to define a new right -side vector, does it help us find the LS solution? Yes, as Theorem 4 of [38] shows. Our goal in this section is to explore the possibility of using the same sort of feedback in the RBI methods. Some results in this direction are in [38]; we review those now.

15.2   Feedback in RBI methods

One issue that makes the KL methods more complicated than the ART is the support of the limit vectors, meaning the set of indices j for which the entries of the vector are positive. In [34] it was shown that when the system $Ax = b$ has no nonnegative solutions and A has the *full rank property* there is a subset S of $\{j = 1, ..., J\}$ with cardinality at most $I - 1$, such that every nonnegative minimizer of $KL(Ax, b)$ has zero for its jth entry whenever j is not in S. It follows that the minimizer is unique. The same result holds for the EMML, although it has not been proven that the set S is the same set as in the SMART case. The same result holds for the vectors of the LC for both RBI-SMART and RBI-EMML.

A simple, yet helpful, example to refer to as we proceed is the following.

$$A = \begin{bmatrix} 1 & .5 \\ 0 & .5 \end{bmatrix}, \qquad b = \begin{bmatrix} .5 \\ 1 \end{bmatrix}.$$

There is no nonnegative solution to this system of equations and the support set S for SMART, EMML, and the RBI methods is $S = \{j = 2\}$.

15.2.1 The RBI-SMART

Our analysis of the SMART and EMML methods has shown that the theory for SMART is somewhat nicer than that for EMML and the resulting theorems for SMART are a bit stronger. The same is true for RBI-SMART, compared to RBI-EMML. For that reason we begin with RBI-SMART.

Recall that the iterative step for RBI-SMART is

$$x_j^{k+1} = x_j^k \exp(m_n^{-1} s_j^{-1} \sum_{i \in B_n} A_{ij} \log(b_i/(Ax^k)_i)),$$

where $n = k (\mathrm{mod}\, N) + 1$, $s_j = \sum_{i=1}^{I} A_{ij}$, $s_{nj} = \sum_{i \in B_n} A_{ij}$, and $m_n = \max\{s_{nj}/s_j, j = 1, ..., J\}$.

For each n let

$$G_n(x,z) = \sum_{j=1}^{J} s_j KL(x_j, z_j) - m_n^{-1} \sum_{i \in B_n} KL((Ax)_i, (Az)_i)$$

$$+ m_n^{-1} \sum_{i \in B_n} KL((Ax)_i, b_i).$$

Lemma 15.1. *For each nonnegative x and z,*

$$\sum_{j=1}^{J} s_j KL(x_j, z_j) - m_n^{-1} \sum_{i \in B_n} KL((Ax)_i, (Az)_i) \geq 0,$$

so that $G_n(x,z) \geq 0$.

Lemma 15.2. *For each nonnegative x and z,*

$$G_n(x,z) = G_n(z',z) + \sum_{j=1}^{J} s_j KL(x_j, z_j'),$$

where

$$z_j' = z_j \exp(m_n^{-1} s_j^{-1} \sum_{i \in B_n} A_{ij} \log(b_i/(Az)_i)).$$

We assume that there are no nonnegative solutions to the nonnegative system $Ax = b$. We apply the RBI-SMART and get the limit cycle $\{z^1, ..., z^N\}$, where N is the number of blocks. We also let $z^0 = z^N$ and for each i let $c_i = (Az^{n-1})_i$ where $i \in B_n$, the nth block. Prompted by what we learned concerning the ART, we ask if the nonnegative minimizers of $KL(Ax, b)$ and $KL(Ax, c)$ are the same. This would be the correct question to ask if we were using the slower unrescaled block-iterative SMART, in which the m_n are replaced by one. For the rescaled case it turns out that the proper question to ask is: Are the nonnegative minimizers of the functions

$$\sum_{n=1}^{N} m_n^{-1} \sum_{i \in B_n} KL((Ax)_i, b_i)$$

and

$$\sum_{n=1}^{N} m_n^{-1} \sum_{i \in B_n} KL((Ax)_i, c_i)$$

the same? The answer is "Yes, probably." The difficulty has to do with the support of these minimizers; specifically, are the supports of both minimizers the same as the support of the LC vectors? If so, then we can prove that the two minimizers are identical. This is our motivation for the feedback approach.

The feedback approach is the following: beginning with $b^0 = b$ we apply the RBI-SMART and obtain the LC, from which we extract the vector c, which we also call c^0. We then let $b^1 = c^0$ and apply the RBI-SMART to the system $b^1 = Ax$. From the resulting LC we extract $c^1 = b^2$, and so on. In this way we obtain an infinite sequence of *data vectors* $\{b^k\}$. We denote by $\{z^{k,1}, ..., z^{k,N}\}$ the LC we obtain from the system $b^k = Ax$, so that

$$b_i^{k+1} = (Az^{k,n})_i, \text{ for } i \in B_n.$$

One issue we must confront is how we use the support sets. At the first step of feedback we apply RBI-SMART to the system $b = b^0 = Ax$, beginning with a positive vector x^0. The resulting limit cycle vectors are supported on a set S^0 with cardinality less than I. At the next step we apply the RBI-SMART to the system $b^1 = Ax$. Should we begin with a positive vector (not necessarily the same x^0 as before) or should our starting vector be supported on S^0?

Lemma 15.3. *The RBI-SMART sequence $\{x^k\}$ is bounded.*

Proof: For each j let $M_j = \max\{b_i/A_{ij}, |A_{ij} > 0\}$ and let $C_j = \max\{x_j^0, M_j\}$. Then $x_j^k \leq C_j$ for all k. □

Lemma 15.4. *Let S be the support of the LC vectors. Then*

$$\sum_{n=1}^{N} m_n^{-1} \sum_{i \in B_n} A_{ij} \log(b_i/c_i) \leq 0 \tag{15.1}$$

for all j, with equality for those $j \in S$. Therefore,

$$\sum_{n=1}^{N} m_n^{-1} \sum_{i \in B_n} KL((Ax)_i, b_i) - \sum_{n=1}^{N} m_n^{-1} \sum_{i \in B_n} KL((Ax)_i, c_i) \geq$$

$$\sum_{n=1}^{N} m_n^{-1} \sum_{i \in B_n} (b_i - c_i),$$

with equality if the support of the vector x lies within the set S.

Proof: For $j \in S$ consider $\log(z_j^n / z_j^{n-1})$ and sum over the index n, using the fact that $z^N = z^0$. For general j assume there is a j for which the inequality does not hold. Then there is M and $\epsilon > 0$ such that for $m \geq M$

$$\log(x_j^{(m+1)N} / x_j^{mN}) \geq \epsilon.$$

Therefore, the sequence $\{x_j^{mN}\}$ is unbounded.   □

Lemma 15.5. *We have*

$$\sum_{n=1}^{N} G_n(z^{k,n}, z^{k,n-1}) = \sum_{n=1}^{N} m_n^{-1} \sum_{i \in B_n} (b_i^k - b_i^{k+1}),$$

so that the sequence $\{\sum_{n=1}^{N} m_n^{-1}(\sum_{i \in B_n} b_i^k)\}$ is decreasing and that the sequence $\{\sum_{n=1}^{N} G_n(z^{k,n}, z^{k,n-1})\} \to 0$ as $k \to \infty$.

Proof: Calculate $G_n(z^{k,n}, z^{k,n-1})$ using Lemma 15.2.   □

Lemma 15.6. *For all vectors $x \geq 0$, the sequence*

$$\left\{ \sum_{n=1}^{N} m_n^{-1} \sum_{i \in B_n} KL((Ax)_i, b_i^k) \right\}$$

is decreasing and the sequence

$$\sum_{n=1}^{N} m_n^{-1} \sum_{i \in B_n} (b_i^k - b_i^{k+1}) \to 0,$$

as $k \to \infty$.

Proof: Calculate

$$\left\{ \sum_{n=1}^{N} m_n^{-1} \sum_{i \in B_n} KL((Ax)_i, b_i^k) \right\} - \left\{ \sum_{n=1}^{N} m_n^{-1} \sum_{i \in B_n} KL((Ax)_i, b_i^{k+1}) \right\}$$

and use the previous lemma.   □

Lemma 15.7. *For each fixed n, the sequence $\{z^{k,n}\}$ is bounded.*

Since the sequence $\{z^{k,0}\}$ is bounded there is a subsequence $\{z^{k_t,0}\}$ converging to a limit vector $z^{*,0}$. Since the sequence $\{z^{k_t,1}\}$ is bounded there is subsequence converging to some vector $z^{*,1}$. Proceeding in this way we find subsequences $\{z^{k_m,n}\}$ converging to $z^{*,n}$ for each fixed n. Our goal is to show that, with certain restrictions on A, $z^{*,n} = z^*$ for each n. We then show that the sequence $\{b^k\}$ converges to Az^* and that z^* minimizes

$$\sum_{n=1}^{N} m_n^{-1} \sum_{i \in B_n} KL((Ax)_i, b_i).$$

It follows from Lemma 15.5 that

$$\{\sum_{n=1}^{N} G_n(z^{*,n}, z^{*,n-1})\} = 0.$$

Open Question 7. Can we find suitable restrictions on the matrix A that permit us to conclude that $z^{*,n} = z^{*,n-1} = z^*$ for each n.

Lemma 15.8. *The sequence $\{b^k\}$ converges to Az^*.*

Proof: Since the sequence $\{\sum_{n=1}^{N} m_n^{-1} \sum_{i \in B_n} KL((Az^*)_i, b_i^k)\}$ is decreasing and a subsequence converges to zero, it follows that the whole sequence converges to zero. □

Open Question 8. Can we use Lemma 15.4 to obtain conditions that permit us to conclude that the vector z^* is a nonnegative minimizer of the function

$$\sum_{n=1}^{N} m_n^{-1} \sum_{i \in B_n} KL((Ax)_i, b_i)?$$

15.2.2   The RBI-EMML

We turn now to the RBI-EMML method, having the iterative step

$$x_j^{k+1} = (1 - m_n^{-1} s_j^{-1} s_{nj}) x_j^k + m_n^{-1} s_j^{-1} x_j^k \sum_{i \in B_n} A_{ij} b_i / (Ax^k)_i,$$

with $n = k \pmod{N} + 1$. As we warned earlier, developing the theory for feedback with respect to the RBI-EMML algorithm appears to be more difficult than in the RBI-SMART case.

Applying the RBI-EMML algorithm to the system of equations $Ax = b$ having no nonnegative solution, we obtain the LC $\{z^1, ..., z^N\}$. As before, for each i we let $c_i = (Az^{n-1})_i$ where $i \in B_n$. There is a subset S of $\{j = 1, ..., J\}$ with cardinality less than I such that for all n we have $z_j^n = 0$ if j is not in S.

The first question that we ask is: Are the nonnegative minimizers of the functions

$$\sum_{n=1}^{N} m_n^{-1} \sum_{i \in B_n} KL(b_i, (Ax)_i)$$

and

$$\sum_{n=1}^{N} m_n^{-1} \sum_{i \in B_n} KL(c_i, (Ax)_i)$$

the same?

As before, the feedback approach involves setting $b^0 = b$, $c^0 = c = b^1$ and for each k defining $b^{k+1} = c^k$, where c^k is extracted from the limit cycle

$$LC(k) = \{z^{k,1}, ..., z^{k,N} = z^{k,0}\}$$

obtained from the system $b^k = Ax$ as $c_i^k = (Az^{k,n-1})_i$ where n is such that $i \in B_n$. Again, we must confront the issue of how we use the support sets. At the first step of feedback we apply RBI-EMML to the system $b = b^0 = Ax$, beginning with a positive vector x^0. The resulting limit cycle vectors are supported on a set S^0 with cardinality less than I. At the next step we apply the RBI-EMML to the system $b^1 = Ax$. Should we begin with a positive vector (not necessarily the same x^0 as before) or should our starting vector be supported on S^0? One approach could be to assume first that $J < I$ and that $S = \{j = 1, ..., J\}$ always and then see what can be discovered.

Some conjectures. Our conjectures, subject to restrictions involving the support sets, are as follows:

(1) the sequence $\{b^k\}$ converges to a limit vector b^∞;

(2) the system $b^\infty = Ax$ has a nonnegative solution, say x^∞;

(3) the LC obtained for each k converge to the singleton x^∞;

(4) the vector x^∞ minimizes the function

$$\sum_{n=1}^{N} m_n^{-1} \sum_{i \in B_n} KL(b_i, (Ax)_i)$$

over nonnegative x.

Some results concerning feedback for RBI-EMML were presented in [38]. We sketch those results now. We have that

$$\sum_{j=1}^{J} s_j \sum_{n=1}^{N} (z_j^{k,n} - z_j^{k,n-1}) = 0.$$

We then rewrite it in terms of b^k and b^{k+1}, and conclude that the quantity

$$\sum_{n=1}^{N} m_n^{-1} \sum_{i \in B_n} b_i^k$$

is the same for $k = 0, 1, \ldots$. There is a constant $B > 0$ such that $z_j^{k,n} \leq B$ for all k, n, and j.

We use the convexity of the log function and the fact that the terms $1 - m_n^{-1} s_{nj}$ and $m_n^{-1} A_{ij}$, $i \in B_n$ sum to one, to show that

$$s_j \log(z_j^{k,n-1}/z_j^{k,n}) \leq m_n^{-1} \sum_{i \in B_n} A_{ij} \log(b_i^{k+1}/b_i^k).$$

It follows that the sequence

$$\left\{ \sum_{n=1}^{N} m_n^{-1} \sum_{i \in B_n} KL((Ax)_i, b_i^k) \right\}$$

is decreasing for each nonnegative vector x and the sequence

$$\left\{ \sum_{n=1}^{N} m_n^{-1} \sum_{i \in B_n} A_{ij} \log(b_i^k) \right\}$$

is increasing.

VII | Optimization

16 | Iterative Optimization

Optimization means finding a maximum or minimum value of a real-valued function of one or several variables. Constrained optimization means that the acceptable solutions must satisfy some additional restrictions, such as being nonnegative. Even if we know equations that optimal points must satisfy, solving these equations is often difficult and usually cannot be done algebraically. In this chapter we sketch the conditions that must hold in order for a point to be an optimum point, and then use those conditions to motivate iterative algorithms for finding the optimum points. We shall consider only minimization problems, since any maximization problem can be converted into a minimization problem by changing the sign of the function involved.

16.1 Functions of a Single Real Variable

If $f(x)$ is a continuous, real-valued function of a real variable x and we want to find an x for which the function takes on its minimum value, then we need only examine those places where the derivative, $f'(x)$, is zero, and those places where $f'(x)$ does not exist; of course, without further assumptions, there is no guarantee that a minimum exists. Therefore, if $f(x)$ is differentiable at all x, and if its minimum value occurs at x^*, then $f'(x^*) = 0$. If the problem is a *constrained minimization*, that is, if the allowable x lie within some interval, say, $[a, b]$, then we must also examine the endpoints, $x = a$ and $x = b$. If the constrained minimum occurs at $x^* = a$ and $f'(a)$ exists, then $f'(a)$ need not be zero; however, we must have $f'(a) \geq 0$, since, if $f'(a) < 0$, we could select $x = c$ slightly to the right of $x = a$ with $f(c) < f(a)$. Similarly, if the minimum occurs at $x = b$, and $f'(b)$ exists, we must have $f'(b) \leq 0$. We can combine these endpoint conditions by saying that if the minimum occurs at one of the two endpoints, moving away from the minimizing point into the interval $[a, b]$ cannot result in the function growing smaller. For functions of several variables similar conditions hold, involving the partial derivatives of the function.

16.2   Functions of Several Real Variables

Suppose, from now on, that $f(x) = f(x_1, ..., x_N)$ is a continuous, real-valued function of the N real variables $x_1, ..., x_N$ and that $x = (x_1, ..., x_N)^T$ is the column vector of unknowns, lying in the N-dimensional space R^N. When the problem is to find a minimum (or a maximum) of $f(x)$, we call $f(x)$ the *objective function*. As in the case of one variable, without additional assumptions, there is no guarantee that a minimum (or a maximum) exists.

16.2.1   Cauchy's Inequality for the Dot Product

For any two vectors v and w in R^N the dot product is defined to be

$$v \cdot w = \sum_{n=1}^{N} v_n w_n.$$

Cauchy's Inequality tells us that $|v \cdot w| \le ||v||_2 ||w||_2$, with equality if and only if $w = \alpha v$ for some real number α. In the multivariable case we speak of the derivative of a function at a point, in the direction of a given vector; these are the *directional derivatives* and their definition involves the dot product.

16.2.2   Directional Derivatives

If $f : D \subseteq R^N \to R$ and, for some z in the interior of D and some h, the limit

$$D_h f(z) = \lim_{t \to 0} \frac{1}{t}(f(z + th) - f(z))$$

exists, then $D_h f(z)$ is the *Gateaux differential* of f, at z, with respect to h [133]. The partial derivatives of f at the point z, denoted $\frac{\partial f}{\partial x_n}(z)$, at z are the Gateaux differentials with respect to the unit vectors in the coordinate directions. If $D_h f(z)$ is linear in the vector h, which happens, for example, if the first partial derivatives of f at z are continuous, then f is said to be *Gateaux differentiable* at z. In that case, for any unit vector, that is, for any vector $u = (u_1, ..., u_N)^T$ with its Euclidean norm

$$||u||_2 = \sqrt{u_1^2 + ... + u_N^2},$$

equal to one, $D_u f(z)$ is the *directional derivative* of f, at the point $x = z$, in the direction of u, and

$$D_u f(z) = \frac{\partial f}{\partial x_1}(z)u_1 + \ldots + \frac{\partial f}{\partial x_N}(z)u_N.$$

Notice that this directional derivative is the dot product of u with the *gradient* of $f(x)$ at $x = z$, defined by

$$\nabla f(z) = (\frac{\partial f}{\partial x_1}(z), \ldots, \frac{\partial f}{\partial x_N}(z))^T.$$

According to Cauchy's Inequality, the dot product $\nabla f(z) \cdot u$ will take on its maximum value when u is a positive multiple of $\nabla f(z)$, and therefore, its minimum value when u is a negative multiple of $\nabla f(z)$. Consequently, the gradient of f at $x = z$ points in the direction, from $x = z$, of the greatest increase in the function f. This suggests that, if we are trying to minimize f, and we are currently at $x = z$, we should consider moving in the direction of $-\nabla f(z)$; this leads to Cauchy's iterative method of *steepest descent*, which we shall discuss in more detail later.

If the minimum value of $f(x)$ occurs at $x = x^*$, then either all the directional derivatives are zero at $x = x^*$, in which case $\nabla f(z) = 0$, or at least one directional derivative does not exist. But, what happens when the problem is a constrained minimization?

16.2.3   Constrained Minimization

Unlike the single-variable case, in which constraining the variable simply meant requiring that it lie within some interval, in the multivariable case constraints can take many forms. For example, we can require that each of the entries x_n be nonnegative, or that each x_n lie within an interval $[a_n, b_n]$ that depends on n, or that the norm of x, defined by $||x||_2 = \sqrt{x_1^2 + \ldots + x_N^2}$, which measures the distance from x to the origin, does not exceed some bound. In fact, for any set C in N-dimensional space, we can pose the problem of minimizing $f(x)$, subject to the restriction that x be a member of the set C. In place of endpoints, we have what are called boundary-points of C, which are those points in C that are not entirely surrounded by other points in C. For example, in the one-dimensional case, the points $x = a$ and $x = b$ are the boundary-points of the set $C = [a, b]$. If $C = R_+^N$ is the subset of N-dimensional space consisting of all the vectors x whose entries are nonnegative, then the boundary-points of C are all nonnegative vectors x having at least one zero entry.

Suppose that C is arbitrary in R^N and the point $x = x^*$ is the solution to the problem of minimizing $f(x)$ over all x in the set C. Assume also that all the directional derivatives of $f(x)$ exist at each x. If x^* is not a boundary-point of C, then all the directional derivatives of $f(x)$, at the point $x = x^*$, must be nonnegative, in which case they must all be zero, so that we must have $\nabla f(z) = 0$. On the other hand, speaking somewhat loosely, if x^* is a boundary-point of C, then it is necessary only that the directional derivatives of $f(x)$, at the point $x = x^*$, in directions that point back into the set C, be nonnegative.

16.2.4   An Example

To illustrate these concepts, consider the problem of minimizing the function of two variables, $f(x_1, x_2) = x_1 + 3x_2$, subject to the constraint that $x = (x_1, x_2)$ lie within the unit ball $C = \{x = (x_1, x_2) | x_1^2 + x_2^2 \leq 1\}$. With the help of simple diagrams we discover that the minimizing point $x^* = (x_1^*, x_2^*)$ is a boundary-point of C, and that the line $x_1 + 3x_2 = x_1^* + 3x_2^*$ is tangent to the unit circle at x^*. The gradient of $f(x)$, at $x = z$, is $\nabla f(z) = (1, 3)^T$, for all z, and is perpendicular to this tangent line. But, since the point x^* lies on the unit circle, the vector $(x_1^*, x_2^*)^T$ is also perpendicular to the line tangent to the circle at x^*. Consequently, we know that $(x_1^*, x_2^*)^T = \alpha(1, 3)^T$, for some real α. From $x_1^2 + x_2^2 = 1$, it follows that $|\alpha| = \sqrt{10}$. This gives us two choices for x^*: either $x^* = (\sqrt{10}, 3\sqrt{10})$, or $x^* = (-\sqrt{10}, -3\sqrt{10})$. Evaluating $f(x)$ at both points reveals that $f(x)$ attains its maximum at the first, and its minimum at the second.

Every direction vector u can be written in the form $u = \beta(1, 3)^T + \gamma(-3, 1)^T$, for some β and γ. The directional derivative of $f(x)$, at $x = x^*$, in any direction that points from $x = x^*$ back into C, must be nonnegative. Such directions must have a nonnegative dot product with the vector $(-x_1^*, -x_2^*)^T$, which tells us that

$$0 \leq \beta(1, 3)^T \cdot (-x_1^*, -x_2^*)^T + \gamma(-3, 1)^T \cdot (-x_1^*, x_2^*)^T,$$

or

$$0 \leq (3\gamma - \beta)x_1^* + (-3\beta - \gamma)x_2^*.$$

Consequently, the gradient $(1, 3)^T$ must have a nonnegative dot product with every direction vector u that has a nonnegative dot product with $(-x_1^*, -x_2^*)^T$. For the dot product of $(1, 3)^T$ with any u to be nonnegative we need $\beta \geq 0$. So we conclude that $\beta \geq 0$ for all β and γ for which

$$0 \leq (3\gamma - \beta)x_1^* + (-3\beta - \gamma)x_2^*.$$

Saying this another way, if $\beta < 0$, then

$$(3\gamma - \beta)x_1^* + (-3\beta - \gamma)x_2^* < 0,$$

for all γ. Taking the limit, as $\beta \to 0$ from the left, it follows that

$$3\gamma x_1^* - \gamma x_2^* \leq 0,$$

for all γ. The only way this can happen is if $3x_1^* - x_2^* = 0$. Therefore, our optimum point must satisfy the equation $x_2^* = 3x_1^*$, which is what we found previously.

We have just seen the conditions necessary for x^* to minimize $f(x)$, subject to constraints, be used to determine the point x^* algebraically. In more complicated problems we will not be able to solve for x^* merely by performing simple algebra. But we may still be able to find x^* using iterative optimization methods.

16.3   Gradient Descent Optimization

Suppose that we want to minimize $f(x)$, over all x, without constraints. Begin with an arbitrary initial guess, $x - x^0$. Having proceeded to x^k, we show how to move to x^{k+1}. At the point $x = x^k$, the direction of greatest rate of decrease of $f(x)$ is $u = -\nabla f(x^k)$. Therefore, it makes sense to move from x^k in the direction of $-\nabla f(x^k)$, and to continue in that direction until the function stops decreasing. In other words, we let

$$x^{k+1} = x^k - \alpha_k \nabla f(x^k),$$

where $\alpha_k \geq 0$ is the *step size*, determined by the condition

$$f(x^k - \alpha_k \nabla f(x^k)) \leq f(x^k - \alpha \nabla f(x^k)),$$

for all $\alpha \geq 0$. This iterative procedure is Cauchy's *steepest descent* method. To establish the convergence of this algorithm to a solution requires additional restrictions on the function f; we shall not consider these issues further. Our purpose here is merely to illustrate an iterative minimization philosophy that we shall recall in various contexts.

If the problem is a constrained minimization, then we must proceed more carefully. One method, known as *interior-point* iteration, begins with x^0 within the constraint set C and each subsequent step is designed to produce another member of C; if the algorithm converges, the limit is then guaranteed to be in C. For example, if $C = R_+^N$, the nonnegative cone in R^N, we could modify the steepest descent method so that, first, x^0 is a nonnegative vector, and second, the step from x^k in C is restricted so that we stop before x^{k+1} ceases to be nonnegative. A somewhat different modification of the steepest descent method would be to take the full step from x^k to x^{k+1}, but then to take as the true x^{k+1} that vector in C nearest to what would have been x^{k+1}, according to the original steepest descent algorithm; this new iterative scheme is the *projected steepest descent* algorithm. It is not necessary, of course, that every intermediate vector x^k be in C; all we want is that the limit be in C. However, in applications, iterative methods must always be stopped before reaching their limit point, so, if we must have a member of C for our (approximate) answer, then we would need x^k in C when we stop the iteration.

16.4   The Newton-Raphson Approach

The Newton-Raphson approach to minimizing a real-valued function f : $R^J \to R$ involves finding x^* such that $\nabla f(x^*) = 0$.

16.4.1   Functions of a Single Variable

We begin with the problem of finding a root of a function $g : R \to R$. If x^0 is not a root, compute the line tangent to the graph of g at $x = x^0$ and let x^1 be the point at which this line intersects the horizontal axis; that is,

$$x^1 = x^0 - g(x^0)/g'(x^0).$$

Continuing in this fashion, we have

$$x^{k+1} = x^k - g(x^k)/g'(x^k).$$

This is the *Newton-Raphson algorithm* for finding roots. Convergence, when it occurs, is usually more rapid than gradient descent, but requires that x^0 be sufficiently close to the solution.

Now suppose that $f : R \to R$ is a real-valued function that we wish to minimize by solving $f'(x) = 0$. Letting $g(x) = f'(x)$ and applying the Newton-Raphson algorithm to $g(x)$ gives the iterative step

$$x^{k+1} = x^k - f'(x^k)/f''(x^k).$$

This is the Newton-Raphson optimization algorithm. Now we extend these results to functions of several variables.

16.4.2   Functions of Several Variables

The Newton-Raphson algorithm for finding roots of functions $g : R^J \to R^J$ has the iterative step

$$x^{k+1} = x^k - [\mathcal{J}(g)(x^k)]^{-1}g(x^k),$$

where $\mathcal{J}(g)(x)$ is the Jacobian matrix of first partial derivatives, $\frac{\partial g_m}{\partial x_j}(x^k)$, for $g(x) = (g_1(x), ..., g_J(x))^T$.

To minimize a function $f : R^J \rightarrow R$, we let $g(x) = \nabla f(x)$ and find a root of g. Then the Newton-Raphson iterative step becomes

$$x^{k+1} = x^k - [\nabla^2 f(x^k)]^{-1} \nabla f(x^k),$$

where $\nabla^2 f(x) = \mathcal{J}(g)(x)$ is the Hessian matrix of second partial derivatives of f.

16.5 Rates of Convergence

In this section we illustrate the concept of *rate of convergence* [22] by considering the fixed-point iteration $x_{k+1} = g(x_k)$, for the twice continuously differentiable function $g : R \rightarrow R$. We suppose that $g(z) = z$ and we are interested in the distance $|x_k - z|$.

16.5.1 Basic Definitions

Definition 16.1. Suppose the sequence $\{x_k\}$ converges to z. If there are positive constants λ and α such that

$$\lim_{n \to \infty} \frac{|x_{k+1} - z|}{|x_k - z|^\alpha} = \lambda,$$

then $\{x_k\}$ is said to converge to z with order α and asymptotic error constant λ. If $\alpha = 1$, the convergence is said to be linear; if $\alpha = 2$, the convergence is said to be quadratic.

16.5.2 Illustrating Quadratic Convergence

According to the Mean-Value Theorem,

$$g(x) = g(z) + g'(z)(x - z) + \frac{1}{2} g''(c)(x - z)^2, \qquad (16.1)$$

for some c between x and z. Suppose now that $g'(z) = 0$. Then we have

$$x_{k+1} = g(x_k) = z + \frac{1}{2} g''(c_k)(x_k - z)^2, \qquad (16.2)$$

for some c_k between x_k and z. Therefore,

$$|x_{k+1} - z| = \frac{1}{2} |g''(c_k)| \, |x_k - z|^2,$$

and the convergence is quadratic, with $\lambda = \frac{1}{2} |g''(z)|$.

16.5.3 Motivating the Newton-Raphson Method

Suppose that we are seeking a root z of the function $f : R \to R$. We define

$$g(x) = x - h(x)f(x),$$

for some function $h(x)$ to be determined. Then $f(z) = 0$ implies that $g(z) = z$. In order to have quadratic convergence of the iterative sequence $x_{k+1} = g(x_k)$, we want $g'(z) = 0$. From

$$g'(x) = 1 - h'(x)f(x) - h(x)f'(x),$$

it follows that we want

$$h(z) = 1/f'(z).$$

Therefore, we choose

$$h(x) = 1/f'(x),$$

so that

$$g(x) = x - f(x)/f'(x).$$

The iteration then takes the form

$$x_{k+1} = g(x_k) = x_k - f(x_k)/f'(x_k),$$

which is the Newton-Raphson iteration.

16.6   Other Approaches

Choosing the negative of the gradient as the next direction makes good sense in minimization problems, but it is not the only, or even the best, way to proceed. For least-squares problems the method of conjugate directions is a popular choice (see [48]). Other modifications of the gradient can also be used, as, for example, in the EMML algorithm.

17 | Convex Sets and Convex Functions

Generally, if a function has a local minimum, it need not be a global minimum. For convex functions, however, a local minimum is necessarily a global one. For that reason, convex functions play an important role in optimization. Convex sets are often used to describe additional conditions placed on the variables. In this chapter we consider several algorithms pertaining to convex sets and convex functions, whose convergence is a consequence of the KM Theorem 5.16.

17.1 Optimizing Functions of a Single Real Variable

Let $f : R \to R$ be a differentiable function. From the Mean-Value Theorem we know that

$$f(b) = f(a) + f'(c)(b - a),$$

for some c between a and b. If there is a constant L with $|f'(x)| \leq L$ for all x, that is, the derivative is bounded, then we have

$$|f(b) - f(a)| \leq L|b - a|, \tag{17.1}$$

for all a and b; functions that satisfy Equation (17.1) are said to be L-*Lipschitz*.

Suppose $g : R \to R$ is differentiable and attains its minimum value. We want to minimize the function $g(x)$. Solving $g'(x) = 0$ to find the optimal $x = x^*$ may not be easy, so we may turn to an iterative algorithm for finding roots of $g'(x)$, or one that minimizes $g(x)$ directly. In the latter case, we may consider a steepest descent algorithm of the form

$$x^{k+1} = x^k - \gamma g'(x^k),$$

for some $\gamma > 0$. We denote by T the operator

$$Tx = x - \gamma g'(x).$$

Then, using $g'(x^*) = 0$, we find that

$$|x^* - x^{k+1}| = |Tx^* - Tx^k|.$$

We would like to know if there are choices for γ that make T an averaged operator. For functions $g(x)$ that are *convex*, the answer is yes.

17.1.1   The Convex Case

A function $g : R \to R$ is called *convex* if, for each pair of distinct real numbers a and b, the line segment connecting the two points $A = (a, g(a))$ and $B = (b, g(b))$ is on or above the graph of $g(x)$. The function $g(x) = x^2$ is a simple example of a convex function.

Proposition 17.1. *The following are equivalent:*

(1) $g(x)$ is convex;

(2) for all points $a < x < b$

$$g(x) \leq \frac{g(b) - g(a)}{b - a}(x - a) + g(a);$$

(3) for all points $a < x < b$

$$g(x) \leq \frac{g(b) - g(a)}{b - a}(x - b) + g(b);$$

(4) for all points a and b and for all α in the interval $(0, 1)$

$$g((1 - \alpha)a + \alpha b) \leq (1 - \alpha)g(a) + \alpha g(b).$$

It follows from Proposition 17.1 that, if $g(x)$ is convex, then, for every triple of points $a < x < b$, we have

$$\frac{g(x) - g(a)}{x - a} \leq \frac{g(b) - g(a)}{b - a} \leq \frac{g(b) - g(x)}{b - x}. \tag{17.2}$$

If $g(x)$ is a differentiable function, then convexity can be expressed in terms of properties of the derivative, $g'(x)$; for every triple of points $a < x < b$, we have

$$g'(a) \leq \frac{g(b) - g(a)}{b - a} \leq g'(b). \tag{17.3}$$

If $g(x)$ is differentiable and convex, then $g'(x)$ is an increasing function. In fact, the converse is also true, as we shall see shortly.

Recall that the line tangent to the graph of $g(x)$ at the point $x = a$ has the equation

$$y = g'(a)(x - a) + g(a).$$

Theorem 17.2. *For the differentiable function $g(x)$, the following are equivalent:*

(1) $g(x)$ is convex;

(2) for all a and x we have

$$g(x) \geq g(a) + g'(a)(x - a); \tag{17.4}$$

(3) the derivative, $g'(x)$, is an increasing function, or, equivalently,

$$(g'(x) - g'(a))(x - a) \geq 0, \tag{17.5}$$

for all a and x.

Proof: Assume that $g(x)$ is convex. If $x > a$, then

$$g'(a) \leq \frac{g(x) - g(a)}{x - a},$$

while, if $x < a$, then

$$\frac{g(a) - g(x)}{a - x} \leq g'(a).$$

In either case, the inequality in (17.4) holds. Now, assume that the inequality in (17.4) holds. Then

$$g(x) \geq g'(a)(x - a) + g(a),$$

and

$$g(a) \geq g'(x)(a - x) + g(x).$$

Adding the two inequalities, we obtain

$$g(a) + g(x) \geq (g'(x) - g'(a))(a - x) + g(a) + g(x),$$

from which we conclude that

$$(g'(x) - g'(a))(x - a) \geq 0.$$

So $g'(x)$ is increasing. Finally, we assume the derivative is increasing and show that $g(x)$ is convex. If $g(x)$ is not convex, then there are points $a < b$ such that, for all x in (a, b),

$$\frac{g(x) - g(a)}{x - a} > \frac{g(b) - g(a)}{b - a}.$$

By the Mean-Value Theorem there is c in (a, b) with

$$g'(c) = \frac{g(b) - g(a)}{b - a}.$$

Select x in the interval (a, c). Then there is d in (a, x) with

$$g'(d) = \frac{g(x) - g(a)}{x - a}.$$

Then $g'(d) > g'(c)$, which contradicts the assumption that $g'(x)$ is increasing. This concludes the proof.   □

If $g(x)$ is twice differentiable, we can say more.

Theorem 17.3. *If $g(x)$ is twice differentiable, then $g(x)$ is convex if and only if $g''(x) \geq 0$, for all x.*

Proof: According to the Mean-Value Theorem, as applied to the function $g'(x)$, for any points $a < b$ there is c in (a, b) with $g'(b) - g'(a) = g''(c)(b - a)$. If $g''(x) \geq 0$, the right side of this equation is nonnegative, so the left side is also. Now assume that $g(x)$ is convex, which implies that $g'(x)$ is an increasing function. Since $g'(x + h) - g'(x) \geq 0$ for all $h > 0$, it follows that $g''(x) \geq 0$.   □

Suppose that $g(x)$ is convex and the function $f(x) = g'(x)$ is L-Lipschitz. If $g(x)$ is twice differentiable, this would be the case if

$$0 \leq g''(x) \leq L,$$

for all x. As we shall see, if γ is in the interval $(0, \frac{2}{L})$, then T is an averaged operator and the iterative sequence converges to a minimizer of $g(x)$. In this regard, we have the following result.

Theorem 17.4. *Let $h(x)$ be convex and differentiable and its derivative, $h'(x)$, nonexpansive, that is,*

$$|h'(b) - h'(a)| \leq |b - a|,$$

for all a and b. Then $h'(x)$ is firmly nonexpansive, which means that

$$(h'(b) - h'(a))(b - a) \geq (h'(b) - h'(a))^2.$$

Proof: Since $h(x)$ is convex and differentiable, the derivative, $h'(x)$, must be increasing. Therefore, if $b > a$, then $|b - a| = b - a$ and

$$|h'(b) - h(a)| = h'(b) - h'(a). \qquad \square$$

If $g(x)$ is convex and $f(x) = g'(x)$ is L-Lipschitz, then $\frac{1}{L}g'(x)$ is nonexpansive, so that $\frac{1}{L}g'(x)$ is firmly nonexpansive and $g'(x)$ is $\frac{1}{L}$-ism. Then, for $\gamma > 0$, $\gamma g'(x)$ is $\frac{1}{\gamma L}$-ism, which tells us that the operator

$$Tx = x - \gamma g'(x)$$

is averaged, whenever $0 < \gamma < \frac{2}{L}$. It follows from the KM Theorem 5.16 that the iterative sequence $x^{k+1} = Tx^k = x^k - \gamma g'(x^k)$ converges to a minimizer of $g(x)$.

In the next section we extend these results to functions of several variables.

17.2   Optimizing Functions of Several Real Variables

Let $F : R^J \to R^N$ be a R^N-valued function of J real variables. The function $F(x)$ is said to be *differentiable* at the point x^0 if there is an N by J matrix $F'(x^0)$ such that

$$\lim_{h \to 0} \frac{1}{||h||_2}[F(x^0 + h) - F(x^0) - F'(x^0)h] = 0.$$

It can be shown that, if F is differentiable at $x = x^0$, then F is continuous there as well [88].

If $f : R^J \to R$ is differentiable, then $f'(x^0) = \nabla f(x^0)$, the gradient of f at x^0. The function $f(x)$ is differentiable if each of its first partial derivatives is continuous. If the derivative $f' : R^J \to R^J$ is, itself, differentiable, then $f'' : R^J \to R^J$, and $f''(x) = H(x) = \nabla^2 f(x)$, the Hessian matrix whose entries are the second partial derivatives of f. The function $f(x)$ will be twice differentiable if each of the second partial derivatives is continuous. In that case, the mixed second partial derivatives are independent of the order of the variables, the Hessian matrix is symmetric, and the chain rule applies.

Let $f : R^J \to R$ be a differentiable function. From the Mean-Value Theorem ([88], p. 41) we know that, for any two points a and b, there is α in $(0, 1)$ such that

$$f(b) = f(a) + \langle \nabla f((1 - \alpha)a + \alpha b), b - a \rangle.$$

If there is a constant L with $||\nabla f(x)||_2 \leq L$ for all x, that is, the gradient is bounded in norm, then we have

$$|f(b) - f(a)| \leq L||b - a||_2, \tag{17.6}$$

for all a and b; functions that satisfy Equation (17.6) are said to be L-$Lipschitz$.

In addition to real-valued functions $f : R^J \rightarrow R$, we shall also be interested in functions $F : R^J \rightarrow R^J$, such as $F(x) = \nabla f(x)$, whose range is R^J, not R. We say that $F : R^J \rightarrow R^J$ is L-Lipschitz if there is $L > 0$ such that

$$||F(b) - F(a)||_2 \leq L||b - a||_2,$$

for all a and b.

Suppose $g : R^J \rightarrow R$ is differentiable and attains its minimum value. We want to minimize the function $g(x)$. Solving $\nabla g(x) = 0$ to find the optimal $x = x^*$ may not be easy, so we may turn to an iterative algorithm for finding roots of $\nabla g(x)$, or one that minimizes $g(x)$ directly. In the latter case, we may again consider a steepest descent algorithm of the form

$$x^{k+1} = x^k - \gamma \nabla g(x^k),$$

for some $\gamma > 0$. We denote by T the operator

$$Tx = x - \gamma \nabla g(x).$$

Then, using $\nabla g(x^*) = 0$, we find that

$$||x^* - x^{k+1}||_2 = ||Tx^* - Tx^k||_2.$$

We would like to know if there are choices for γ that make T an averaged operator. As in the case of functions of a single variable, for functions $g(x)$ that are $convex$, the answer is yes.

17.2.1   The Convex Case

We begin with some definitions.

Definition 17.5. The function $g(x) : R^J \rightarrow R$ is said to be convex if, for each pair of distinct vectors a and b and for every α in the interval $(0, 1)$ we have

$$g((1 - \alpha)a + \alpha b) \leq (1 - \alpha)g(a) + \alpha g(b).$$

The function $g(x)$ is convex if and only if, for every x and z in R^J and real t, the function $f(t) = g(x + tz)$ is a convex function of t. Therefore, the theorems for the multivariable case can also be obtained from previous results for the single-variable case.

Definition 17.6. A convex function $g : R^J \to [-\infty, +\infty]$ is proper if there is no x with $g(x) = -\infty$ and some x with $g(x) < +\infty$.

Definition 17.7. The effective domain of g is $\mathrm{dom}(g) = \{x | g(x) < +\infty\}$.

Definition 17.8. A proper convex function g is closed if it is lower semi-continuous, that is, if $g(x) = \lim \inf g(y)$, as $y \to x$.

Definition 17.9. The subdifferential of g at x is the set

$$\partial g(x) = \{x^* | \langle x^*, z - x \rangle \le g(z) - g(x), \text{for all } z\}.$$

The domain of ∂g is the set $\mathrm{dom}\, \partial g = \{x | \partial g(x) \ne \emptyset\}$.

If g is differentiable, then the subdifferential contains only the gradient, that is,

$$\partial g(x) = \{\nabla g(x)\}.$$

In this chapter we shall focus on the optimization of differentiable functions g, leaving to a later chapter the nondifferentiable, or nonsmooth, case. If $g(x)$ is a differentiable function, then convexity can be expressed in terms of properties of the derivative, $\nabla g(x)$. Note that, by the chain rule, $f'(t) = \nabla g(x + tz) \cdot z$.

Theorem 17.10. *For the differentiable function $g(x)$, the following are equivalent:*

(1) $g(x)$ is convex;

(2) for all a and b we have

$$g(b) \ge g(a) + \langle \nabla g(a), b - a \rangle ; \tag{17.7}$$

(3) for all a and b we have

$$\langle \nabla g(b) - \nabla g(a), b - a \rangle \ge 0. \tag{17.8}$$

As in the case of functions of a single variable, we can say more when the function $g(x)$ is twice differentiable. To guarantee that the second derivative matrix is symmetric, we assume that the second partial derivatives are continuous. Note that, by the chain rule again, $f''(t) = z^T \nabla^2 g(x + tz) z$.

Theorem 17.11. *Let each of the second partial derivatives of $g(x)$ be continuous, so that $g(x)$ is twice continuously differentiable. Then $g(x)$ is convex if and only if the second derivative matrix $\nabla^2 g(x)$ is nonnegative definite, for each x.*

Suppose that $g(x) : R^J \to R$ is convex and the function $F(x) = \nabla g(x)$ is L-Lipschitz. As we shall see, if γ is in the interval $(0, \frac{2}{L})$, then the operator $T = I - \gamma F$ defined by

$$Tx = x - \gamma \nabla g(x)$$

is an averaged operator and the iterative sequence converges to a minimizer of $g(x)$. In this regard, we have the following analog of Theorem 17.4.

Theorem 17.12. *Let $h(x)$ be convex and differentiable and its derivative, $\nabla h(x)$, nonexpansive, that is,*

$$||\nabla h(b) - \nabla h(a)||_2 \le ||b - a||_2,$$

for all a and b. Then $\nabla h(x)$ is firmly nonexpansive, which means that

$$\langle \nabla h(b) - \nabla h(a), b - a \rangle \ge ||\nabla h(b) - \nabla h(a)||_2^2.$$

Unlike the proof of Theorem 17.4, the proof of this theorem is not trivial. In [93] Golshtein and Tretyakov prove the following theorem, from which Theorem 17.12 follows immediately.

Theorem 17.13. *Let $g : R^J \to R$ be convex and differentiable. The following are equivalent:*

$$||\nabla g(x) - \nabla g(y)||_2 \le ||x - y||_2; \tag{17.9}$$

$$g(x) \ge g(y) + \langle \nabla g(y), x - y \rangle + \frac{1}{2}||\nabla g(x) - \nabla g(y)||_2^2; \tag{17.10}$$

$$\langle \nabla g(x) - \nabla g(y), x - y \rangle \ge ||\nabla g(x) - \nabla g(y)||_2^2. \tag{17.11}$$

Proof: The only difficult step in the proof is showing that Inequality (17.9) implies Inequality (17.10). To prove this part, let $x(t) = (1 - t)y + tx$, for $0 \le t \le 1$. Then

$$g'(x(t)) = \langle \nabla g(x(t)), x - y \rangle,$$

so that

$$\int_0^1 \langle \nabla g(x(t)) - \nabla g(y), x - y \rangle dt = g(x) - g(y) - \langle \nabla g(y), x - y \rangle.$$

Therefore,

$$g(x) - g(y) - \langle \nabla g(y), x - y \rangle \leq \int_0^1 ||\nabla g(x(t)) - \nabla g(y)||_2 ||x(t) - y||_2 dt$$

$$\leq \int_0^1 ||x(t) - y||_2^2 dt$$

$$= \int_0^1 ||t(x - y)||_2^2 dt = \frac{1}{2} ||x - y||_2^2,$$

according to Inequality (17.9). Therefore,

$$g(x) \leq g(y) + \langle \nabla g(y), x - y \rangle + \frac{1}{2} ||x - y||_2^2.$$

Now let $x = y - \nabla g(y)$, so that

$$g(y - \nabla g(y)) \leq g(y) + \langle \nabla g(y), \nabla g(y) \rangle + \frac{1}{2} ||\nabla g(y)||_2^2.$$

Consequently,

$$g(y - \nabla g(y)) \leq g(y) - \frac{1}{2} ||\nabla g(y)||_2^2.$$

Therefore,

$$\inf g(x) \leq g(y) - \frac{1}{2} ||\nabla g(y)||_2^2,$$

or

$$g(y) \geq \inf g(x) + \frac{1}{2} ||\nabla g(y)||_2^2. \tag{17.12}$$

Now fix y and define the function $h(x)$ by

$$h(x) = g(x) - g(y) - \langle \nabla g(y), x - y \rangle.$$

Then $h(x)$ is convex, differentiable, and nonnegative,

$$\nabla h(x) = \nabla g(x) - \nabla g(y),$$

and $h(y) = 0$, so that $h(x)$ attains its minimum at $x = y$. Applying Inequality (17.12) to the function $h(x)$, with z in the role of x and x in the role of y, we find that

$$\inf h(z) = 0 \leq h(x) - \frac{1}{2} ||\nabla h(x)||_2^2.$$

From the definition of $h(x)$, it follows that

$$0 \leq g(x) - g(y) - \langle \nabla g(y), x - y \rangle - \frac{1}{2} ||\nabla g(x) - \nabla g(y)||_2^2. \qquad \square$$

This completes the proof of the implication.

If $g(x)$ is convex and $f(x) = \nabla g(x)$ is L-Lipschitz, then $\frac{1}{L}\nabla g(x)$ is nonexpansive, so that $\frac{1}{L}\nabla g(x)$ is firmly nonexpansive and $\nabla g(x)$ is $\frac{1}{L}$-ism. Then for $\gamma > 0$, $\gamma\nabla g(x)$ is $\frac{1}{\gamma L}$-ism, which tells us that the operator

$$Tx = x - \gamma\nabla g(x)$$

is averaged, whenever $0 < \gamma < \frac{2}{L}$. It follows from the KM Theorem 5.16 that the iterative sequence $x^{k+1} = Tx^k = x^k - \gamma\nabla g(x^k)$ converges to a minimizer of $g(x)$, whenever minimizers exist.

17.3 Convex Feasibility

The *convex feasibility problem* (CFP) is to find a point in the nonempty intersection C of finitely many closed, convex sets C_i in R^J. The *successive orthogonal projection* (SOP) method [96] is the following. Begin with an arbitrary x^0. For $k = 0, 1, ...$, and $i = k(\bmod I) + 1$, let

$$x^{k+1} = P_i x^k,$$

where $P_i x$ denotes the orthogonal projection of x onto the set C_i. Since each of the operators P_i is firmly nonexpansive, the product

$$T = P_I P_{I-1} \cdots P_2 P_1$$

is averaged. Since C is not empty, T has fixed points. By the KM Theorem 5.16, the sequence $\{x^k\}$ converges to a member of C. It is useful to note that the limit of this sequence will not generally be the point in C closest to x^0; it is if the C_i are hyperplanes, however.

17.3.1 The CFP in Linear Programming

Following Rockafellar [139], we define a *real interval* to be any nonempty closed convex subset of the real line. Let $I_1, ..., I_J$ be real intervals and L a nonempty subspace of R^J. Is there a vector $x = (x_1, ..., x_J)^T$ in L, with x_j in I_j, for each j? This is an example of a CFP. To see this, let $C = \prod_{j=1}^J I_j$ be the set of all x in R^J with x_j in I_j, for each j. Then C is a nonempty closed convex set. The problem then is to find x in $C \cap L$. According to Theorem 22.6 of [139], there will be such an x unless there is $z = (z_1, ..., z_J)^T$ in $L^\perp$ with $\sum_{j=1}^J z_j c_j > 0$, for all c_j in I_j, for $j = 1, ..., J$.

17.3.2 The SOP for Hyperplanes

For any x, $P_i x$, the orthogonal projection of x onto the closed, convex set C_i, is the unique member of C_i for which

$$\langle P_i x - x, y - P_i x \rangle \geq 0,$$

for every y in C_i. It follows from this characterization that

$$||y - P_i x||_2^2 + ||P_i x - x||_2^2 \leq ||y - x||_2^2,$$

for all x and for all y in C_i.

When the C_i are hyperplanes, we can say more.

Lemma 17.14. *If C_i is a hyperplane, then*

$$\langle P_i x - x, y - P_i x \rangle = 0,$$

for all y in C_i.

Since both $P_i x$ and y are in C_i, so is $P_i x + t(y - P_i x)$, for every real t. We can use Lemma 17.14 to show that

$$||y - P_i x||_2^2 + ||P_i x - x||_2^2 = ||y - x||_2^2,$$

for every y in the hyperplane C_i.

Theorem 17.15. *When the C_i are hyperplanes, the SOP algorithm converges to the member of the intersection that is closest to x^0.*

Proof: Let the C_i be hyperplanes with C their nonempty intersection. Let $\hat{c}$ be in C. For $x^{k+1} = P_i x^k$, where $i = k \pmod{I} + 1$, we have

$$||\hat{c} - x^k||_2^2 - ||\hat{c} - x^{k+1}||_2^2 = ||x^k - x^{k+1}||_2^2. \tag{17.13}$$

It follows that the sequence $\{||\hat{c} - x^k||_2\}$ is decreasing and that the sequence $\{||x^k - x^{k+1}||_2^2\}$ converges to zero. Therefore, the sequence $\{x^k\}$ is bounded, so it has a cluster point, x^*, and the cluster point must be in C. Therefore, replacing $\hat{c}$ with x^*, we find that the sequence $\{||x^* - x^k||_2^2\}$ converges to zero, which means that $\{x^k\}$ converges to x^*. Summing over k on both sides of Equation (17.13), we get

$$||\hat{c} - x^*||_2^2 - ||\hat{c} - x^0||_2^2$$

on the left side, while on the right side we get a quantity that does not depend on which $\hat{c}$ in C we have selected. It follows that minimizing $||\hat{c} - x^0||_2^2$ over $\hat{c}$ in C is equivalent to minimizing $||\hat{c} - x^*||_2^2$ over $\hat{c}$ in C; the minimizer of the latter problem is clearly $\hat{c} = x^*$. $\qquad \square$

Note that the SOP is the ART algorithm, for the case of hyperplanes.

17.3.3   The SOP for Half-Spaces

If the C_i are half-spaces, that is, there is some I by J matrix A and vector b so that

$$C_i = \{x | (Ax)_i \geq b_i\},$$

then the SOP becomes the Agmon-Motzkin-Schoenberg (AMS) Algorithm. When the intersection is nonempty, the algorithm converges, by the KM Theorem, to a member of that intersection. When the intersection is empty, we get subsequential convergence to a limit cycle.

17.3.4   The SOP when C is empty

When the intersection C of the sets C_i, $i = 1, ..., I$ is empty, the SOP cannot converge. Drawing on our experience with two special cases of the SOP, the ART and the AMS Algorithms, we conjecture that, for each $i = 1, ..., I$, the subsequences $\{x^{nI+i}\}$ converge to $c^{*,i}$ in C_i, with $P_i c^{*,i-1} = c^{*,i}$ for $i = 2, 3, ..., I$, and $P_1 c^{*,I} = c^{*,1}$; the set $\{c^{*,i}\}$ is then a limit cycle. For the special case of $I = 2$ we can prove this.

Theorem 17.16. *Let C_1 and C_2 be nonempty, closed convex sets in $\mathcal{X}$, with $C_1 \cap C_2 = \emptyset$. Assume that there is a unique $\hat{c}_2$ in C_2 minimizing the function $f(x) = ||c_2 - P_1 c_2||_2$, over all c_2 in C_2. Let $\hat{c}_1 = P_1 \hat{c}_2$. Then $P_2 \hat{c}_1 = \hat{c}_2$. Let z^0 be arbitrary and, for $n = 0, 1, ...,$ let*

$$z^{2n+1} = P_1 z^{2n},$$
$$z^{2n+2} = P_2 z^{2n+1}.$$

Then

$$\{z^{2n+1}\} \rightarrow \hat{c}_1,$$
$$\{z^{2n}\} \rightarrow \hat{c}_2.$$

Proof: We apply the CQ algorithm, with the iterative step given by Equation (6.4), with $C = C_2$, $Q = C_1$, and the matrix $A = I$, the identity matrix. The CQ iterative step is now

$$x^{k+1} = P_2(x^k + \gamma(P_1 - I)x^k).$$

Using the acceptable choice of $\gamma = 1$, we have

$$x^{k+1} = P_2 P_1 x^k.$$

This CQ iterative sequence then converges to $\hat{c}_2$, the minimizer of the function $f(x)$. Since $z^{2n} = x^n$, we have $\{z^{2n}\} \rightarrow \hat{c}_2$. Because

$$||P_2 \hat{c}_1 - \hat{c}_1||_2 \leq ||\hat{c}_2 - \hat{c}_1||_2,$$

it follows from the uniqueness of $\hat{c}_2$ that $P_2\hat{c}_1 = \hat{c}_2$. This completes the proof. $\qquad\qquad\qquad\qquad\qquad\qquad\qquad\qquad\qquad\qquad\qquad\qquad\qquad\square$

The paper of De Pierro and Iusem includes related results [77].

17.4 Optimization over a Convex Set

Suppose now that $g : R^J \rightarrow R$ is a convex, differentiable function and we want to find a minimizer of $g(x)$ over a closed, convex set C, if such minimizers exists. We saw earlier that, if $\nabla g(x)$ is L-Lipschitz, and γ is in the interval $(0, 2/L)$, then the operator $Tx = x - \gamma \nabla g(x)$ is averaged. Since P_C, the orthogonal projection onto C, is also averaged, their product, $S = P_C T$, is averaged. Therefore, by the KM Theorem 5.16, the sequence $\{x^{k+1} = Sx^k\}$ converges to a fixed point of S, whenever such fixed points exist. Note that $\hat{x}$ is a fixed point of S if and only if $\hat{x}$ minimizes $g(x)$ over x in C.

17.4.1 Linear Optimization over a Convex Set

Suppose we take $g(x) = d^T x$, for some fixed vector d. Then $\nabla g(x) = d$ for all x, and $\nabla g(x)$ is L-Lipschitz for every $L > 0$. Therefore, the operator $Tx - x - \gamma d$ is averaged, for any positive γ. Since P_C is also averaged, the product, $S = P_C T$ is averaged and the iterative sequence $x^{k+1} = Sx^k$ converges to a minimizer of $g(x) = d^T x$ over C, whenever minimizers exist.

For example, suppose that C is the closed, convex region in the plane bounded by the coordinate axes and the line $x + y = 1$. Let $d^T = (1, -1)$. The problem then is to minimize the function $g(x, y) = x - y$ over C. Let $\gamma = 1$ and begin with $x^0 = (1, 1)^T$. Then $x^0 - d = (0, 2)^T$ and $x^1 = P_C(0, 2)^T = (0, 1)^T$, which is the solution.

For this algorithm to be practical, $P_C x$ must be easy to calculate. In those cases in which the set C is more complicated than in the example, other algorithms, such as the simplex algorithm, will be preferred. We consider these ideas further, when we discuss the linear programming problem.

17.5 Geometry of Convex Sets

Definition 17.17. A point x in a convex set C is said to be an *extreme point* of C if the set obtained by removing x from C remains convex.

Said another way, $x \in C$ is an extreme point of C if x cannot be written as

$$x = (1 - \alpha)y + \alpha z,$$

for $y, z \neq x$ and $\alpha \in (0, 1)$. For example, the point $x = 1$ is an extreme point of the convex set $C = [0, 1]$. Every point on the boundary of a sphere in R^J is an extreme point of the sphere. The set of all extreme points of a convex set is denoted $\text{Ext}(C)$.

A nonzero vector d is said to be a *direction of unboundedness* of a convex set C if, for all x in C and all $\gamma \geq 0$, the vector $x + \gamma d$ is in C. For example, if C is the nonnegative orthant in R^J, then any nonnegative vector d is a direction of unboundedness.

The fundamental problem in linear programming is to minimize the function

$$f(x) = c^T x,$$

over the *feasible set* F, that is, the convex set of all $x \geq 0$ with $Ax = b$. In Chapter 22 we present an algebraic description of the extreme points of the feasible set F, in terms of *basic feasible solutions*, show that there are at most finitely many extreme points of F, and show that every member of F can be written as a convex combination of the extreme points, plus a direction of unboundedness. These results will be used to prove the basic theorems about the primal and dual linear programming problems and to describe the simplex algorithm.

17.6   Projecting onto Convex Level Sets

Suppose that $f : R^J \to R$ is a convex function and $C = \{x | f(x) \leq 0\}$. Then C is a convex set. A vector t is said to be a *subgradient* of f at x if, for all z, we have

$$f(z) - f(x) \geq \langle t, z - x \rangle.$$

Such subgradients always exist, for convex functions. If f is differentiable at x, then f has a unique subgradient, namely, its gradient, $t = \nabla f(x)$.

Unless f is a linear function, calculating the orthogonal projection, $P_C z$, of z onto C requires the solution of an optimization problem. For that reason, closed-form approximations of $P_C z$ are often used. One such approximation occurs in the *cyclic subgradient projection* (CSP) method. Given x not in C, let

$$\Pi_C x = x - \alpha t,$$

where t is any subgradient of f at x and $\alpha = \frac{f(x)}{||t||^2} > 0$.

Proposition 17.18. *For any c in C, $||c - \Pi_C x||_2^2 < ||c - x||_2^2$.*

Proof: Since x is not in C, we know that $f(x) > 0$. Then,

$$||c - \Pi_C x||_2^2 = ||c - x + \alpha t||_2^2$$
$$= ||c - x||_2^2 + 2\alpha \langle c - x, t \rangle + \alpha f(x). \qquad \square$$

Since t is a subgradient, we know that

$$\langle c - x, t \rangle \leq f(c) - f(x),$$

so that

$$||c - \Pi_C x||_2^2 - ||c - x||_2^2 \leq 2\alpha(f(c) - f(x)) + \alpha f(x) < 0.$$

The CSP method is a variant of the SOP method, in which P_{C_i} is replaced with Π_{C_i}.

17.7 Projecting onto the Intersection of Convex Sets

As we saw previously, the SOP algorithm need not converge to the point in the intersection closest to the starting point. To obtain the point closest to x^0 in the intersection of the convex sets C_i, we can use *Dykstra's Algorithm*, a modification of the SOP method [83]. For simplicity, we shall discuss only the case of $C = A \cap B$, the intersection of two closed, convex sets.

17.7.1 A Motivating Lemma

The following lemma will help motivate Dykstra's Algorithm.

Lemma 17.19. *If* $x = c + p + q$, *where* $c = P_A(c + p)$ *and* $c = P_B(c + q)$, *then* $c = P_C x$.

Proof: Let d be arbitrary in C. Then

$$\langle c - (c + p), d - c \rangle \geq 0,$$

since d is in A, and

$$\langle c - (c + q), d - c \rangle \geq 0,$$

since d is in B. Adding the two inequalities, we get

$$\langle -p - q, d - c \rangle \geq 0.$$

But

$$-p - q = c - x,$$

so

$$\langle c - x, d - c \rangle \geq 0,$$

for all d in C. Therefore, $c = P_C x$. $\square$

17.7.2   Dykstra's Algorithm

Algorithm 17.20 (Dykstra). Let $b_0 = x$, and $p_0 = q_0 = 0$. Then let

$$a_n = P_A(b_{n-1} + p_{n-1}),$$
$$b_n = P_B(a_n + q_{n-1}),$$

and define p_n and q_n by

$$x = a_n + p_n + q_{n-1} = b_n + p_n + q_n.$$

Using the algorithm, we construct two sequences, $\{a_n\}$ and $\{b_n\}$, both converging to $c = P_C x$, along with two other sequences, $\{p_n\}$ and $\{q_n\}$. Usually, but not always, $\{p_n\}$ converges to p and $\{q_n\}$ converges to q, so that

$$x = c + p + q,$$

with

$$c = P_A(c + p) = P_B(c + q).$$

Generally, however, $\{p_n + q_n\}$ converges to $x - c$.

In [17], Bregman considers the problem of minimizing a convex function $f : R^J \rightarrow R$ over the intersection of half-spaces, that is, over the set of points x for which $Ax => b$. His approach is a *primal-dual* algorithm involving the notion of projecting onto a convex set, with respect to a generalized distance constructed from f. Such generalized projections have come to be called *Bregman projections*. In [58], Censor and Reich extend Dykstra's Algorithm to Bregman projections, and, in [18], Bregman, Censor, and Reich show that the extended Dykstra Algorithm of [58] is the natural extension of Bregman's primal-dual algorithm to the case of intersecting convex sets. We shall consider these results in more detail in a subsequent chapter.

17.7.3   The Halpern-Lions-Wittmann-Bauschke Algorithm

There is yet another approach to finding the orthogonal projection of the vector x onto the nonempty intersection C of finitely many closed, convex sets C_i, $i = 1, ..., I$.

Algorithm 17.21 (HLWB). Let x^0 be arbitrary. Then let

$$x^{k+1} = t_k x + (1 - t_k)P_{C_i} x^k,$$

where P_{C_i} denotes the orthogonal projection onto C_i, t_k is in the interval $(0, 1)$, and $i = k(\mod I) + 1$.

Several authors have proved convergence of the sequence $\{x^k\}$ to $P_C x$, with various conditions imposed on the parameters $\{t_k\}$. As a result, the algorithm is known as the Halpern-Lions-Wittmann-Bauschke (HLWB) Algorithm, after the names of several who have contributed to the evolution of the theorem; see also Corollary 2 in Reich's paper [137]. The conditions imposed by Bauschke [7] are $\{t_k\} \to 0$, $\sum t_k = \infty$, and $\sum |t_k - t_{k+1}| < +\infty$. The HLWB algorithm has been extended by Deutsch and Yamada [79] to minimize certain (possibly nonquadratic) functions over the intersection of fixed point sets of operators more general than P_{C_i}.

18 | Generalized Projections onto Convex Sets

The *convex feasibility problem* (CFP) is to find a member of the nonempty set $C = \bigcap_{i=1}^{I} C_i$, where the C_i are closed convex subsets of R^J. In most applications the sets C_i are more easily described than the set C and algorithms are sought whereby a member of C is obtained as the limit of an iterative procedure involving (exact or approximate) orthogonal or generalized projections onto the individual sets C_i.

In his often cited paper [17] Bregman generalizes the SOP algorithm for the convex feasibility problem to include projections with respect to a generalized distance, and uses this *successive generalized projections* (SGP) method to obtain a *primal-dual algorithm* to minimize a convex function $f : R^J \to R$ over the intersection of half-spaces, that is, over x with $Ax \geq b$. The generalized distance is built from the function f, which then must exhibit additional properties, beyond convexity, to guarantee convergence of the algorithm

18.1 Bregman Functions and Bregman Distances

The class of functions f that are used to define the generalized distance have come to be called *Bregman functions*; the associated generalized distances are then *Bregman distances*, which are used to define generalized projections onto closed convex sets (see the book by Censor and Zenios [61] for details). In [10] Bauschke and Borwein introduce the related class of *Bregman-Legendre functions* and show that these functions provide an appropriate setting in which to study Bregman distances and generalized projections associated with such distances. For further details concerning Bregman and Bregman-Legendre functions, see the appendix.

Bregman's successive generalized projection method uses projections with respect to Bregman distances to solve the convex feasibility problem.

209

Let $f : R^J \rightarrow (-\infty, +\infty]$ be a closed, proper convex function, with effective domain $D = \mathrm{dom} f = \{x | f(x) < +\infty\}$ and $\emptyset \neq \mathrm{int}\, D$. Denote by $D_f(\cdot, \cdot) :$ $D \times \mathrm{int}\, D \rightarrow [0, +\infty)$ the Bregman distance, given by

$$D_f(x, z) = f(x) - f(z) - \langle \nabla f(z), x - z \rangle$$

and by $P_{C_i}^f$ the Bregman projection operator associated with the convex function f and the convex set C_i; that is

$$P_{C_i}^f z = \arg\min_{x \in C_i \cap D} D_f(x, z).$$

The Bregman projection of x onto C is characterized by *Bregman's Inequality*:

$$\langle \nabla f(P_C^f x) - \nabla f(x), c - P_C^f \rangle \geq 0, \tag{18.1}$$

for all c in C.

18.2 The Successive Generalized Projections Algorithm

Bregman considers the following generalization of the SOP algorithm:

Algorithm 18.1. [Bregman's SGP Method] Beginning with $x^0 \in \mathrm{int}\,\mathrm{dom} f$, for $k = 0, 1, ...,$ let $i = i(k) := k(\mathrm{mod}\, I) + 1$ and

$$x^{k+1} = P_{C_{i(k)}}^f(x^k). \tag{18.2}$$

He proves that the sequence $\{x^k\}$ given by Equation (18.2) converges to a member of $C \cap \mathrm{dom} f$, whenever this set is nonempty and the function f is what came to be called a Bregman function ([17]). Bauschke and Borwein [10] prove that Bregman's SGP method converges to a member of C provided that one of the following holds:

(1) f is Bregman-Legendre;

(2) $C \cap \mathrm{int} D \neq \emptyset$ and $\mathrm{dom}\, f^*$ is open; or

(3) $\mathrm{dom}\, f$ and $\mathrm{dom}\, f^*$ are both open, with f^* the function conjugate to f.

In [17] Bregman goes on to use the SGP to find a minimizer of a Bregman function $f(x)$ over the set of x such that $Ax = b$. Each hyperplane associated with a single equation is a closed, convex set. The SGP finds the Bregman projection of the starting vector onto the intersection of the hyperplanes. If the starting vector has the form $x^0 = A^T d$, for some vector d, then this Bregman projection also minimizes $f(x)$ over x in the intersection. Alternating Bregman projections also appears in Reich's paper [138].

18.3 Bregman's Primal-Dual Algorithm

The problem is to minimize $f : R^J \rightarrow R$ over the set of all x for which $Ax \geq b$. Begin with x^0 such that $x^0 = A^T u^0$, for some $u^0 \geq 0$. For $k = 0, 1, ...,$ let $i = k(\mod I) + 1$. Having calculated x^k, there are three possibilities:

(1) if $(Ax^k)_i < b_i$, then let x^{k+1} be the Bregman projection onto the hyperplane $H_i = \{x | (Ax)_i = b_i\}$, so that

$$\nabla f(x^{k+1}) = \nabla f(x^k) + \lambda_k a^i,$$

where a^i is the ith column of A^T. With $\nabla f(x^k) = A^T u^k$, for $u^k \geq 0$, update u^k by

$$u_i^{k+1} = u_i^k + \lambda_k,$$

and

$$u_m^{k+1} = u_m^k,$$

for $m \neq i$;

(2) if $(Ax^k)_i = b_i$, or $(Ax^k)_i > b_i$ and $u_i^k = 0$, then $x^{k+1} = x^k$, and $u^{k+1} = u^k$;

(3) if $(Ax^k)_i > b_i$ and $u_i^k > 0$, then let μ_k be the smaller of the numbers μ_k' and μ_k'', where

$$\nabla f(y) = \nabla f(x^k) - \mu_k' a^i$$

puts y in H_i, and

$$\mu_k'' = u_i^k.$$

Then take x^{k+1} with

$$\nabla f(x^{k+1}) = \nabla f(x^k) - \mu_k a^i.$$

With appropriate assumptions made about the function f, the sequence $\{x^k\}$ so defined converges to a minimizer of $f(x)$ over the set of x with $Ax \geq b$. For a detailed proof of this result, see [61].

Bregman also suggests that this primal-dual algorithm be used to find approximate solutions for linear programming problems, where the problem is to minimize a linear function $c^T x$, subject to constraints. His idea is to replace the function $c^T x$ with $h(x) = c^T x + \epsilon f(x)$, and then apply his primal-dual method to $h(x)$.

18.4   Dykstra's Algorithm for Bregman Projections

We are concerned now with finding the Bregman projection of x onto the intersection C of finitely many closed convex sets, C_i. The problem can be solved by extending Dykstra's Algorithm to include Bregman projections.

18.4.1   A Helpful Lemma

The following lemma helps to motivate the extension of Dykstra's Algorithm.

Lemma 18.2. *Suppose that*

$$\nabla f(c) - \nabla f(x) = \nabla f(c) - \nabla f(c+p) + \nabla f(c) - \nabla f(c+q),$$

with $c = P_A^f(c+p)$ and $c = P_B^f(c+q)$. Then $c = P_C^f x$.

Proof: Let d be arbitrary in C. We have

$$\langle \nabla f(c) - \nabla f(c+p), d-c \rangle \geq 0,$$

and

$$\langle \nabla f(c) - \nabla f(c+q), d-c \rangle \geq 0.$$

Adding, we obtain

$$\langle \nabla f(c) - \nabla f(x), d-c \rangle \geq 0. \qquad \square$$

This suggests the following algorithm for finding $c = P_C^f x$, which turns out to be the extension of Dykstra's Algorithm to Bregman projections.

Algorithm 18.3 (Bregman-Dykstra). Begin with $b^0 = x$, $p_0 = q_0 = 0$. Define

$$b_{n-1} + p_{n-1} = \nabla f^{-1}(\nabla f(b_{n-1}) + r_{n-1}),$$
$$a_n = P_A^f(b_{n-1} + p_{n-1}),$$
$$r_n = \nabla f(b_{n-1}) + r_{n-1} - \nabla f(a_n),$$
$$\nabla f(a_n + q_{n-1}) = \nabla f(a_n) + s_{n-1},$$
$$b_n = P_B^f(a_n + q_{n-1}),$$
$$s_n = \nabla f(a_n) + s_{n-1} - \nabla f(b_n).$$

In place of

$$\nabla f(c+p) - \nabla f(c) + \nabla f(c+q) - \nabla f(c),$$

we have

$$r_{n-1} + s_{n-1} = [\nabla f(b_{n-1}) + r_{n-1}] - \nabla f(b_{n-1})$$
$$+ [\nabla f(a_n) + s_{n-1}] - \nabla f(a_n),$$

and also

$$r_n + s_{n-1} = [\nabla f(a_n) + s_{n-1}] - \nabla f(a_n)$$
$$+ [\nabla f(b_n) + r_n] - \nabla f(b_n).$$

But we also have

$$r_{n-1} + s_{n-1} = \nabla f(x) - \nabla f(b_{n-1}),$$

and

$$r_n + s_{n-1} = \nabla f(x) - \nabla f(a_n).$$

Then the sequences $\{a_n\}$ and $\{b_n\}$ converge to c. For further details, see the papers of Censor and Reich [58] and Bauschke and Lewis [12].

In [18] Bregman, Censor, and Reich show that the extension of Dykstra's Algorithm to Bregman projections can be viewed as an extension of Bregman's primal-dual algorithm to the case in which the intersection of half-spaces is replaced by the intersection of closed convex sets.

19 | The Split Feasibility Problem

The *split feasibility problem* (SFP) [54] is to find $c \in C$ with $Ac \in Q$, if such points exist, where A is a real I by J matrix and C and Q are nonempty, closed convex sets in R^J and R^I, respectively. In this chapter we discuss the CQ algorithm for solving the SFP, as well as recent extensions and applications.

19.1   The CQ Algorithm

In [45] the CQ algorithm for solving the SFP was presented, for the real case. It has the iterative step

$$x^{k+1} = P_C(x^k - \gamma A^T(I - P_Q)Ax^k), \qquad (19.1)$$

where I is the identity operator and $\gamma \in (0, 2/\rho(A^T A))$, for $\rho(A^T A)$ the spectral radius of the matrix $A^T A$, which is also its largest eigenvalue. The CQ algorithm can be extended to the complex case, in which the matrix A has complex entries, and the sets C and Q are in C^J and C^I, respectively. The iterative step of the extended CQ algorithm is then

$$x^{k+1} = P_C(x^k - \gamma A^\dagger(I - P_Q)Ax^k). \qquad (19.2)$$

The CQ algorithm converges to a solution of the SFP, for any starting vector x^0, whenever the SFP has solutions. When the SFP has no solutions, the CQ algorithm converges to a minimizer of the function

$$f(x) = \frac{1}{2}\|P_Q Ax - Ax\|_2^2$$

over the set C, provided such constrained minimizers exist. Therefore the CQ algorithm is an iterative constrained optimization method. As shown in [46], convergence of the CQ algorithm is a consequence of Theorem 5.16.

The function $f(x)$ is convex and differentiable on R^J and its derivative is the operator

$$\nabla f(x) = A^T(I - P_Q)Ax;$$

see [3].

Lemma 19.1. *The derivative operator ∇f is λ-Lipschitz continuous for $\lambda = \rho(A^T A)$, therefore it is ν-ism for $\nu = \frac{1}{\lambda}$.*

Proof: We have

$$\|\nabla f(x) - \nabla f(y)\|_2^2 = \|A^T(I - P_Q)Ax - A^T(I - P_Q)Ay\|_2^2$$
$$\leq \lambda \|(I - P_Q)Ax - (I - P_Q)Ay\|_2^2.$$

Also

$$\|(I - P_Q)Ax - (I - P_Q)Ay\|_2^2 = \|Ax - Ay\|_2^2 + \|P_Q Ax - P_Q Ay\|_2^2$$
$$- 2\langle P_Q Ax - P_Q Ay, Ax - Ay \rangle$$

and, since P_Q is firmly nonexpansive,

$$\langle P_Q Ax - P_Q Ay, Ax - Ay \rangle \geq \|P_Q Ax - P_Q Ay\|_2^2.$$

Therefore,

$$\|\nabla f(x) - \nabla f(y)\|_2^2 \leq \lambda(\|Ax - Ay\|_2^2 - \|P_Q Ax - P_Q Ay\|_2^2)$$
$$\leq \lambda \|Ax - Ay\|_2^2 \leq \lambda^2 \|x - y\|_2^2.$$

This completes the proof.   □

If $\gamma \in (0, 2/\lambda)$, then $B = P_C(I - \gamma A^T(I - P_Q)A)$ is averaged and, by the KM Theorem 5.16, the orbit sequence $\{B^k x\}$ converges to a fixed point of B, whenever such points exist. If z is a fixed point of B, then $z = P_C(z - \gamma A^T(I - P_Q)Az)$. Therefore, for any c in C we have

$$\langle c - z, z - (z - \gamma A^T(I - P_Q)Az) \rangle \geq 0.$$

This tells us that

$$\langle c - z, A^T(I - P_Q)Az \rangle \geq 0,$$

which means that z minimizes $f(x)$ relative to the set C.

The CQ algorithm employs the relaxation parameter γ in the interval $(0, 2/L)$, where L is the largest eigenvalue of the matrix $A^T A$. Choosing the best relaxation parameter in any algorithm is a nontrivial procedure. Generally speaking, we want to select γ near to $1/L$. We saw a simple estimate for L in our discussion of singular values of sparse matrices: if A is normalized so that each row has length one, then the spectral radius of $A^T A$ does not exceed the maximum number of nonzero elements in any column of A. A similar upper bound on $\rho(A^T A)$ was obtained for nonnormalized, ϵ-sparse A.

19.2   Particular Cases of the CQ Algorithm

It is easy to find important examples of the SFP: if $C \subseteq R^J$ and $Q = \{b\}$, then solving the SFP amounts to solving the linear system of equations $Ax = b$; if C is a proper subset of R^J, such as the nonnegative cone, then we seek solutions of $Ax = b$ that lie within C, if there are any. Generally, we cannot solve the SFP in closed form and iterative methods are needed.

A number of well-known iterative algorithms, such as Landweber's Algorithm [115] and the Projected Landweber's Algorithm (see [13]), are particular cases of the CQ algorithm.

19.2.1   Landweber's Algorithm

With x^0 arbitrary and $k = 0, 1, \ldots$ let

$$x^{k+1} = x^k + \gamma A^T (b - Ax^k). \tag{19.3}$$

This is Landweber's Algorithm.

19.2.2   The Projected Landweber's Algorithm

For a general nonempty closed convex C, x^0 arbitrary, and $k = 0, 1, \ldots$, the Projected Landweber's Algorithm for finding a solution of $Ax = b$ in C has the iterative step

$$x^{k+1} = P_C(x^k + \gamma A^T (b - Ax^k)). \tag{19.4}$$

19.2.3   Convergence of Landweber Algorithms

From the convergence theorem for the CQ algorithm it follows that Landweber's Algorithm converges to a solution of $Ax = b$ and the Projected Landweber's Algorithm converges to a solution of $Ax = b$ in C, whenever such solutions exist. When there are no solutions of the desired type, Landweber's Algorithm converges to a least-squares approximate solution of $Ax = b$, while the Projected Landweber's Algorithm converges to a minimizer, over the set C, of the function $||b - Ax||_2$, whenever such a minimizer exists.

19.2.4   The Simultaneous ART (SART)

Another example of the CQ algorithm is the *simultaneous algebraic reconstruction technique* (SART) of Anderson and Kak for solving $Ax = b$, for nonnegative matrix A [2]. Let A be an I by J matrix with nonnegative entries. Let $A_{i+} > 0$ be the sum of the entries in the ith row of A and $A_{+j} > 0$ be the sum of the entries in the jth column of A. Consider the (possibly inconsistent) system $Ax = b$.

Algorithm 19.2 (SART). Let x^0 be arbitrary. Then let

$$x_j^{k+1} = x_j^k + \frac{1}{A_{+j}} \sum_{i=1}^{I} A_{ij}(b_i - (Ax^k)_i)/A_{i+}.$$

We make the following changes of variables:

$$B_{ij} = A_{ij}/(A_{i+})^{1/2}(A_{+j})^{1/2},$$
$$z_j = x_j(A_{+j})^{1/2},$$
$$c_i = b_i/(A_{i+})^{1/2}.$$

Then the SART iterative step can be written as

$$z^{k+1} = z^k + B^T(c - Bz^k).$$

This is a particular case of Landweber's Algorithm, with $\gamma = 1$. The convergence of SART follows from the KM Theorem 5.16, once we know that the largest eigenvalue of $B^T B$ is less than two; in fact, we show that it is one [45].

If $B^T B$ had an eigenvalue greater than one and some of the entries of A are zero, then, replacing these zero entries with very small positive entries, we could obtain a new A whose associated $B^T B$ also had an eigenvalue greater than one. Therefore, we assume, without loss of generality, that A has all positive entries. Since the new $B^T B$ also has only positive entries, this matrix is irreducible and the Perron-Frobenius Theorem applies. We shall use this to complete the proof.

Let $u = (u_1, ..., u_J)^T$ with $u_j = (A_{+j})^{1/2}$ and $v = (v_1, ..., v_I)^T$, with $v_i = (A_{i+})^{1/2}$. Then we have $Bu = v$ and $B^T v = u$; that is, u is an eigenvector of $B^T B$ with associated eigenvalue equal to one, and all the entries of u are positive, by assumption. The Perron-Frobenius Theorem applies and tells us that the eigenvector associated with the largest eigenvalue has all positive entries. Since the matrix $B^T B$ is symmetric, its eigenvectors are orthogonal; therefore u itself must be an eigenvector associated with the largest eigenvalue of $B^T B$. The convergence of SART follows.

19.2.5   Application of the CQ Algorithm in Dynamic ET

To illustrate how an image reconstruction problem can be formulated as a SFP, we consider briefly *emission tomography* (ET) image reconstruction. The objective in ET is to reconstruct the internal spatial distribution of intensity of a radionuclide from counts of photons detected outside the patient. In static ET the intensity distribution is assumed constant over the scanning time. Our data are photon counts at the detectors, forming the positive vector b and we have a matrix A of detection probabilities; our model is $Ax = b$, for x a nonnegative vector. We could then take $Q = \{b\}$ and $C = R_+^N$, the nonnegative cone in R^N.

In *dynamic* ET [86] the intensity levels at each voxel may vary with time. The observation time is subdivided into, say, T intervals and one static image, call it x^t, is associated with the time interval denoted by t, for $t = 1, ..., T$. The vector x is the concatenation of these T image vectors x^t. The discrete time interval at which each data value is collected is also recorded and the problem is to reconstruct this succession of images.

Because the data associated with a single time interval is insufficient, by itself, to generate a useful image, one often uses prior information concerning the time history at each fixed voxel to devise a model of the behavior of the intensity levels at each voxel, as functions of time. One may, for example, assume that the radionuclide intensities at a fixed voxel are increasing with time, or are concave (or convex) with time. The problem then is to find $x \geq 0$ with $Ax = b$ and $Dx \geq 0$, where D is a matrix chosen to describe this additional prior information. For example, we may wish to require that, for each fixed voxel, the intensity is an increasing function of (discrete) time; then we want

$$x_j^{t+1} - x_j^t \geq 0,$$

for each t and each voxel index j. Or, we may wish to require that the intensity at each voxel describes a concave function of time, in which case nonnegative second differences would be imposed:

$$(x_j^{t+1} - x_j^t) - (x_j^{t+2} - x_j^{t+1}) \geq 0.$$

In either case, the matrix D can be selected to include the left sides of these inequalities, while the set Q can include the nonnegative cone as one factor.

19.2.6 Related Methods

One of the obvious drawbacks to the use of the CQ algorithm is that we would need the projections P_C and P_Q to be easily calculated. Several authors have offered remedies for that problem, using approximations of the convex sets by the intersection of hyperplanes and orthogonal projections onto those hyperplanes [156].

In a recent paper [55] Censor et al. discuss the application of the CQ algorithm to the problem of intensity-modulated radiation therapy treatment planning. Details concerning this application are in a later chapter.

The split feasibility problem can be formulated as an optimization problem, namely, to minimize

$$h(x) = \iota_C(x) + \iota_Q(Ax),$$

where $\iota_C(x)$ is the indicator function of the set C. The CQ algorithm solves the more general problem of minimizing the function

$$f(x) = \iota_C(x) + ||P_Q Ax - Ax||_2^2.$$

The second term in $f(x)$ is differentiable, allowing us to apply the forward-backward splitting method of Combettes and Wajs [69], to be discussed in a subsequent chapter. The CQ algorithm is then a special case of their method.

20 | Nonsmooth Optimization

In this chapter we consider the problem of optimizing functions f that are convex, but possibly nondifferentiable.

Let $f : R^J \to (-\infty, +\infty]$ be a closed, proper, convex function. When f is differentiable, we can find minimizers of f using techniques such as gradient descent. When f is not necessarily differentiable, the minimization problem is more difficult. One approach is to augment the function f and to convert the problem into one of minimizing a differentiable function. Moreau's approach uses Euclidean distances to augment f, leading to the definition of *proximal operators* [139], or *proximity operators* [69]. More general methods, using Bregman distances to augment f, have been considered by Teboulle [150] and by Censor and Zenios [60].

The *interior-point algorithm* (IPA) is an iterative method for minimizing a convex function $f : R^J \to (-\infty, +\infty]$ over the set $\overline{D}$, the closure of the essential domain of a second convex function $h : R^J \to (-\infty, +\infty]$, where D is the set of all x for which $h(x)$ is finite. The IPA is an interior-point algorithm, in the sense that each iterate lies within the interior of D. The IPA generalizes the PMD algorithm of Censor and Zenios [60] and is related to the proximity operators of Moreau and to the entropic proximal mappings of Teboulle [150].

20.1 Moreau's Proximity Operators

The Moreau envelope of the function f is the function

$$m_f(z) = \inf_x \{ f(x) + \frac{1}{2} ||x - z||_2^2 \}, \tag{20.1}$$

which is also the infimal convolution of the functions $f(x)$ and $\frac{1}{2}||x||_2^2$. It can be shown that the infimum is uniquely attained at the point denoted $x = \text{prox}_f z$ (see [139]).

Proposition 20.1. *The infimum of $m_f(z)$, over all z, is the same as the infimum of $f(x)$, over all x.*

Proof: We have

$$\inf_z m_f(z) = \inf_z \inf_x \{f(x) + \frac{1}{2}||x - z||_2^2\}$$

$$= \inf_x \inf_z \{f(x) + \frac{1}{2}||x - z||_2^2\}$$

$$= \inf_x \{f(x) + \frac{1}{2}\inf_z ||x - z||_2^2\} = \inf_x f(x). \qquad \square$$

Later, we shall show that the minimizers of $m_f(z)$ and $f(x)$ are the same, as well.

The function $m_f(z)$ is differentiable and $\nabla m_f(z) = z - \text{prox}_f z$. The point $x = \text{prox}_f z$ is characterized by the property $z - x \in \partial f(x)$. Consequently, x is a global minimizer of f if and only if $x = \text{prox}_f x$.

For example, consider the indicator function of the convex set C, $f(x) = \iota_C(x)$ that is zero if x is in the closed convex set C and $+\infty$ otherwise. Then $m_f z$ is the minimum of $\frac{1}{2}||x - z||_2^2$ over all x in C, and $\text{prox}_f z = P_C z$, the orthogonal projection of z onto the set C.

If $f : R \to R$ is $f(t) = \omega|t|$, then

$$\text{prox}_f(t) = t - \frac{t}{|t|}\omega,$$

for $|t| \leq \omega$, and equals zero, otherwise.

The operators $\text{prox}_f : z \to \text{prox}_f z$ are proximal operators. These operators generalize the projections onto convex sets, and, like those operators, they are firmly nonexpansive [69].

The conjugate function associated with f is the function $f^*(x^*) = \sup_x(\langle x^*, x \rangle - f(x))$. In similar fashion, we can define $m_{f^*}z$ and $\text{prox}_{f^*}z$. Both m_f and m_{f^*} are convex and differentiable.

The support function of the convex set C is $\sigma_C(x) = \sup_{u \in C}\langle x, u \rangle$. It is easy to see that $\sigma_C = \iota_C^*$. For $f^*(z) = \sigma_C(z)$, we can find $m_{f^*}z$ using Moreau's Theorem ([139], p. 338).

Moreau's Theorem generalizes the decomposition of members of R^J with respect to a subspace.

Theorem 20.2. (Moreau's Theorem) *Let f be a closed, proper, convex function. Then*

$$m_f z + m_{f^*}z = \frac{1}{2}||z||_2^2,$$

$$\text{prox}_f z + \text{prox}_{f^*}z = z.$$

In addition, we have

$$\text{prox}_{f*}z \in \partial f(\text{prox}_f z),$$
$$\text{prox}_{f*}z = \nabla m_f(z),$$
$$\text{prox}_f z = \nabla m_{f*}(z).$$

Since $\sigma_C = \iota_C^*$, we have

$$\text{prox}_{\sigma_C} z = z - \text{prox}_{\iota_C} z = z - P_C z.$$

The following proposition illustrates the usefulness of these concepts.

Proposition 20.3. *The minimizers of m_f and the minimizers of f are the same.*

Proof: From Moreau's Theorem we know that

$$\nabla m_f(z) = \text{prox}_{f*}z = z - \text{prox}_f z,$$

so $\nabla m_f z = 0$ is equivalent to $z = \text{prox}_f z$. $\qquad\square$

Because the minimizers of m_f are also minimizers of f, we can find global minimizers of f using gradient descent iterative methods on m_f.

Let x^0 be arbitrary. For $k = 0, 1, ...$, let

$$x^{k+1} = x^k - \gamma_k \nabla m_f(x^k). \tag{20.2}$$

We know from Moreau's Theorem that

$$\nabla m_f z = \text{prox}_{f*}z = z - \text{prox}_f z,$$

so that Equation (20.2) can be written as

$$x^{k+1} = x^k - \gamma_k(x^k - \text{prox}_f x^k),$$

which leads to the Proximal Minimization Algorithm:

Algorithm 20.4 (Proximal Minimization). Let x^0 be arbitrary. For $k = 0, 1, ...$, let

$$x^{k+1} = (1 - \gamma_k)x^k + \gamma_k \text{prox}_f x^k. \tag{20.3}$$

Because

$$x^k - \text{prox}_f x^k \in \partial f(\text{prox}_f x^k),$$

the iteration in Equation (20.3) has the increment

$$x^{k+1} - x^k \in -\gamma_k \partial f(x^{k+1}),$$

in contrast to what we would have with the usual gradient descent method for differentiable f,

$$x^{k+1} - x^k = -\gamma_k \nabla f(x^k).$$

It follows from the definition of $\partial f(x^{k+1})$ that $f(x^k) \geq f(x^{k+1})$ for the iteration in Equation (20.3).

20.2   Forward-Backward Splitting

In [69] Combettes and Wajs consider the problem of minimizing the function $f = f_1 + f_2$, where f_2 is differentiable and its gradient is λ-Lipschitz continuous. The function f is minimized at the point x if and only if

$$0 \in \partial f(x) = \partial f_1(x) + \nabla f_2(x),$$

so we have

$$-\gamma \nabla f_2(x) \in \gamma \partial f_1(x),$$

for any $\gamma > 0$. Therefore

$$x - \gamma \nabla f_2(x) - x \in \gamma \partial f_1(x). \tag{20.4}$$

From Equation (20.4) we conclude that

$$x = \text{prox}_{\gamma f_1}(x - \gamma \nabla f_2(x)).$$

This suggests an algorithm, called the *forward-backward splitting* for minimizing the function $f(x)$.

Algorithm 20.5 (Forward-Backward Splitting). Beginning with an arbitrary x^0, and having calculated x^k, we let

$$x^{k+1} = \text{prox}_{\gamma f_1}(x^k - \gamma \nabla f_2(x^k)),$$

with γ chosen to lie in the interval $(0, 2/\lambda)$.

The operator $I - \gamma \nabla f_2$ is then averaged. Since the operator $\text{prox}_{\gamma f_1}$ is firmly nonexpansive, we know from the KM Theorem 5.16 that the sequence $\{x^k\}$ converges to a minimizer of the function $f(x)$, whenever minimizers exist. It is also possible to allow γ to vary with the k.

20.2.1   The CQ Algorithm as Forward-Backward Splitting

Recall that the split feasibility problem (SFP) is to find x in C with Ax in Q. The CQ algorithm minimizes the function

$$g(x) = ||P_Q Ax - Ax||_2^2,$$

over $x \in C$, whenever such minimizers exist, and so solves the SFP whenever it has solutions. The CQ algorithm minimizes the function

$$f(x) = \iota_C(x) + g(x),$$

where ι_C is the indicator function of the set C. With $f_1(x) = \iota_C(x)$ and $f_2(x) = g(x)$, the function $f(x)$ has the form considered by Combettes and Wajs, and the CQ algorithm becomes a special case of their forward-backward splitting method.

20.3   Proximity Operators using Bregman Distances

Several authors have extended Moreau's results by replacing the Euclidean squared distance with a Bregman distance. Let h be a closed proper convex function that is differentiable on the nonempty set $\text{int}D$. The corresponding *Bregman distance* $D_h(x, z)$ is defined for $x \in R^J$ and $z \in \text{int}D$ by

$$D_h(x, z) = h(x) - h(z) - \langle \nabla h(z), x - z \rangle.$$

Note that $D_h(x, z) \geq 0$ always and that $D_h(x, z) = +\infty$ is possible. If h is essentially strictly convex, then $D_h(x, z) = 0$ implies that $x = z$.

Teboulle [150] considers the function

$$R(x, z) = f(x) + D_h(x, z), \tag{20.5}$$

and shows that, with certain restrictions on f and h, the function $R(\cdot, z)$ attains its minimum value at a unique $x = E_h(f, z)$. The operator $E_h(f, \cdot)$ is then shown to have properties analogous to the proximity operators $\text{prox}_f(\cdot)$. He then shows that several nonlinear programming problems can be formulated using such functions $R(x, z)$.

Censor and Zenios [60] also consider $R(x, z)$. They are less interested in the properties of the operator $E_h(f, \cdot)$ and more interested in the behavior of their PMD iterative algorithm:

Algorithm 20.6 (PMD). Let x^0 be in the zone of the Bregman function h. Then let

$$x^{k+1} = \operatorname{argmin} \left(f(x) + D_h(x, x^k) \right).$$

In their work, the function h is a Bregman function with zone S. They show that, subject to certain assumptions, if the function f has a minimizer within the closure of S, then the PMD iterates converge to such a minimizer. It is true that their method and results are somewhat more general. in that they consider also the minimizers of $R(x, z)$ over another closed convex set X; however, this set X is unrelated to the function h.

The interior-point algorithm presented in this chapter has the same iterative step as the PMD method of Censor and Zenios. However, the assumptions about f and h are different, and our theorem asserts convergence of the iterates to a constrained minimizer of f over $\overline{D}$, whenever such a minimizer exists. In other words, we solve a constrained minimization problem, whereas Censor and Zenios solve the unconstrained minimization problem, under a restrictive assumption on the location of minimizers of f.

Algorithm 20.7 (IPA). For each k, the next IPA iterate, x^{k+1}, minimizes the function $R(x, x^k)$, as given in Equation (20.5).

Then x^{k+1} satisfies the inclusion

$$\nabla h(x^k) - \nabla h(x^{k+1}) \in \partial f(x^{k+1}),$$

where $\partial f(x)$ is the subdifferential of f at x. In order to prove convergence of the IPA, we restrict h to the class of Bregman-Legendre functions.

20.4   The Interior-Point Algorithm (IPA)

The objective is to minimize the convex function $f : R^J \to R$, over $\overline{D}$, the closure of the effective domain D of the Bregman-Legendre function $h : R^J \to R$. We assume throughout this section that there is $\hat{x}$ in $\overline{D}$ with $f(\hat{x}) \le f(x)$ for all x in $\overline{D}$.

In order for the iterative scheme to be well defined, we need to assume that x^{k+1} is again in $\text{int}(D)$ for each k. It follows then that

$$\nabla h(x^k) - \nabla h(x^{k+1}) \in \partial f(x^{k+1}). \tag{20.6}$$

Since

$$f(x^{k+1}) + D_h(x^{k+1}, x^k) \le f(x^k),$$

it follows immediately that the sequence $\{f(x^k)\}$ is decreasing. Since the sequence $\{f(x^k)\}$ is bounded below by $f(\hat{x})$, the sequence $\{D_h(x^{k+1}, x^k)\}$ converges to zero and the sequence $\{f(x^k)\}$ converges to some value $\hat{f} \ge f(\hat{x})$.

Proposition 20.8. *For every initial vector x^0, $\hat{f} = f(\hat{x})$.*

Proof: Suppose not; let $\hat{f} - f(\hat{x}) = \delta > 0$. Since $\hat{x} \in \overline{D}$, there is $z \in D$ with $f(z) < f(\hat{x}) + \delta/2$. Then

$$D_h(z, x^k) - D_h(z, x^{k+1}) = D_h(x^{k+1}, x^k) + \langle \nabla h(x^k)$$
$$- \nabla h(x^{k+1}), x^{k+1} - z \rangle.$$

Using Equation (20.6) and the definition of the subdifferential, we find that

$$D_h(z, x^k) - D_h(z, x^{k+1}) \ge D_h(x^{k+1}, x^k) + f(x^{k+1}) - f(z)$$
$$\ge D_h(x^{k+1}, x^k) + \delta/2. \tag{20.7}$$

Consequently, the sequence $\{D_h(z, x^k)\}$ is decreasing, and (20.7) must converge to zero. But, this cannot happen, unless $\delta = 0$. $\square$

In the discussion that follows we shall indicate the properties of the function $h(x)$ needed at each step.

Suppose now that $\hat{x}$ is a unique minimizer of $f(x)$ over $x \in \overline{D}$, then the function $g(x) = f(x) + \iota_{\overline{D}}(x)$ is a closed, proper convex function and the level set

$$\{x|g(x) \leq g(\hat{x})\} = \{\hat{x}\}$$

is nonempty and bounded. It follows from Corollary 8.7.1 of [139] that every set of the form

$$\{x|g(x) \leq \alpha\}$$

is then bounded. We conclude that the sequence $\{x^k\}$ is bounded, and, furthermore, that it converges to $\hat{x}$.

If $\hat{x}$ is not unique, but can be chosen in D, then, mimicking the proof of the proposition, we can show that the sequence $\{D_h(\hat{x}, x^k)\}$ is decreasing. If, in addition, the function $D_h(\hat{x}, \cdot)$ has bounded level sets, then, once again, we can conclude that the sequence $\{x^k\}$ is bounded, has a subsequence $\{x^{k_n}\}$ converging to x^*, $f(x^*) = f(\hat{x})$, and $\{D_h(x^*, x^k)\}$ is decreasing.

Finally, if h is a Bregman-Legendre function, $\{D_h(x^*, x^{k_n})\} \to 0$. Since $\{D_h(x^*, x^k)\}$ is decreasing, it follows that $\{D_h(x^*, x^k)\} \to 0$. From this, we can conclude that $x^k \to x^*$.

In summary, we have the following theorem.

Theorem 20.9. *Let h be a Bregman-Legendre function and let $\hat{x}$ be a minimizer of $f(x)$, over x in $\overline{D}$. If $\hat{x}$ is unique, then the sequence $\{x^k\}$ converges to $\hat{x}$. If $\hat{x}$ is not unique, but can be chosen in D, then $\{x^k\}$ converges to x^* in D, with $f(x^*) = f(\hat{x})$.*

20.5 Computing the Iterates

As we have seen, the point x^{k+1} has the property that

$$\nabla h(x^k) - \nabla h(x^{k+1}) \in \partial f(x^{k+1}).$$

Even when f is differentiable, and so

$$\nabla h(x^k) - \nabla h(x^{k+1}) = \nabla f(x^{k+1}), \qquad (20.8)$$

it is not obvious how we might calculate x^{k+1} efficiently. In this section we consider a trick that is sometimes helpful.

The function h is chosen because we are interested in D, not specifically in h itself. When f is differentiable on R^J, the functions $F(x) = f(x) + h(x)$ and $h(x)$ will have the same D, and

$$D_F(x, z) \geq D_f(x, z),$$

for all x in D and z in $\text{int}(D)$. We can rewrite Equation (20.8) as

$$\nabla F(x^{k+1}) = \nabla F(x^k) - \nabla f(x^k). \tag{20.9}$$

Our approach is then the following: having selected D, we attempt to find a function $F(x)$ with $D = \text{dom}(F)$ and $D_F(x,z) \geq D_f(x,z)$, and for which Equation (20.9) can be solved easily for x^{k+1}. Since we start with f and D and then select F, we do not have an explicit description of h. It was for this reason that we introduced the properties of h as needed, in the previous section. We do know that $h = F - f$ will be convex, and if f is differentiable over $\overline{D}$, then F and h will have the same essential domain.

In the next section we give several examples of this approach.

20.6   Some Examples

A useful property of the KL distance is given by the following lemma.

Lemma 20.10. *For any $c > 0$, with $a \geq c$ and $b \geq c$, we have $KL(a - c, b - c) \geq KL(a,b)$.*

Proof: Let $g(c) = KL(a - c, b - c)$ and differentiate with respect to c, to obtain

$$g'(c) = \frac{a-c}{b-c} - 1 - \log(\frac{a-c}{b-c}) \geq 0. \qquad \square$$

We see then that the function $g(c)$ is increasing with c.

In these examples, we seek to minimize the function $KL(Ax, b)$, where $b = (b_1, ..., b_I)^T$ is a vector with positive entries, $A = (A_{ij})$ is an I by J matrix with $A_{ij} \geq 0$, and $s_j = \sum_{i=1}^{I} A_{ij} > 0$, for each j, and the vector x is in $\mathcal{P}$, where $\mathcal{P}$ is the set of all vectors x such that $(Ax)_i = \sum_{j=1}^{J} A_{ij}x_j > 0$, for each i.

20.6.1   Minimizing $KL(Ax, b)$ over $x \geq 0$

In our first example, we seek to minimize $f(x) = KL(Ax, b)$ over vectors x with nonnegative entries. We take $F(x)$ to be the function

$$F(x) = \sum_{j=1}^{J} t_j x_j \log x_j,$$

with $t_j \geq s_j$.

Then

$$D_F(x,z) = \sum_{j=1}^{J} t_j KL(x_j, z_j),$$

$$D_f(x,z) = KL(Ax, Az).$$

Lemma 20.11. $D_F(x,z) \geq D_f(x,z)$.

Proof: We have

$$D_F(x,z) \geq \sum_{j=1}^{J} s_j KL(x_j, z_j) \geq \sum_{j=1}^{J}\sum_{i=1}^{I} KL(A_{ij}x_j, A_{ij}z_j)$$

$$\geq \sum_{i=1}^{I} KL((Ax)_i, (Az)_i) = KL(Ax, Az). \qquad \square$$

The gradient of $F(x)$ has entries

$$\nabla F(x)_j - t_j \log x_j,$$

and the gradient of $f(x)$ has entries

$$\nabla f(x)_j = \sum_{i=1}^{I} A_{ij} \log((Ax)_i/b_i).$$

Solving Equation (20.9) we find

$$t_j \log x_j^{k+1} = t_j \log x_j^k + \sum_{i=1}^{I} A_{ij} \log(b_i/(Ax^k)_i),$$

so that

$$x_j^{k+1} = x_j^k \exp\left(t_j^{-1}\sum_{i=1}^{I} A_{ij} \log(b_i/(Ax^k)_i)\right).$$

If $t_j = s_j$, we get the SMART iterative algorithm for minimizing $KL(Ax, b)$ over nonnegative x [34, 47].

20.6.2 Minimizing $KL(Ax, b)$ with bounds on x

Let $u_j < v_j$, for each j. Let $\mathcal{P}_{uv}$ be the set of all vectors x such that $u_j \leq x_j \leq v_j$, for each j. Now, we seek to minimize $f(x) = KL(Ax, b)$, over all vectors x in $\mathcal{P} \cap \mathcal{P}_{uv}$. We let

$$F(x) = \sum_{j=1}^{J} t_j\Big((x_j - u_j)\log(x_j - u_j) + (v_j - x_j)\log(v_j - x_j)\Big).$$

Then we have

$$D_F(x, z) = \sum_{j=1}^{J} t_j \Big(KL(x_j - u_j, z_j - u_j) + KL(v_j - x_j, v_j - z_j) \Big),$$

and, as before,

$$D_f(x, z) = KL(Ax, Az).$$

As a corollary of Lemma 20.10, we have

Lemma 20.12. Let $u = (u_1, ..., u_J)^T$, and x and z in $\mathcal{P}$ with $(Ax)_i \geq (Au)_i$, $(Az)_i \geq (Au)_i$, for each i. Then $KL(Ax, Az) \leq KL(Ax - Au, Az - Au)$.

Lemma 20.13. $D_F(x, z) \geq D_f(x, z)$.

Proof: We can easily show that $D_F(x, z) \geq KL(Ax - Au, Az - Au) + KL(Av - Ax, Av - Az)$, along the lines used previously. Then, from Lemma 20.12, we have $KL(Ax - Au, Az - Au) \geq KL(Ax, Az) = D_f(x, z)$.

The iterative step of the algorithm is obtained by solving for x_j^{k+1} in Equation 20.9.

Algorithm 20.14 (IPA-UV). Let x^0 be an arbitrary vector with $u_j \leq x_j^0 \leq v_j$, for each j. Then let

$$x_j^{k+1} = \alpha_j^k u_j + (1 - \alpha_j^k) v_j,$$

where

$$(\alpha_j^k)^{-1} = 1 + \Big(\frac{x_j^k - u_j}{v_j - x_j^k} \Big) \exp \Big(\sum_{i=1}^{I} A_{ij} \log(b_i/(Ax^k)_i) \Big).$$

This algorithm is closely related to those presented in [40] and discussed in the next chapter.

In this chapter, the IPA was presented as an extension of the PMD method, and in the context of proximal minimization. In the next chapter, we present the original derivation of the IPA, as a special case of the multiple-distance successive generalized projection (MSGP) algorithm.

21

An Interior-Point Optimization Method

Investigations in [37] into several well-known iterative algorithms, including the expectation maximization maximum likelihood (EMML) method, the multiplicative algebraic reconstruction technique (MART), as well as block-iterative and simultaneous versions of MART, revealed that the iterative step of each algorithm involved weighted arithmetic or geometric means of Bregman projections onto hyperplanes; interestingly, the projections involved were associated with Bregman distances that differed from one hyperplane to the next. This representation of the EMML algorithm as a weighted arithmetic mean of Bregman projections provided the key step in obtaining block-iterative and row-action versions of EMML. Because it is well known that convergence is not guaranteed if one simply extends Bregman's algorithm to multiple distances by replacing the single distance D_f in Equation (18.2) with multiple distances D_{f_i}, the appearance of distinct distances in these algorithms suggested that a somewhat more sophisticated algorithm employing multiple Bregman distances might be possible.

21.1 Multiple-Distance Successive Generalized Projection

In [41] such an iterative multiprojection method for solving the CFP, called the *multiple-distance successive generalized projection* method (MSGP), was presented in the context of Bregman functions, and subsequently, in the framework of Bregman-Legendre functions [43]; see the appendix on Bregman functions for definitions and details concerning these functions. The MSGP extends Bregman's SGP method by allowing the Bregman projection onto each set C_i to be performed with respect to a Bregman distance D_{f_i} derived from a Bregman-Legendre function f_i. The MSGP method depends on the selection of a super-coercive Bregman-

Legendre function h whose Bregman distance D_h satisfies the inequality $D_h(x, z) \geq D_{f_i}(x, z)$ for all $x \in \operatorname{dom} h \subseteq \cap_{i=1}^{I} \operatorname{dom} f_i$ and all $z \in \operatorname{int} \operatorname{dom} h$, where $\operatorname{dom} h = \{x | h(x) < +\infty\}$. By using different Bregman distances for different convex sets, we found that we can sometimes calculate the desired Bregman projections in closed form, thereby obtaining computationally tractable iterative algorithms (see [37]).

21.2   An Interior-Point Algorithm (IPA)

Consideration of a special case of the MSGP, involving only a single convex set C_1, leads us to an interior-point optimization method. If $I = 1$ and $f := f_1$ has a unique minimizer $\hat{x}$ in $\operatorname{int} \operatorname{dom} h$, then the MSGP iteration using $C_1 = \{\hat{x}\}$ is

$$\nabla h(x^{k+1}) = \nabla h(x^k) - \nabla f(x^k).$$

This suggests an *interior-point algorithm* (IPA) that could be applied more broadly to minimize a convex function f over the closure of $\operatorname{dom} h$. This is the IPA method discussed previously. In this chapter, we present its original derivation, as suggested by the MSGP.

First, we present the MSGP method and prove convergence, in the context of Bregman-Legendre functions. Then we investigate the IPA suggested by the MSGP algorithm.

21.3   The MSGP Algorithm

We begin by setting out the assumptions we shall make and the notation we shall use in this section.

21.3.1   Assumptions and Notation

We make the following assumptions throughout this section. Let $C = \cap_{i=1}^{I} C_i$ be the nonempty intersection of closed convex sets C_i. The function h is super-coercive and Bregman-Legendre with essential domain $D = \operatorname{dom} h$ and $C \cap \operatorname{dom} h \neq \emptyset$. For $i = 1, 2, ..., I$ the function f_i is also Bregman-Legendre, with $D \subseteq \operatorname{dom} f_i$, so that $\operatorname{int} D \subseteq \operatorname{int} \operatorname{dom} f_i$; also $C_i \cap \operatorname{int} \operatorname{dom} f_i \neq \emptyset$. For all $x \in \operatorname{dom} h$ and $z \in \operatorname{int} \operatorname{dom} h$ we have $D_h(x, z) \geq D_{f_i}(x, z)$, for each i.

21.3.2   The MSGP Algorithm

Algorithm 21.1. [MSGP] Let $x^0 \in \operatorname{int} \operatorname{dom} h$ be arbitrary. For $k = 0, 1, ...$ and $i(k) := k(\operatorname{mod} I) + 1$ let

$$x^{k+1} = \nabla h^{-1}\left(\nabla h(x^k) - \nabla f_{i(k)}(x^k) + \nabla f_{i(k)}(P_{C_{i(k)}}^{f_{i(k)}}(x^k))\right). \tag{21.1}$$

21.3.3 A Preliminary Result

For each $k = 0, 1, \ldots$ define the function $G^k(\cdot) : \operatorname{dom} h \to [0, +\infty)$ by

$$G^k(x) = D_h(x, x^k) - D_{f_{i(k)}}(x, x^k) + D_{f_{i(k)}}(x, P_{C_{i(k)}}^{f_{i(k)}}(x^k)). \tag{21.2}$$

The next proposition provides a useful identity, which can be viewed as an analogue of Pythagoras' theorem. The proof is not difficult and we omit it.

Proposition 21.2. *For each $x \in \operatorname{dom} h$, each $k = 0, 1, \ldots$, and x^{k+1} given by Equation (21.1) we have*

$$G^k(x) = G^k(x^{k+1}) + D_h(x, x^{k+1}). \tag{21.3}$$

Consequently, x^{k+1} is the unique minimizer of the function $G^k(\cdot)$.

This identity (21.3) is the key ingredient in the convergence proof for the MSGP algorithm.

21.3.4 The MSGP Convergence Theorem

We shall prove the following convergence theorem:

Theorem 21.3. *Let $x^0 \in \operatorname{int} \operatorname{dom} h$ be arbitrary. Any sequence x^k obtained from the iterative scheme given by Algorithm 21.1 converges to $x^\infty \in C \cap \operatorname{dom} h$. If the sets C_i are hyperplanes, then x^∞ minimizes the function $D_h(x, x^0)$ over all $x \in C \cap \operatorname{dom} h$; if, in addition, x^0 is the global minimizer of h, then x^∞ minimizes $h(x)$ over all $x \in C \cap \operatorname{dom} h$.*

Proof: All details concerning Bregman functions (including the properties and results referred to below) are in Appendix A. Let c be a member of $C \cap \operatorname{dom} h$. From the Pythagorean identity (21.3) it follows that

$$G^k(c) = G^k(x^{k+1}) + D_h(c, x^{k+1}).$$

Using the definition of $G^k(\cdot)$, we write

$$G^k(c) = D_h(c, x^k) - D_{f_{i(k)}}(c, x^k) + D_{f_{i(k)}}(c, P_{C_{i(k)}}^{f_{i(k)}}(x^k)).$$

From Bregman's Inequality (18.1) we have that

$$D_{f_{i(k)}}(c, x^k) - D_{f_{i(k)}}(c, P_{C_{i(k)}}^{f_{i(k)}}(x^k)) \geq D_{f_{i(k)}}(P_{C_{i(k)}}^{f_{i(k)}}(x^k), x^k).$$

Consequently, we know that

$$D_h(c, x^k) - D_h(c, x^{k+1}) \geq G^k(x^{k+1}) + D_{f_{i(k)}}(P_{C_{i(k)}}^{f_{i(k)}}(x^k), x^k) \geq 0.$$

It follows that $\{D_h(c, x^k)\}$ is decreasing and finite and the sequence $\{x^k\}$ is bounded. Therefore,

$$\{D_{f_{i(k)}}(P_{C_{i(k)}}^{f_{i(k)}}(x^k), x^k)\} \to 0$$

and $\{G^k(x^{k+1})\} \to 0$; from the definition of $G^k(x)$ it follows that

$$\{D_{f_{i(k)}}(x^{k+1}, P_{C_{i(k)}}^{f_{i(k)}}(x^k))\} \to 0$$

as well. Using the Bregman inequality we obtain the inequality

$$D_h(c, x^k) \geq D_{f_{i(k)}}(c, x^k) \geq D_{f_{i(k)}}(c, P_{C_{i(k)}}^{f_{i(k)}}(x^k)),$$

which tells us that the sequence $\{P_{C_{i(k)}}^{f_{i(k)}}(x^k)\}$ is also bounded. Let x^* be an arbitrary cluster point of the sequence $\{x^k\}$ and let $\{x^{k_n}\}$ be a subsequence of the sequence $\{x^k\}$ converging to x^*.

We first show that $x^* \in \text{dom}\, h$ and $\{D_h(x^*, x^k)\} \to 0$. If x^* is in int dom h, then our claim is verified, so suppose that x^* is in bdry dom h. If c is in dom h but not in int dom h, then, applying Property 2 of Bregman-Legendre functions, we conclude that $x^* \in \text{dom}\, h$ and $\{D_h(x^*, x^k)\} \to 0$. If, on the other hand, c is in int dom h, then by Result 2 of Bregman-Legendre functions, x^* would have to be in int dom h also. It follows that $x^* \in \text{dom}\, h$ and $\{D_h(x^*, x^k)\} \to 0$. Now we show that x^* is in C.

Label $x^* = x_0^*$. Since there must be at least one index i that occurs infinitely often as $i(k)$, we assume, without loss of generality, that the subsequence $\{x^{k_n}\}$ has been selected so that $i(k) = 1$ for all $n = 1, 2, \ldots$. Passing to subsequences as needed, we assume that, for each $m = 0, 1, 2, \ldots, I - 1$, the subsequence $\{x^{k_n+m}\}$ converges to a cluster point x_m^*, which is in dom h, according to the same argument we used in the previous paragraph. For each m the sequence $\{D_{f_m}(c, P_{C_m}^{f_m}(x^{k_n+m-1}))\}$ is bounded, so, again, by passing to subsequences as needed, we assume that the subsequence $\{P_{C_m}^{f_m}(x^{k_n+m-1})\}$ converges to $c_m^* \in C_m \cap \overline{\text{dom}\, f_m}$.

Since the sequence $\{D_{f_m}(c, P_{C_m}^{f_m}(x^{k_n+m-1}))\}$ is bounded and $c \in \text{dom}\, f_m$, it follows, from either Property 2 or Result 2, that $c_m^* \in \text{dom}\, f_m$. We know that

$$\{D_{f_m}(P_{C_m}^{f_m}(x^{k_n+m-1}), x^{k_n+m-1})\} \to 0$$

and both $P_{C_m}^{f_m}(x^{k_n+m-1})$ and x^{k_n+m-1} are in int dom f_m. Applying Result 1, Property 3, or Result 3, depending on the assumed locations of c_m^* and x_{m-1}^*, we conclude that $c_m^* = x_{m-1}^*$.

We also know that

$$\{D_{f_m}(x^{k_n+m}, P_{C_m}^{f_m}(x^{k_n+m-1}))\} \to 0,$$

from which it follows, using the same arguments, that $x_m^* = c_m^*$. Therefore, we have $x^* = x_m^* = c_m^*$ for all m; so $x^* \in C$.

Since $x^* \in C \cap \operatorname{dom} h$, we may now use x^* in place of the generic c, to obtain that the sequence $\{D_h(x^*, x^k)\}$ is decreasing. However, we also know that the sequence $\{D_h(x^*, x^{k_n})\} \to 0$. So we have $\{D_h(x^*, x^k)\} \to 0$. Applying Result 5, we conclude that $\{x^k\} \to x^*$.

If the sets C_i are hyperplanes, then we get equality in Bregman's Inequality (18.1) and so

$$D_h(c, x^k) - D_h(c, x^{k+1}) = G^k(x^{k+1}) + D_{f_{i(k)}}(P_{C_{i(k)}}^{f_{i(k)}}(x^k), x^k).$$

Since the right side of this equation is independent of which c we have chosen in the set $C \cap \operatorname{dom} h$, the left side is also independent of this choice. This implies that

$$D_h(c, x^0) - D_h(c, x^M) = D_h(x^*, x^0) - D_h(x^*, x^M),$$

for any positive integer M and any $c \in C \cap \operatorname{dom} h$. Therefore

$$D_h(c, x^0) - D_h(x^*, x^0) = D_h(c, x^M) - D_h(x^*, x^M).$$

Since $\{D_h(x^*, x^M)\} \to 0$ as $M \to +\infty$ and $\{D_h(c, x^M)\} \to \alpha \geq 0$, we have that $D_h(c, x^0) - D_h(x^*, x^0) \geq 0$. This completes the proof. $\square$

21.4   An Interior-Point Algorithm for Iterative Optimization

We consider now an interior-point algorithm for iterative optimization. This algorithm was first presented in [42] and applied to transmission tomography in [129]. The IPA is suggested by a special case of the MSGP, involving functions h and $f := f_1$.

21.4.1   Assumptions

We assume, for the remainder of this section, that h is a super-coercive Legendre function with essential domain $D = \operatorname{dom} h$. We also assume that f is continuous on the set $\overline{D}$, takes the value $+\infty$ outside this set and is differentiable in $\operatorname{int} D$. Thus, f is a closed, proper convex function on R^J. We assume also that $\hat{x} = \operatorname{argmin}_{x \in \overline{D}} f(x)$ exists, but not that it is unique. As in the previous section, we assume that $D_h(x, z) \geq D_f(x, z)$ for all $x \in \operatorname{dom} h$ and $z \in \operatorname{int} \operatorname{dom} h$. As before, we denote by h^* the function conjugate to h.

21.4.2   The IPA

The IPA is an iterative procedure that, under conditions to be described shortly, minimizes the function f over the closure of the essential domain of h, provided that such a minimizer exists.

Algorithm 21.4 (IPA). Let x^0 be chosen arbitrarily in int D. For $k = 0, 1, \ldots$ let x^{k+1} be the unique solution of the equation

$$\nabla h(x^{k+1}) = \nabla h(x^k) - \nabla f(x^k). \tag{21.4}$$

Note that Equation (21.4) can also be written as

$$x^{k+1} = \nabla h^{-1}(\nabla h(x^k) - \nabla f(x^k)) = \nabla h^*(\nabla h(x^k) - \nabla f(x^k)). \tag{21.5}$$

21.4.3   Motivating the IPA

As already noted, the IPA was originally suggested by consideration of a special case of the MSGP. Suppose that $\overline{x} \in \text{dom}\, h$ is the unique global minimizer of the function f, and that $\nabla f(\overline{x}) = 0$. Take $I = 1$ and $C = C_1 = \{\overline{x}\}$. Then $P^f_{C_1}(x^k) = \overline{x}$ always and the iterative MSGP step becomes that of the IPA. Since we are assuming that $\overline{x}$ is in dom h, the convergence theorem for the MSGP tells us that the iterative sequence $\{x^k\}$ converges to $\overline{x}$.

In most cases, the global minimizer of f will not lie within the essential domain of the function h and we are interested in the minimum value of f on the set $\overline{D}$, where $D = \text{dom}\, h$; that is, we want $\hat{x} = \text{argmin}_{x \in \overline{D}}\, f(x)$, whenever such a minimum exists. As we shall see, the IPA can be used to advantage even when the specific conditions of the MSGP do not hold. Two aspects of the IPA suggest strongly that it may converge under more general conditions than those required for convergence of the MSGP. The sequence $\{x^k\}$ defined by Equation (21.4) is entirely within the interior of dom h. In addition, as we showed previously, the sequence $\{f(x^k)\}$ is decreasing.

22 | Linear and Convex Programming

The term *linear programming* (LP) refers to the problem of optimizing a linear function of several variables over linear equality or inequality constraints. In this chapter we present the problem and establish the basic facts. For a much more detailed discussion, consult [130].

22.1 Primal and Dual Problems

Associated with the basic problem in LP, called the *primary problem*, there is a second problem, the *dual problem*. Both of these problems can be written in two equivalent ways, the canonical form and the standard form.

22.1.1 Canonical and Standard Forms

Let b and c be fixed vectors and A a fixed matrix. The problem

$$\text{minimize } z = c^T x, \text{ subject to } Ax \geq b, x \geq 0 \text{ (PC)}$$

is the so-called primary problem of LP, in *canonical form*. The dual problem in canonical form is

$$\text{maximize } w = b^T y, \text{ subject to } A^T y \leq c, y \geq 0. \text{ (DC)}$$

The primary problem, in *standard form*, is

$$\text{minimize } z = c^T x, \text{ subject to } Ax = b, x \geq 0 \text{ (PS)}$$

with the dual problem in standard form given by

$$\text{maximize } w = b^T y, \text{ subject to } A^T y \leq c. \text{ (DS)}$$

Notice that the dual problem in standard form does not require that y be nonnegative. Note also that the standard problems make sense only if the system $Ax = b$ is under-determined and has infinitely many solutions. For that reason, we shall assume, for the standard problems, that the I by J matrix A has more columns than rows, so $J > I$, and it has full row rank.

If we are given the primary problem in canonical form, we can convert it to standard form by augmenting the variables, that is, by defining

$$u_i = (Ax)_i - b_i,$$

for $i = 1, ..., I$, and rewriting $Ax \geq b$ as

$$\tilde{A}\tilde{x} = b,$$

for $\tilde{A} = \begin{bmatrix} A & -I \end{bmatrix}$ and $\tilde{x} = [x^T u^T]^T$.

22.1.2   Weak Duality

Consider the problems (PS) and (DS). Say that x is *feasible* if $x \geq 0$ and $Ax = b$. Let F be the set of feasible x. Say that y is *feasible* if $A^T y \leq c$.

Theorem 22.1. (Weak Duality) *Let x and y be feasible vectors. Then*

$$z = c^T x \geq b^T y = w.$$

Corollary 22.2. *If z is not bounded below, then there are no feasible y.*

Corollary 22.3. *If x and y are both feasible, and $z = w$, then both x and y are optimal for their respective problems.*

The proof of the theorem and its corollaries are left as exercises.

The nonnegative quantity $c^T x - b^T y$ is called the *duality gap*. The *complementary slackness condition* says that, for optimal x and y, we have

$$x_j(c_j - (A^T y)_j) = 0,$$

for each j, which says that the duality gap is zero. Primal-dual algorithms for solving linear programming problems are based on finding sequences $\{x^k\}$ and $\{y^k\}$ that drive the duality gap down to zero [130].

22.1.3   Strong Duality

The *Strong Duality Theorem* makes a stronger statement.

Theorem 22.4. (Strong Duality) *If one of the problems (PS) or (DS) has an optimal solution, then so does the other and $z = w$ for the optimal vectors.*

Before we consider the proof of the theorem, we need a few preliminary results.

Definition 22.5. A point x in F is said to be a *basic feasible solution* if the columns of A corresponding to positive entries of x are linearly independent.

Denote by B an invertible matrix obtained by deleting from A columns associated with zero entries of x. The entries of an arbitrary x corresponding to the columns not deleted are called the *basic variables*. Then, assuming that the columns of B are the first I columns of A, we write $x^T = (x_B^T, x_N^T)$, and

$$A = \begin{bmatrix} B & N \end{bmatrix},$$

so that $Ax = Bx_B = b$, and $x_B = B^{-1}b$. The following theorems are taken from the book by Nash and Sofer [130]. We begin with a characterization of the extreme points of F (recall Definition 17.17).

Theorem 22.6. *A point x is in Ext(F) if and only if x is a basic feasible solution.*

Proof: Suppose that x is a basic feasible solution, and we write $x^T = (x_B^T, 0^T)$, $A = \begin{bmatrix} B & N \end{bmatrix}$. If x is not an extreme point of F, then there are $y \neq x$ and $z \neq x$ in F, and α in $(0, 1)$, with

$$x = (1 - \alpha)y + \alpha z.$$

Then $y^T = (y_B^T, y_N^T)$, $z^T = (z_B^T, z_N^T)$, and $y_N \geq 0$, $z_N \geq 0$. From

$$0 = x_N = (1 - \alpha)y_N + (\alpha)z_N$$

it follows that

$$y_N = z_N = 0,$$

and $b = By_B = Bz_B = Bx_B$. But, since B is invertible, we have $x_B = y_B = z_B$. This is a contradiction, so x must be in Ext(F).

Conversely, suppose that x is in Ext(F). Since it is in F, we know that $Ax = b$ and $x \geq 0$. By reordering the variables if necessary, we may assume that $x^T = (x_B^T, x_N^T)$, with $x_B > 0$ and $x_N = 0$; we do not know that x_B is a vector of length I, however, so when we write $A = \begin{bmatrix} B & N \end{bmatrix}$, we do not know that B is square. If B is invertible, then x is a basic feasible solution. If not, we shall construct $y \neq x$ and $z \neq x$ in F, such that

$$x = \frac{1}{2}y + \frac{1}{2}z.$$

If $\{B_1, B_2, ..., B_K\}$ are the columns of B and are linearly dependent, then there are constants $p_1, p_2, ..., p_K$, not all zero, with

$$p_1 B_1 + ... + p_K B_K = 0.$$

With $p^T = (p_1, ..., p_K)$, we have

$$B(x_B + \alpha p) = B(x_B - \alpha p) = B x_B = b,$$

for all $\alpha \in (0, 1)$. We then select α so small that both $x_B + \alpha p > 0$ and $x_B - \alpha p > 0$. Let

$$y^T = (x_B^T + \alpha p^T, x_N^T)$$

and

$$z^T = (x_B^T - \alpha p^T, x_N^T).$$

This completes the proof. $\square$

Lemma 22.7. *There are at most finitely many basic feasible solutions, so there are at most finitely many members of Ext(F).*

Theorem 22.8. *If F is not empty, then Ext(F) is not empty. In that case, let $\{v^1, ..., v^M\}$ be the members of Ext(F). Every x in F can be written as*

$$x = d + \alpha_1 v^1 + ... + \alpha_M v^M,$$

for some $\alpha_m \geq 0$, with $\sum_{m=1}^{M} \alpha_m = 1$, and some direction of unboundedness, d.

Proof: We consider only the case in which F is bounded, so there is no direction of unboundedness; the unbounded case is similar. Let x be a feasible point. If x is an extreme point, fine. If not, then x is not a basic feasible solution. The columns of A that correspond to the positive entries of x are not linearly independent. Then we can find a vector p such that $Ap = 0$ and $p_j = 0$ if $x_j = 0$. If $|\epsilon|$ is small, $x + \epsilon p \geq 0$ and $(x + \epsilon p)_j = 0$ if $x_j = 0$, then $x + \epsilon p$ is in F. We can alter ϵ in such a way that eventually $y = x + \epsilon p$ has one more zero entry than x has, and so does $z = x - \epsilon p$. Both y and z are in F and x is the average of these points. If y and z are not basic, repeat the argument on y and z, each time reducing the number of positive entries. Eventually, we will arrive at the case where the number of nonzero entries is I, and so we will have a basic feasible solution. $\square$

Proof (of the Strong Duality Theorem): Suppose now that x_* is a solution of the problem (PS) and $z_* = c^T x_*$. Without loss of generality, we may assume that x_* is a basic feasible solution, hence an extreme point of F. Then we can write

$$x_*^T = ((B^{-1}b)^T, 0^T),$$

$$c^T = (c_B^T, c_N^T),$$

and $A = \begin{bmatrix} B & N \end{bmatrix}$. Every feasible solution has the form

$$x^T = ((B^{-1}b)^T, 0^T) + ((B^{-1}Nv)^T, v^T),$$

for some $v \geq 0$. From $c^T x \geq c^T x_*$ we find that

$$(c_N^T - c_B^T B^{-1} N)(v) \geq 0,$$

for all $v \geq 0$. It follows that

$$c_N^T - c_B^T B^{-1} N = 0.$$

Nw let $y_* = (B^{-1})^T c_B$, or $y_*^T = c_B^T B^{-1}$. We show that y_* is feasible for (DS); that is, we show that

$$A^T y_* \leq c^T.$$

Since

$$y_*^T A = (y_*^T B, y_*^T N) = (c_B^T, y_*^T N) = (c_B^T, c_B^T B^{-1} N)$$

and

$$c_N^T \geq c_B^T B^{-1} N,$$

we have

$$y_*^T A \leq c^T,$$

so y_* is feasible for (DS). Finally, we show that

$$c^T x_* = y_*^T b.$$

We have

$$y_*^T b = c_B^T B^{-1} b = c^T x_*.$$

This completes the proof. $\square$

22.2   The Simplex Method

In this section we sketch the main ideas of the simplex method. For further details see [130].

Begin with a basic feasible solution of (PS), say

$$x^T = (\hat{b}^T, 0^T) = ((B^{-1}b)^T, 0^T).$$

Compute the vector $y^T = c_B^T B^{-1}$. If

$$\hat{c}_N^T = c_N^T - y^T N \geq 0,$$

then x is optimal. Otherwise, select an *entering variable* x_j such that

$$(\hat{c}_N)_j < 0.$$

Compute $\hat{a}^j = B^{-1}a^j$, where a^j is the jth column of A. Find an index s such that

$$\frac{\hat{b}_s}{(\hat{a}_j)_s} = \min_{1 \leq i \leq I} \left\{ \frac{\hat{b}_i}{(\hat{a}_j)_i} : (\hat{a}_j)_i > 0 \right\}.$$

If there are no such positive denominators, the problem is unbounded. Then x_s is the *leaving variable*, replacing x_j. Redefine B and the basic variables x_B accordingly.

22.3   Convex Programming

Let f and g_i, $i = 1, ..., I$, be convex functions defined on C, a nonempty closed, convex subset of R^J. The *primal problem* in *convex programming* is the following:

> minimize $f(x)$, subject to $g_i(x) \leq 0$, for $i = 1, ..., I$.   (P)

The Lagrangian is

$$L(x, \lambda) = f(x) + \sum_{i=1}^{I} \lambda_i g_i(x).$$

The corresponding dual problem is

> maximize $h(\lambda) = \inf_{x \in C} L(x, \lambda)$, for $\lambda \geq 0$.   (D)

22.3.1 An Example

Let $f(x) = \frac{1}{2}||x||_2^2$. The primary problem is to minimize $f(x)$ over all x for which $Ax \geq b$. Then $g_i = b_i - (Ax)_i$, for $i = 1, ..., I$, and the set C is all of R^J. The Lagrangian is then

$$L(x, \lambda) = \frac{1}{2}||x||_2^2 - \lambda^T Ax + \lambda^T b.$$

The infimum over x occurs when $x = A^T\lambda$ and so

$$h(\lambda) = \lambda^T b - \frac{1}{2}||A^T\lambda||_2^2.$$

For any x satisfying $Ax \geq b$ and any $\lambda \geq 0$ we have $h(\lambda) \leq f(x)$. If x^* is the unique solution of the primal problem and λ^* any solution of the dual problem, we have $f(x^*) = h(\lambda^*)$. The point here is that the constraints in the dual problem are easier to implement in an iterative algorithm, so solving the dual problem is the simpler task.

22.3.2 An Iterative Algorithm for the Dual Problem

In [120] Lent and Censor present the following sequential iterative algorithm for solving the dual problem above. At each step only one entry of the current λ is altered.

Algorithm 22.9 (Lent-Censor). Let a_i denote the ith row of the matrix A. Having calculated x^k and $\lambda^k > 0$, let $i = k(\bmod I) + 1$. Then let

$$\theta = (b_i - (a_i)^T x^k)/a_i^T a_i,$$
$$\delta = \max\{-\lambda_i^k, \omega\theta\},$$

and set

$$\lambda_i^{k+1} = \lambda_i^k + \delta,$$
$$x^{k+1} = x^k + \delta a_i.$$

23 | Systems of Linear Inequalities

Designing linear discriminants for pattern classification involves the problem of solving a system of linear inequalities $Ax \geq b$. In this chapter we discuss the iterative Agmon-Motzkin-Schoenberg (AMS) Algorithm [1,128] for solving such problems. We prove convergence of the AMS algorithm, for both the consistent and inconsistent cases, by mimicking the proof for the ART algorithm. Both algorithms are examples of the method of projection onto convex sets. The AMS algorithm is a special case of the cyclic subgradient projection (CSP) method, so that convergence of the AMS, in the consistent case, follows from the convergence theorem for the CSP algorithm.

23.1 Projection onto Convex Sets

In [158] Youla suggests that problems in image restoration might be viewed geometrically and the method of projection onto convex sets (POCS) employed to solve such inverse problems. In the survey paper [157] he examines the POCS method as a particular case of iterative algorithms for finding fixed points of nonexpansive mappings. This point of view is increasingly important in applications such as medical imaging and a number of recent papers have addressed the theoretical and practical issues involved [9], [11], [8], [41], [45], [51], [65], [66], [68].

In this geometric approach the restored image is a solution of the *convex feasibility problem* (CFP), that is, it lies within the intersection of finitely many closed nonempty convex sets $C_i, i = 1, ..., I$, in R^J (or sometimes, in infinite dimensional Hilbert space). For any nonempty closed convex set C, the *metric projection* of x onto C, denoted $P_C x$, is the unique member of C closest to x. The iterative methods used to solve the CFP employ these metric projections. Algorithms for solving the CFP are discussed in the papers cited above, as well as in the books by Censor and Zenios [61], Stark and Yang [148], and Borwein and Lewis [15].

The simplest example of the CFP is the solving of a system of linear equations $Ax = b$. Let A be an I by J real matrix and for $i = 1, ..., I$ let $B_i = \{x | (Ax)_i = b_i\}$, where b_i denotes the ith entry of the vector b. Now let $C_i = B_i$. Any solution of $Ax = b$ lies in the intersection of the C_i; if the system is inconsistent, then the intersection is empty. The Kaczmarz algorithm [110] for solving the system of linear equations $Ax = b$ has the iterative step

$$x_j^{k+1} = x_j^k + A_{i(k)j}(b_{i(k)} - (Ax^k)_{i(k)}), \tag{23.1}$$

for $j = 1, ..., J$, $k = 0, 1, ...$, and $i(k) = k(\mathrm{mod}\, I) + 1$. This algorithm was rediscovered by Gordon, Bender, and Herman [94], who called it the *algebraic reconstruction technique* (ART). This algorithm is an example of the method of *successive orthogonal projections* (SOP) [96] whereby we generate the sequence $\{x^k\}$ by taking x^{k+1} to be the point in $C_{i(k)}$ closest to x^k. Kaczmarz's algorithm can also be viewed as a method for constrained optimization: whenever $Ax = b$ has solutions, the limit of the sequence generated by Equation (23.1) minimizes the function $||x - x^0||_2$ over all solutions of $Ax = b$.

In the example just discussed the sets C_i are hyperplanes in R^J; suppose now that we take the C_i to be half-spaces and consider the problem of finding x such that $Ax \geq b$. For each i let H_i be the half-space $H_i = \{x | (Ax)_i \geq b_i\}$. Then x will be in the intersection of the sets $C_i = H_i$ if and only if $Ax \geq b$. Methods for solving this CFP, such as Hildreth's algorithm, are discussed in the book by Censor and Zenios [61]. Of particular interest for us here is the behavior of the Agmon-Motzkin-Schoenberg (AMS) Algorithm [1, 128] for solving such systems of inequalities $Ax \geq b$.

Algorithm 23.1 (Agmon-Motzkin-Schoenberg). Let x^0 be arbitrary. Having found x^k, define

$$x_j^{k+1} = x_j^k + A_{i(k)j}(b_{i(k)} - (Ax^k)_{i(k)})_+. \tag{23.2}$$

The AMS algorithm converges to a solution of $Ax \geq b$, if there are solutions. If there are no solutions, the AMS algorithm converges cyclically, that is, subsequences associated with the same m converge, as has been shown by De Pierro and Iusem [77], and by Bauschke, Borwein, and Lewis [11]. We present an elementary proof of this result in this chapter.

Algorithms for solving the CFP fall into two classes: those that employ all the sets C_i at each step of the iteration (the so-called *simultaneous methods*) and those that do not (the *row-action algorithms* or, more generally, *block-iterative methods*).

In the consistent case, in which the intersection of the convex sets C_i is nonempty, all reasonable algorithms are expected to converge to a member of that intersection; the limit may or may not be the member of the intersection closest to the starting vector x^0.

In the inconsistent case, in which the intersection of the C_i is empty, simultaneous methods typically converge to a minimizer of a *proximity function* [51], such as

$$ f(x) = \sum_{i=1}^{I} ||x - P_{C_i} x||_2^2, $$

if a minimizer exists.

Methods that are not simultaneous cannot converge in the inconsistent case, since the limit would then be a member of the (empty) intersection. Such methods often exhibit what is called *cyclic convergence*; that is, subsequences converge to finitely many distinct limits comprising a limit cycle. Once a member of this limit cycle is reached, further application of the algorithm results in passing from one member of the limit cycle to the next. Proving the existence of these limit cycles seems to be a difficult problem.

Tanabe [149] showed the existence of a limit cycle for Kaczmarz's algorithm (see also [74]), in which the convex sets are hyperplanes. The SOP method may fail to have a limit cycle for certain choices of the convex sets. For example, if, in R^2, we take C_1 to be the lower half-plane and $C_2 = \{(x, y) | x > 0, y \geq 1/x\}$, then the SOP algorithm fails to produce a limit cycle. However, Gubin, Polyak, and Riak [96] prove weak convergence to a limit cycle for the method of SOP in Hilbert space, under the assumption that at least one of the C_i is bounded, hence weakly compact. In [11] Bauschke, Borwein, and Lewis present a wide variety of results on the existence of limit cycles. In particular, they prove that if each of the convex sets C_i in Hilbert space is a convex polyhedron, that is, the intersection of finitely many half-spaces, then there is a limit cycle and the subsequential convergence is in norm. This result includes the case in which each C_i is a half-space, so implies the existence of a limit cycle for the AMS algorithm. In this chapter we give a proof of existence of a limit cycle for the AMS algorithm using a modification of our proof for the ART.

In the next section we consider the behavior of the ART for solving $Ax = b$. The proofs given by Tanabe and Dax of the existence of a limit cycle for this algorithm rely heavily on aspects of the theory of linear algebra, as did the proof given in an earlier chapter here. Our goal now is to obtain a more direct proof that can be easily modified to apply to the AMS Algorithm.

We assume throughout this chapter that the real I by J matrix A has full rank and its rows have Euclidean length one.

23.2   Solving $Ax = b$

For $k = 0, 1, ...$, let $i(k) = k(\bmod I) + 1$, and for $i = 1, 2, ..., I$, let $K_i = \{x|(Ax)_i = 0\}$, $B_i = \{x|(Ax)_i = b_i\}$ and p^i be the metric projection of $x = 0$ onto B_i. Let $v_i^r = (Ax^{rI+i-1})_i$ and $v^r = (v_1^r, ..., v_I^r)^T$, for $r = 0, 1, ...$ We begin with some basic lemmas concerning the ART.

Lemma 23.2. *For each* $k = 0, 1, ...,,$

$$||x^k||_2^2 - ||x^{k+1}||_2^2 = (A(x^k)_{i(k)})^2 - (b_{i(k)})^2.$$

Lemma 23.3. *For each* $r = 0, 1, ...,$

$$||x^{rI}||_2^2 - ||x^{(r+1)I}||_2^2 = ||v^r||_2^2 - ||b||_2^2.$$

Lemma 23.4. *For each* $k = 0, 1, ...,$

$$||x^k - x^{k+1}||_2^2 = ((Ax^k)_{i(k)} - b_{i(k)})^2.$$

Lemma 23.5. *There exists a constant* $B > 0$ *such that, for all* $r = 0, 1, ...,$ *if* $||v^r||_2 \le ||b||_2$, *then* $||x^{rI}||_2 \ge ||x^{(r+1)I}||_2 - B$.

Lemma 23.6. *Let* x^0 *and* y^0 *be arbitrary and* $\{x^k\}$ *and* $\{y^k\}$ *the sequences generated by applying the ART, starting with* x^0 *and* y^0, *respectively. Then*

$$||x^0 - y^0||_2^2 - ||x^I - y^I||_2^2 = \sum_{i=1}^{I} ((Ax^{i-1})_i - (Ay^{i-1})_i)^2.$$

23.2.1   When the System $Ax = b$ Is Consistent

In this subsection we give a proof of the following result.

Theorem 23.7. *Let* $A\hat{x} = b$ *and let* x^0 *be arbitrary. Let* $\{x^k\}$ *be generated by Equation (23.1). Then the sequence* $\{||\hat{x} - x^k||_2\}$ *is decreasing and* $\{x^k\}$ *converges to the solution of* $Ax = b$ *closest to* x^0.

Proof: Let $A\hat{x} = b$. It follows from Lemma 23.6 that the sequence $\{||\hat{x} - x^{rI}||_2\}$ is decreasing and the sequence $\{v^r - b\} \to 0$. So $\{x^{rI}\}$ is bounded; let $x^{*,0}$ be a cluster point. Then, for $i = 1, 2, ..., I$ let $x^{*,i}$ be the successor of $x^{*,i-1}$ using the ART. It follows that $(Ax^{*,i-1})_i = b_i$ for each i, from which we conclude that $x^{*,0} = x^{*,i}$ for all i and that $Ax^{*,0} = b$. Using $x^{*,0}$ in place of $\hat{x}$, we have that $\{||x^{*,0} - x^k||_2\}$ is decreasing. But a subsequence converges to zero, so $\{x^k\}$ converges to $x^{*,0}$. By Lemma 23.6, the difference $||\hat{x} - x^k||_2^2 - ||\hat{x} - x^{k+1}||_2^2$ is independent of which solution $\hat{x}$ we pick; consequently, so is $||\hat{x} - x^0||_2^2 - ||\hat{x} - x^{*,0}||_2^2$. It follows that $x^{*,0}$ is the solution closest to x^0. This completes the proof. □

23.2.2 When the System $Ax = b$ Is Inconsistent

In the inconsistent case the sequence $\{x^k\}$ will not converge, since any limit would be a solution. However, for each fixed $i \in \{1, 2, ..., I\}$, the subsequence $\{x^{rI+i}\}$ converges [149], [74]; in this subsection we prove this result and then, in the next section, we extend the proof to get cyclic convergence for the AMS algorithm. We start by showing that the sequence $\{x^{rI}\}$ is bounded. We assume that $I > J$ and A has full rank.

Proposition 23.8. *The sequence $\{x^{rI}\}$ is bounded.*

Proof: Assume that the sequence $\{x^{rI}\}$ is unbounded. We first show that we can select a subsequence $\{x^{r_t I}\}$ with the properties $||x^{r_t I}||_2 \geq t$ and $||v^{r_t}||_2 < ||b||_2$, for $t = 1, 2, ...$.

Assume that we have selected $x^{r_t I}$, with the properties $||x^{r_t I}||_2 \geq t$ and $||v^{r_t}||_2 < ||b||_2$; we show how to select $x^{r_{t+1} I}$. Pick integer $s > 0$ such that

$$||x^{sI}||_2 \geq ||x^{r_t I}||_2 + B + 1,$$

where $B > 0$ is as in Lemma 23.5. With $n + r_t = s$ let $m \geq 0$ be the smallest integer for which

$$||x^{(r_t+n-m-1)I}||_2 < ||x^{sI}||_2 \leq ||x^{(r_t+n-i)I}||_2.$$

Then $||v^{r_t+n-m-1}||_2 < ||b||_2$. Let $x^{r_{t+1}I} = x^{(r_t+n-m-1)I}$. Then we have

$$||x^{r_{t+1}I}||_2 \geq ||x^{(r_t+n-m)I}||_2 - B$$
$$\geq ||x^{sI}||_2 - B$$
$$\geq ||x^{r_t I}||_2 + B + 1 - B \geq t + 1.$$

This gives us the desired subsequence.

For every $k = 0, 1, ...$ let $z^{k+1} = x^{k+1} - p^{i(k)}$. Then $z^{k+1} \in K_{i(k)}$. For $z^{k+1} \neq 0$ let $u^{k+1} = z^{k+1}/||z^{k+1}||_2$. Since the subsequence $\{x^{r_t I}\}$ is unbounded, so is $\{z^{r_t I}\}$, so for sufficiently large t the vectors $u^{r_t I}$ are defined and on the unit sphere. Let $u^{*,0}$ be a cluster point of $\{u^{r_t I}\}$; replacing $\{x^{r_t I}\}$ with a subsequence if necessary, assume that the sequence $\{u^{r_t I}\}$ converges to $u^{*,0}$. Then let $u^{*,1}$ be a subsequence of $u^{r_t I+1}\}$; again, assume the sequence $\{u^{r_t I+1}\}$ converges to $u^{*,1}$. Continuing in this manner, we have $\{u^{r_t I+\tau}\}$ converging to $u^{*,\tau}$ for $\tau = 0, 1, 2, ...$. We know that $\{z^{r_t I}\}$ is unbounded and since $||v^{r_t}||_2 < ||b||_2$, we have, by Lemma 23.4, that $\{z^{r_t I+i-1} - z^{r_t I+i}\}$ is bounded for each i. Consequently $\{z^{r_t I+i}\}$ is unbounded for each i.

Now we have

$$||z^{r_t I+i-1} - z^{r_t I+i}||_2 \geq ||z^{r_t I+i-1}||_2 ||u^{r_t I+i-1} - \langle u^{r_t I+i-1}, u^{r_t I+i} \rangle u^{r_t I+i}||_2.$$

Since the left side is bounded and $||z^{r_t I + i - 1}||_2$ has no infinite bounded subsequence, we conclude that

$$||u^{r_t I + i - 1} - \langle u^{r_t I + i - 1}, u^{r_j + I + i} \rangle u^{r_t I + i}||_2 \to 0.$$

It follows that $u^{*,0} = u^{*,i}$ or $u^{*,0} = -u^{*,i}$ for each $i = 1, 2, ..., I$. Therefore $u^{*,0}$ is in K_i for each i; but, since the null space of A contains only zero, this is a contradiction. This completes the proof of the proposition. □

Now we give a proof of the following result.

Theorem 23.9. *Let A be I by J, with $I > J$ and A with full rank. If $Ax = b$ has no solutions, then, for any x^0 and each fixed $i \in \{0, 1, ..., I\}$, the subsequence $\{x^{rI + i}\}$ converges to a limit $x^{*,i}$. Beginning the iteration in Equation (23.1) at $x^{*,0}$, we generate the $x^{*,i}$ in turn, with $x^{*,I} = x^{*,0}$.*

Proof: Let $x^{*,0}$ be a cluster point of $\{x^{rI}\}$. Beginning the ART at $x^{*,0}$ we obtain $x^{*,n}$, for $n = 0, 1, 2, ...$ It is easily seen that

$$||x^{(r-1)I} - x^{rI}||_2^2 - ||x^{rI} - x^{(r+1)I}||_2^2 = \sum_{i=1}^{I} ((Ax^{(r-1)I + i - 1})_i - (Ax^{rI + i - 1})_i)^2.$$

Therefore the sequence $\{||x^{(r-1)I} - x^{rI}||_2\}$ is decreasing and

$$\left\{ \sum_{i=1}^{I} ((Ax^{(r-1)I + i - 1})_i - (Ax^{rI + i - 1})_i)^2 \right\} \to 0.$$

Therefore $(Ax^{*,i-1})_i = (Ax^{*,I + i - 1})_i$ for each i.

For arbitrary x we have

$$||x - x^{*,0}||^2 - ||x - x^{*,I}||_2^2 = \sum_{i=1}^{I} ((Ax)_i - (Ax^{*,i-1})_i)^2$$
$$- \sum_{i=1}^{I} ((Ax)_i - b_i)^2,$$

so that

$$||x - x^{*,0}||_2^2 - ||x - x^{*,I}||_2^2 = ||x - x^{*,I}||_2^2 - ||x - x^{*,2I}||_2^2.$$

Using $x = x^{*,I}$ we have

$$||x^{*,I} - x^{*,0}||_2 = -||x^{*,I} - x^{*,2I}||_2,$$

from which we conclude that $x^{*,0} = x^{*,I}$. From Lemma 23.6, it follows that the sequence $\{||x^{*,0} - x^{rI}||_2\}$ is decreasing; but a subsequence converges to zero, so the entire sequence converges to zero and $\{x^{rI}\}$ converges to $x^{*,0}$. This completes the proof. □

Now we turn to the problem $Ax \geq b$.

23.3 The Agmon-Motzkin-Schoenberg Algorithm

In this section we are concerned with the behavior of the Agmon-Motzkin-Schoenberg (AMS) Algorithm for finding x such that $Ax \geq b$, if such x exist. We begin with some basic facts concerning the AMS Algorithm.

Let $w_i^r = \min\{(Ax^{rI+i-1})_i, b_i\}$ and $w^r = (w_1^r, ..., w_I^r)^T$, for $r = 0, 1, ...$. The following lemmas are easily proved.

Lemma 23.10. *For* $r = 0, 1, ...,$

$$||x^{rI+i-1}||_2^2 - ||x^{rI+i}||_2^2 = (w_i^r)^2 - (b_i)^2.$$

Lemma 23.11. *For* $r = 0, 1, ...,$

$$||x^{rI}||_2^2 - ||x^{(r+1)I}||_2^2 = ||w^r||_2^2 - ||b||_2^2.$$

Lemma 23.12.

$$||x^{rI+i-1} - x^{rI+i}||_2^2 = (w_i^r - b_i)^2.$$

Lemma 23.13. *There exists* $B > 0$ *such that, for all* $r = 0, 1, ...,$ *if* $||w^r||_2 \leq ||b||_2$ *then* $||x^{rI}||_2 \geq ||x^{(r+1)I}||_2 - B$.

Lemma 23.14. *Let* x^0 *and* y^0 *be arbitrary and* $\{x^k\}$ *and* $\{y^k\}$ *the sequences generated by applying the AMS Algorithm, beginning with* x^0 *and* y^0, *respectively. Then*

$$||x^0 - y^0||_2^2 - ||x^I - y^I||_2^2 = \sum_{i=1}^{I}((Ax^{i-1})_i - (Ay^{i-1})_i)^2 -$$

$$\sum_{i=1}^{I}(((Ax^{i-1})_i - b_i)_+ - ((Ay^{i-1})_i - b_i)_+)^2 \geq 0. \qquad (23.3)$$

Consider for a moment the elements of the second sum in the inequality In Lemma 23.14 above. There are three possibilities:

(1) both $(Ax^{i-1})_i - b_i$ and $(Ay^{i-1})_i - b_i$ are nonnegative, in which case this term becomes $((Ax^{i-1})_i - (Ay^{i-1})_i)^2$ and cancels with the same term in the previous sum;

(2) neither $(Ax^{i-1})_i - b_i$ nor $(Ay^{i-1})_i - b_i$ is nonnegative, in which case this term is zero;

(3) precisely one of $(Ax^{i-1})_i - b_i$ and $(Ay^{i-1})_i - b_i$ is nonnegative; say it is $(Ax^{i-1})_i - b_i$, in which case the term becomes $((Ax^{i-1})_i - b_i)^2$.

Since we then have

$$(Ay^{i-1})_i \leq b_i < (Ax^{i-1})_i$$

it follows that

$$((Ax^{i-1})_i - (Ay^{i-1})_i)^2 \geq ((Ax^{i-1})_i - b_i)^2.$$

We conclude that the right side of Equation (23.3) is nonnegative, as claimed.

It will be important in subsequent discussions to know under what conditions the right side of this equation is zero, so we consider that now. We then have

$$((Ax^{i-1})_i - (Ay^{i-1})_i)^2 - (((Ax^{i-1})_i - b_i)_+ - ((Ay^{i-1})_i - b_i)_+)^2 = 0$$

for each m separately, since each of these terms is nonnegative, as we have just seen.

In Case (1) above this difference is already zero, as we just saw. In Case (2) this difference reduces to $((Ax^{i-1})_i - (Ay^{i-1})_i)^2$, which then is zero precisely when $(Ax^{i-1})_i = (Ay^{i-1})_i$. In Case (3) the difference becomes

$$((Ax^{i-1})_i - (Ay^{i-1})_i)^2 - ((Ax^{i-1})_i - b_i)^2,$$

which equals

$$((Ax^{i-1})_i - (Ay^{i-1})_i + (Ax^{i-1})_i - b_i)(b_i - (Ay^{i-1})_i).$$

Since this is zero, it follows that $(Ay^{i-1})_i = b_i$, which contradicts our assumptions in this case. We conclude therefore that the difference of sums in Equation (23.3) is zero if and only if, for all i, either both $(Ax^{i-1})_i \geq b_i$ and $(Ay^{i-1})_i \geq b_i$ or $(Ax^{i-1})_i = (Ay^{i-1})_i < b_i$.

23.3.1   When $Ax \geq b$ Is Consistent

We now prove the following result.

Theorem 23.15. *Let $A\hat{x} \geq b$. Let x^0 be arbitrary and let $\{x^k\}$ be generated by Equation (23.2). Then the sequence $\{\|\hat{x} - x^k\|_2\}$ is decreasing and the sequence $\{x^k\}$ converges to a solution of $Ax \geq b$.*

Proof: Let $A\hat{x} \geq b$. When we apply the AMS Algorithm beginning at $\hat{x}$ we obtain $\hat{x}$ again at each step. Therefore, by Lemma 23.14 and the discussion that followed, with $y^0 = \hat{x}$, we have

$$||x^k - \hat{x}||_2^2 - ||x^{k+1} - \hat{x}||_2^2 = ((Ax^k)_i - (A\hat{x})_i)^2 - (((Ax^k)_i - b_i)_+$$
$$- (A\hat{x})_i + b_i)^2 \geq 0. \qquad (23.4)$$

Therefore the sequence $\{||x^k - \hat{x}||_2\}$ is decreasing and so $\{x^k\}$ is bounded; let $x^{*,0}$ be a cluster point.

The sequence defined by the right side of Equation (23.4) above converges to zero. It follows from the discussion following Lemma 23.14 that $Ax^{*,0} \geq b$. Continuing as in the case of $Ax = b$, we have that the sequence $\{x^k\}$ converges to $x^{*,0}$. In general it is not the case that $x^{*,0}$ is the solution of $Ax \geq b$ closest to x^0. $\qquad \square$

23.3.2 When $Ax \geq b$ Is Inconsistent

In the inconsistent case the sequence $\{x^k\}$ will not converge, since any limit would be a solution. However, we do have the following result.

Theorem 23.16. *Let A be I by J, with $I > J$ and A with full rank. Let x^0 be arbitrary. The sequence $\{x^{rI}\}$ converges to a limit $x^{*,0}$. Beginning the AMS algorithm at $x^{*,0}$ we obtain $x^{*,k}$, for $k = 1, 2, \dots$. For each fixed $i \in \{0, 1, 2, \dots, I\}$, the subsequence $\{x^{rI+i}\}$ converges to $x^{*,i}$ and $x^{*,I} = x^{*,0}$.*

We start by showing that the sequence $\{x^{rI}\}$ is bounded.

Proposition 23.17. *The sequence $\{x^{rI}\}$ is bounded.*

Proof: Assume that the sequence $\{x^{rI}\}$ is unbounded. We first show that we can select a subsequence $\{x^{r_t I}\}$ with the properties $||x^{r_t I}||_2 \geq t$ and $||w^{r_t}||_2 < ||b||_2$, for $t = 1, 2, \dots$.

Assume that we have selected $x^{r_t I}$, with the properties $||x^{r_t I}||_2 \geq t$ and $||w^{r_t}||_2 < ||b||_2$; we show how to select $x^{r_{t+1} I}$. Pick integer $s > 0$ such that

$$||x^{sI}||_2 \geq ||x^{r_t I}||_2 + B + 1,$$

where $B > 0$ is as in Lemma 23.13. With $n + r_t = s$ let $m \geq 0$ be the smallest integer for which

$$||x^{(r_t+n-m-1)I}||_2 < ||x^{sI}||_2 \leq ||x^{(r_t+n-m)I}||_2.$$

Then $||w^{r_t+n-m-1}||_2 < ||b||_2$. Let $x^{r_{t+1} I} = x^{(r_t+n-m-1)I}$. Then we have

$$||x^{r_{t+1} I}||_2 \geq ||x^{(r_t+n-m)I}||_2 - B \geq ||x^{sI}||_2 - B$$
$$\geq ||x^{r_t I}||_2 + B + 1 - B \geq t + 1.$$

This gives us the desired subsequence.

For every $k = 0, 1, \ldots$ let z^{k+1} be the metric projection of x^{k+1} onto the hyperplane $K_{i(k)}$. Then $z^{k+1} = x^{k+1} - p^{i(k)}$ if $(Ax^k)_i \leq b_i$ and $z^{k+1} = x^{k+1} - (Ax^k)_i A^i$ if not; here A^i is the ith column of A^T. Then $z^{k+1} \in K_{i(k)}$. For $z^{k+1} \neq 0$ let $u^{k+1} = z^{k+1}/\|z^{k+1}\|_2$. Let $u^{*,0}$ be a cluster point of $\{u^{r_t I}\}$; replacing $\{x^{r_t I}\}$ with a subsequence if necessary, assume that the sequence $\{u^{r_t I}\}$ converges to $u^{*,0}$. Then let $u^{*,1}$ be a subsequence of $\{u^{r_t I+1}\}$; again, assume the sequence $\{u^{r_t I+1}\}$ converges to $u^{*,1}$. Continuing in this manner, we have $\{u^{r_t I+m}\}$ converging to $u^{*,m}$ for $m = 0, 1, 2, \ldots$. Since $\|w^{r_t}\|_2 < \|b\|_2$, we have, by Lemma 23.12, that $\{z^{r_t I+i-1} - z^{r_t I+i}\}$ is bounded for each i. Now we have

$$\|z^{r_t I+i-1} - z^{r_t I+i}\|_2 \geq \|z^{r_t I+i-1}\|_2 \|u^{r_t I+i-1} - \langle u^{r_t I+i-1}, u^{r_t+I+i} \rangle u^{r_t I+i}\|_2.$$

The left side is bounded. We consider the sequence $\|z^{r_t I+i-1}\|_2$ in two cases: (1) the sequence is unbounded; (2) the sequence is bounded.

In the first case, it follows, as in the case of $Ax = b$, that $u^{*,i-1} = u^{*,i}$ or $u^{*,i-1} = -u^{*,i}$. In the second case we must have $(Ax^{r_t I+i-1})_i > b_i$ for t sufficiently large, so that, from some point on, we have $x^{r_t I+i-1} = x^{r_t I+i}$, in which case we have $u^{*,i-1} = u^{*,i}$. So we conclude that $u^{*,0}$ is in the null space of A, which is a contradiction. This concludes the proof of the proposition. $\square$

Proof (of Theorem 23.16): Let $x^{*,0}$ be a cluster point of $\{x^{rI}\}$. Beginning the AMS iteration (23.2) at $x^{*,0}$ we obtain $x^{*,m}$, for $m = 0, 1, 2, \ldots$. Using Lemma 23.14, it is easily seen that the sequence $\{\|x^{rI} - x^{(r+1)I}\|_2\}$ is decreasing and that the sequence

$$\left\{ \sum_{i=1}^{I} ((Ax^{(r-1)I+i-1})_i - (Ax^{rI+i-1})_i)^2 - \sum_{i=1}^{I} (((Ax^{(r-1)I+i-1})_i - b_i)_+ \right.$$
$$\left. - ((Ax^{rI+i-1})_i - b_i)_+)^2 \right\} \to 0.$$

Again, by the discussion following Lemma 23.14, we conclude one of two things: either Case (1): $(Ax^{*,i-1})_i = (Ax^{*,jI+i-1})_i$ for each $j = 1, 2, \ldots$; or Case (2): $(Ax^{*,i-1})_i > b_i$ and, for each $j = 1, 2, \ldots$, $(Ax^{*,jI+i-1})_i > b_i$. Let A^i denote the ith column of A^T. As the AMS iteration proceeds from $x^{*,0}$ to $x^{*,I}$, from $x^{*,I}$ to $x^{*,2I}$ and, in general, from $x^{*,jI}$ to $x^{*,(j+1)I}$ we have either $x^{*,i-1} - x^{*,i} = 0$ and $x^{*,jI+i-1} - x^{*,jI+i} = 0$, for each $j = 1, 2, \ldots$, which happens in Case (2), or $x^{*,i-1} - x^{*,i} = x^{*,jI+i-1} - x^{*,jI+i} = (b_i - (Ax^{*,i-1})_i)A^i$, for $j = 1, 2, \ldots$, which happens in Case (1). It follows, therefore, that

$$x^{*,0} - x^{*,I} = x^{*,jI} - x^{*,(j+1)I}$$

for $j = 1, 2, \ldots$. Since the original sequence $\{x^{rI}\}$ is bounded, we have

$$||x^{*,0} - x^{*,jI}||_2 \le ||x^{*,0}||_2 + ||x^{*,jI}||_2 \le K$$

for some K and all $j = 1, 2, \ldots$. But we also have

$$||x^{*,0} - x^{*,jI}||_2 = j||x^{*,0} - x^{*,I}||_2.$$

We conclude that $||x^{*,0} - x^{*,I}||_2 = 0$ or $x^{*,0} = x^{*,I}$.

From Lemma 23.14, using $y^0 = x^{*,0}$, it follows that the sequence $\{||x^{*,0} - x^{rI}||_2\}$ is decreasing; but a subsequence converges to zero, so the entire sequence converges to zero and $\{x^{rI}\}$ converges to $x^{*,0}$. This completes the proof of Theorem 23.16. $\qquad\square$

24 | Constrained Iteration Methods

The ART and its simultaneous and block-iterative versions are designed to solve general systems of linear equations $Ax = b$. The SMART, EMML, and RBI methods require that the entries of A be nonnegative, those of b positive and produce nonnegative x. In this chapter we present variations of the SMART and EMML that impose the constraints $u_j \leq x_j \leq v_j$, where the u_j and v_j are selected lower and upper bounds on the individual entries x_j. These algorithms were applied to transmission tomography image reconstruction in [129].

24.1 Modifying the KL Distance

The SMART, EMML, and RBI methods are based on the Kullback-Leibler distance between nonnegative vectors. To impose more general constraints on the entries of x we derive algorithms based on shifted KL distances, also called Fermi-Dirac generalized entropies.

For a fixed real vector u, the shifted KL distance $KL(x - u, z - u)$ is defined for vectors x and z having $x_j \geq u_j$ and $z_j \geq u_j$. Similarly, the shifted distance $KL(v - x, v - z)$ applies only to those vectors x and z for which $x_j \leq v_j$ and $z_j \leq v_j$. For $u_j \leq v_j$, the combined distance

$$KL(x - u, z - u) + KL(v - x, v - z)$$

is restricted to those x and z whose entries x_j and z_j lie in the interval $[u_j, v_j]$. Our objective is to mimic the derivation of the SMART, EMML, and RBI methods, replacing KL distances with shifted KL distances, to obtain algorithms that enforce the constraints $u_j \leq x_j \leq v_j$, for each j. The algorithms that result are the ABMART and ABEMML block-iterative methods. These algorithms were originally presented in [40], in which the vectors u and v were called a and b, hence the names of the algorithms.

Throughout this chapter we shall assume that the entries of the matrix A are nonnegative. We shall denote by B_n, $n = 1, ..., N$ a partition of the index set $\{i = 1, ..., I\}$ into blocks. For $k = 0, 1, ...$ let $n(k) = k(\mathrm{mod}\,N) + 1$.

The Projected Landweber's Algorithm can also be used to impose the restrictions $u_j \leq x_j \leq v_j$; however, the projection step in that algorithm is implemented by clipping, or setting equal to u_j or v_j values of x_j that would otherwise fall outside the desired range. The result is that the values u_j and v_j can occur more frequently than may be desired. One advantage of the AB methods is that the values u_j and v_j represent barriers that can only be reached in the limit and are never taken on at any step of the iteration.

24.2   The ABMART Algorithm

We assume that $(Au)_i \leq b_i \leq (Av)_i$ and seek a solution of $Ax = b$ with $u_j \leq x_j \leq v_j$, for each j.

Algorithm 24.1 (ABMART). Select x^0 satisfying $u_j \leq x_j^0 \leq v_j$, for each j. Having calculated x^k, we take

$$x_j^{k+1} = \alpha_j^k v_j + (1 - \alpha_j^k) u_j, \qquad (24.1)$$

with $n = n(k)$,

$$\alpha_j^k = \frac{c_j^k \prod^n (d_i^k)^{A_{ij}}}{1 + c_j^k \prod^n (d_i^k)^{A_{ij}}},$$

$$c_j^k = \frac{(x_j^k - u_j)}{(v_j - x_j^k)},$$

$$d_j^k = \frac{(b_i - (Au)_i)((Av)_i - (Ax^k)_i)}{((Av)_i - b_i)((Ax^k)_i - (Au)_i)},$$

where $\prod^n$ denotes the product over those indices i in $B_{n(k)}$.

Notice that, at each step of the iteration, x_j^k is a convex combination of the endpoints u_j and v_j, so that x_j^k lies in the interval $[u_j, v_j]$.

We have the following convergence theorem for the ABMART algorithm:

Theorem 24.2. *If there is a solution of the system $Ax = b$ that satisfies the constraints $u_j \leq x_j \leq v_j$ for each j, then, for any N and any choice of the blocks B_n, the ABMART sequence converges to that constrained solution* *of $Ax = b$ for which the Fermi-Dirac generalized entropic distance from x to x^0,*

$$KL(x - u, x^0 - u) + KL(v - x, v - x^0),$$

is minimized. If there is no constrained solution of $Ax = b$, then, for $N = 1$, the ABMART sequence converges to the minimizer of

$$KL(Ax - Au, b - Au) + KL(Av - Ax, Av - b)$$

for which

$$KL(x - u, x^0 - u) + KL(v - x, v - x^0)$$

is minimized.

The proof is similar to that for RBI-SMART and is found in [40].

24.3 The ABEMML Algorithm

We make the same assumptions as in the previous section.

We have the following convergence theorem for the ABEMML algorithm:

Theorem 24.3. *If there is a solution of the system $Ax = b$ that satisfies the constraints $u_j \le x_j \le v_j$ for each j, then, for any N and any choice of the blocks B_n, the ABEMML sequence converges to such a constrained solution of $Ax = b$. If there is no constrained solution of $Ax = b$, then, for $N = 1$, the ABMART sequence converges to a constrained minimizer of*

$$KL(Ax - Au, b - Au) + KL(Av - Ax, Av - b).$$

The proof is similar to that for RBI-EMML and is to be found in [40]. In contrast to the ABMART theorem, this is all we can say about the limits of the ABEMML sequences.

Open Question 9. How does the limit of the ABEMML iterative sequence depend, in the consistent case, on the choice of blocks, and, in general, on the choice of x^0?

Algorithm 24.4. Select x^0 satisfying $u_j \le x_j^0 \le v_j$, for each j. Having calculated x^k, let

$$x_j^{k+1} = \alpha_j^k v_j + (1 - \alpha_j^k) u_j,$$

where

$$\alpha_j^k = \gamma_j^k / d_j^k,$$
$$\gamma_j^k = (x_j^k - u_j) e_j^k,$$

$$\beta_j^k = (v_j - x_j^k)f_j^k,$$
$$d_j^k = \gamma_j^k + \beta_j^k,$$
$$e_j^k = \left(1 - \sum_{i \in B_n} A_{ij}\right) + \sum_{i \in B_n} A_{ij}\left(\frac{b_i - (Au)_i}{(Ax^k)_i - (Au)_i}\right),$$
$$f_j^k = \left(1 - \sum_{i \in B_n} A_{ij}\right) + \sum_{i \in B_n} A_{ij}\left(\frac{(Av)_i - b_i}{(Av)_i - (Ax^k)_i}\right).$$

25 | Fourier Transform Estimation

In many remote-sensing problems, the measured data is related to the function to be imaged by Fourier transformation. In the *Fourier* approach to tomography, the data are often viewed as line integrals through the object of interest. These line integrals can then be converted into values of the Fourier transform of the object function. In magnetic-resonance imaging (MRI), adjustments to the external magnetic field cause the measured data to be Fourier-related to the desired proton-density function. In such applications, the imaging problem becomes a problem of estimating a function from finitely many noisy values of its Fourier transform. To overcome these limitations, one can use iterative and noniterative methods for incorporating prior knowledge and regularization; data-extrapolation algorithms form one class of such methods.

We focus on the use of iterative algorithms for improving resolution through extrapolation of Fourier-transform data. The reader should consult the appendices for brief discussion of some of the applications of these methods.

25.1 The Limited-Fourier-Data Problem

For notational convenience, we shall discuss only the one-dimensional case, involving the estimation of the (possibly complex-valued) function $f(x)$ of the real variable x, from finitely many values $F(\omega_n)$, $n = 1, ..., N$ of its Fourier transform. Here we adopt the definitions

$$F(\omega) = \int f(x)e^{ix\omega} dx,$$

$$f(x) = \frac{1}{2\pi} \int F(\omega)e^{-ix\omega} d\omega.$$

261

Because it is the case in the applications of interest to us here, we shall assume that the object function has bounded support, that is, there is $A > 0$, such that $f(x) = 0$ for $|x| > A$.

The values $\omega = \omega_n$ at which we have measured the function $F(\omega)$ may be structured in some way; they may be equi-spaced along a line, or, in the higher-dimensional case, arranged in a cartesian grid pattern, as in MRI. According to the Central Slice Theorem, the Fourier data in tomography lie along rays through the origin. Nevertheless, in what follows, we shall not assume any special arrangement of these data points.

Because the data are finite, there are infinitely many functions $f(x)$ consistent with the data. We need some guidelines to follow in selecting a best estimate of the true $f(x)$. First, we must remember that the data values are noisy, so we want to avoid over-fitting the estimate to noisy data. This means that we should include regularization in whatever method we adopt. Second, the limited data are often insufficient to provide the desired resolution, so we need to incorporate additional prior knowledge about $f(x)$, such as nonnegativity, upper and lower bounds on its values, its support, its overall shape, and so on. Third, once we have selected prior information to include, we should be conservative in choosing an estimate consistent with that information. This may involve the use of constrained minimum-norm solutions. Fourth, we should not expect our prior information to be perfectly accurate, so our estimate should not be overly sensitive to slight changes in the prior information. Finally, the estimate we use will be one for which there are good algorithms for its calculation.

25.2   Minimum-Norm Estimation

To illustrate the notion of minimum-norm estimation, we begin with the finite-dimensional problem of solving an underdetermined system of linear equations, $Ax = b$, where A is a real I by J matrix with $J > I$ and AA^T is invertible.

25.2.1   The Minimum-Norm Solution of $Ax = b$

Each equation can be written as

$$b_i = (a^i)^T x = \langle x, a^i \rangle,$$

where the vector a^i is the ith column of the matrix A^T and $\langle u, v \rangle$ denoted the inner, or dot product of the vectors u and v.

Lemma 25.1. *Every vector x in R^J can be written as*

$$x = A^T z + w, \tag{25.1}$$

with $Aw = 0$ and

$$||x||_2^2 = ||A^T z||_2^2 + ||w||_2^2.$$

Consequently, $Ax = b$ if and only if $A(A^T z) = b$ and $A^T z$ is the solution having the smallest norm. This minimum-norm solution $\hat{x} = A^T z$ can be found explicitly; it is

$$\hat{x} = A^T z = A^T (AA^T)^{-1} b. \tag{25.2}$$

Proof: Multiply both sides of Equation (25.1) by A and solve for z. □

It follows from Lemma 25.1 that the minimum-norm solution $\hat{x}$ of $Ax = b$ has the form $\hat{x} = A^T z$, which means that $\hat{x}$ is a linear combination of the a^i:

$$\hat{x} = \sum_{i=1}^{I} z_i a^i.$$

25.2.2 Minimum-Weighted-Norm Solution of $Ax = b$

As we shall see later, it is sometimes convenient to introduce a new norm for the vectors. Let Q be a J by J symmetric positive-definite matrix and define

$$||x||_Q^2 = x^T Q x.$$

With $Q = C^T C$, where C is the positive-definite symmetric square-root of Q, we can write

$$||x||_Q^2 = ||y||_2^2,$$

for $y = Cx$. Now suppose that we want to find the solution of $Ax = b$ for which $||x||_Q^2$ is minimum. We write $Ax = b$ as

$$AC^{-1}y = b,$$

so that, from Equation (25.2), we find that the solution y with minimum norm is

$$\hat{y} = (AC^{-1})^T (AC^{-1}(AC^{-1})^T)^{-1} b,$$

or

$$\hat{y} = (AC^{-1})^T (AQ^{-1}A^T)^{-1} b,$$

so that the $\hat{x}_Q$ with minimum weighted norm is

$$\hat{x}_Q = C^{-1}\hat{y} = Q^{-1}A^T(AQ^{-1}A^T)^{-1}b. \tag{25.3}$$

Notice that, writing

$$\langle u, v \rangle_Q = u^T Q v,$$

we find that

$$b_i = \langle Q^{-1}a^i, \hat{x}_Q \rangle_Q,$$

and the minimum-weighted-norm solution of $Ax = b$ is a linear combination of the columns g^i of $Q^{-1}A^T$, that is,

$$\hat{x}_Q = \sum_{i=1}^{I} d_i g^i,$$

where

$$d_i = ((AQ^{-1}A^T)^{-1}b)_i,$$

for each $i = 1, ..., I$.

25.3   Fourier-Transform Data

Returning now to the case in which we have finitely many values of the Fourier transform of $f(x)$, we write

$$F(\omega) = \int f(x)e^{ix\omega}dx = \langle e_\omega, f \rangle,$$

where $e_\omega(x) = e^{-ix\omega}$ and

$$\langle g, h \rangle = \int g(x)h(x)dx.$$

The norm of a function $f(x)$ is then

$$\|f\|_2 = \sqrt{\langle f, f \rangle} = \sqrt{\int |f(x)|^2 dx}.$$

25.3.1   The Minimum-Norm Estimate

Arguing as we did in the finite-dimensional case, we conclude that the minimum-norm solution of the data-consistency equations

$$F(\omega_n) = \langle e_{\omega_n}, f \rangle, n = 1, ..., N,$$

has the form

$$\hat{f}(x) = \sum_{n=1}^{N} a_n e^{-ix\omega_n}.$$

If the integration assumed to extend over the whole real line, the functions $e_\omega(x)$ are mutually orthogonal and so

$$a_n = \frac{1}{2\pi} F(\omega_n).$$ (25.4)

In most applications, however, the function $f(x)$ is known to have finite support.

Lemma 25.2. *If $f(x) = 0$ for x outside the interval $[a, b]$, then the coefficients a_n satisfy the system of linear equations*

$$F(\omega_n) = \sum_{m=1}^{N} G_{nm} a_m,$$

with

$$G_{nm} = \int_{a}^{b} e^{ix(\omega_n - \omega_m)} dx.$$

For example, suppose that $[a, b] = [-\pi, \pi]$ and

$$\omega_n = -\pi + \frac{2\pi}{N} n,$$

for $n = 1, ..., N$

Lemma 25.3. *In this example, $G_{nn} = 2\pi$ and $G_{nm} = 0$, for $n \neq m$. Therefore, for this special case, we again have*

$$a_n = \frac{1}{2\pi} F(\omega_n).$$

25.3.2   Minimum-Weighted-Norm Estimates

Let $p(x) \geq 0$ be a weight function. Let

$$\langle g, h \rangle_p = \int g(x) h(x) p(x)^{-1} dx,$$

with the understanding that $p(x)^{-1} = 0$ outside of the support of $p(x)$. The associated weighted norm is then

$$\|f\|_p = \sqrt{\int |f(x)|^2 p(x)^{-1} dx}.$$

We can then write

$$F(\omega_n) = \langle p e_\omega, f \rangle_p = \int (p(x) e^{-ix\omega}) f(x) p(x)^{-1} dx.$$

It follows that the function consistent with the data and having the minimum weighted norm has the form

$$\hat{f}_p(x) = p(x) \sum_{n=1}^{N} b_n e^{-ix\omega_n}. \tag{25.5}$$

Lemma 25.4. *The coefficients b_n satisfy the system of linear equations*

$$F(\omega_n) = \sum_{m=1}^{N} b_m P_{nm}, \tag{25.6}$$

with

$$P_{nm} = \int p(x) e^{ix(\omega_n - \omega_m)} dx,$$

for $m, n = 1, ..., N$.

Whenever we have prior information about the support of $f(x)$, or about the shape of $|f(x)|$, we can incorporate this information through our choice of the weight function $p(x)$. In this way, the prior information becomes part of the estimate, through the first factor in Equation (25.5), with the second factor providing information gathered from the measurement data. This minimum-weighted-norm estimate of $f(x)$ is called the PDFT, and it is discussed in more detail in [48].

Once we have $\hat{f}_p(x)$, we can take its Fourier transform, $\hat{F}_p(\omega)$, which is then an estimate of $F(\omega)$. Because the coefficients b_n satisfy Equations (25.6), we know that

$$\hat{F}_p(\omega_n) = F(\omega_n),$$

for $n = 1, ..., N$. For other values of ω, the estimate $\hat{F}_p(\omega)$ provides an extrapolation of the data. For this reason, methods such as the PDFT are sometimes called *data-extrapolation methods*. If $f(x)$ is supported on an interval $[a, b]$, then the function $F(\omega)$ is said to be *band-limited*. If $[c, d]$ is an interval containing $[a, b]$ and $p(x) = 1$, for x in $[c, d]$, and $p(x) = 0$ otherwise, then the PDFT estimate is a noniterative version of the Gerchberg-Papoulis band-limited extrapolation estimate of $f(x)$ (see [48]).

25.3.3 Implementing the PDFT

The PDFT can be extended easily to the estimation of functions of several variables. However, there are several difficult steps that can be avoided by iterative implementation. Even in the one-dimensional case, when the values ω_n are not equi-spaced, the calculation of the matrix P can be messy. In the case of higher dimensions, both calculating P and solving for the coefficients can be expensive. In the next section we consider an iterative implementation that solves both of these problems.

25.4 The Discrete PDFT (DPDFT)

The derivation of the PDFT assumes a function $f(x)$ of one or more continuous real variables, with the data obtained from $f(x)$ by integration. The discrete PDFT (DPDFT) begins with $f(x)$ replaced by a finite vector $f = (f_1, ..., f_J)^T$ that is a discretization of $f(x)$; say that $f_j = f(x_j)$ for some point x_j. The integrals that describe the Fourier transform data can be replaced by finite sums,

$$F(\omega_n) = \sum_{j=1}^{J} f_j E_{nj}, \qquad (25.7)$$

where $E_{nj} = e^{ix_j\omega_n}$. We have used a Riemann-sum approximation of the integrals here, but other choices are also available. The problem then is to solve this system of equations for the f_j.

Since the N is fixed, but the J is under our control, we select $J > N$, so that the system becomes under-determined. Now we can use minimum-norm and minimum-weighted-norm solutions of the finite-dimensional problem to obtain an approximate, discretized PDFT solution.

Since the PDFT estimate is a minimum-weighted-norm solution in the continuous-variable formulation, that is, it is the function f, consistent with the data, for which $||f||_p$ is minimum, it is reasonable to let the DPDFT estimate be the corresponding minimum-weighted-norm solution obtained with the positive-definite matrix Q now being a diagonal matrix having for its jth diagonal entry

$$Q_{jj} = 1/p(x_j),$$

if $p(x_j) > 0$, and zero, otherwise. In other words, the DPDFT solution is the vector f, satisfying Equation (25.7), for which the weighted norm $f^T Q f$ is minimum.

25.4.1   Calculating the DPDFT

The DPDFT is a minimum-weighted-norm solution, which can be calculated using, say, the ART algorithm. We know that, in the underdetermined case, the ART provides the solution closest to the starting vector, in the sense of the Euclidean distance. We therefore reformulate the system, so that the minimum-weighted-norm solution becomes a minimum-norm solution, as we did earlier, and then begin the ART iteration with zero. For recent work involving the DPDFT see [144–146].

25.4.2   Regularization

We noted earlier that one of the principles guiding the estimation of $f(x)$ from Fourier transform data should be that we do not want to overfit the estimate to noisy data. In the PDFT, this can be avoided by adding a small positive quantity to the main diagonal of the matrix P. In the DPDFT, implemented using ART, we regularize the ART algorthm, as we discussed earlier.

VIII | Applications

VII

Applications

26 | Tomography

In this chapter we present a brief overview of transmission and emission tomography. These days, the term *tomography* is used by lay people and practitioners alike to describe any sort of scan, from ultrasound to magnetic resonance. It has apparently lost its association with the idea of slicing, as in the expression *three-dimensional tomography*. In this chapter we focus on two important modalities, transmission tomography and emission tomography. An x-ray CAT scan is an example of the first; a positron emission (PET) scan is an example of the second.

26.1 X-Ray Transmission Tomography

Computer-assisted tomography (CAT) scans have revolutionized medical practice. One example of CAT is x-ray transmission tomography. The goal here is to image the spatial distribution of various matter within the body, by estimating the distribution of x-ray attenuation. In the continuous formulation, the data are line integrals of the function of interest.

When an x-ray beam travels along a line segment through the body it becomes progressively weakened by the material it encounters. By comparing the initial strength of the beam as it enters the body with its final strength as it exits the body, we can estimate the integral of the attenuation function, along that line segment. The data in transmission tomography are these line integrals, corresponding to thousands of lines along which the beams have been sent. The image reconstruction problem is to create a discrete approximation of the attenuation function. The inherently three-dimensional problem is usually solved one two-dimensional plane, or slice, at a time, hence the name tomography [99].

The beam attenuation at a given point in the body will depend on the material present at that point; estimating and imaging the attenuation as a function of spatial location will give us a picture of the material within the body. A bone fracture will show up as a place where significant attenuation should be present, but is not.

26.1.1   The Exponential-Decay Model

As an x-ray beam passes through the body, it encounters various types of matter, such as soft tissue, bone, ligaments, air, each weakening the beam to a greater or lesser extent. If the intensity of the beam upon entry is I_{in} and I_{out} is its lower intensity after passing through the body, then

$$I_{\text{out}} = I_{\text{in}} e^{-\int_L f},$$

where $f = f(x, y) \geq 0$ is the *attenuation function* describing the two-dimensional distribution of matter within the slice of the body being scanned and $\int_L f$ is the integral of the function f over the line L along which the x-ray beam has passed. To see why this is the case, imagine the line L parameterized by the variable s and consider the intensity function $I(s)$ as a function of s. For small $\Delta s > 0$, the drop in intensity from the start to the end of the interval $[s, s + \Delta s]$ is approximately proportional to the intensity $I(s)$, to the attenuation $f(s)$ and to Δs, the length of the interval; that is,

$$I(s) - I(s + \Delta s) \approx f(s)I(s)\Delta s.$$

Dividing by Δs and letting Δs approach zero, we get

$$I'(s) = -f(s)I(s).$$

The solution to this differential equation is

$$I(s) = I(0) \exp(-\int_{u=0}^{u=s} f(u)du).$$

From knowledge of I_{in} and I_{out}, we can determine $\int_L f$. If we know $\int_L f$ for every line in the x, y-plane we can reconstruct the attenuation function f. In the real world we know line integrals only approximately and only for finitely many lines. The goal in x-ray transmission tomography is to estimate the attenuation function $f(x, y)$ in the slice, from finitely many noisy measurements of the line integrals. We usually have prior information about the values that $f(x, y)$ can take on. We also expect to find sharp boundaries separating regions where the function $f(x, y)$ varies only slightly. Therefore, we need algorithms capable of providing such images. As we shall see, the line-integral data can be viewed as values of the Fourier transform of the attenuation function.

26.1.2 Reconstruction from Line Integrals

We turn now to the underlying problem of reconstructing such functions from line-integral data. Our goal is to reconstruct the function $f(x, y)$ from line-integral data. Let θ be a fixed angle in the interval $[0, \pi)$. Form the t, s-axis system with the positive t-axis making the angle θ with the positive x-axis. Each point (x, y) in the original coordinate system has coordinates (t, s) in the second system, where the t and s are given by

$$t = x \cos \theta + y \sin \theta,$$
$$s = -x \sin \theta + y \cos \theta.$$

If we have the new coordinates (t, s) of a point, the old coordinates are (x, y) given by

$$x = t \cos \theta - s \sin \theta,$$

and

$$y = t \sin \theta + s \cos \theta.$$

We can then write the function f as a function of the variables t and s. For each fixed value of t, we compute the integral

$$\int_L f(x, y) ds = \int f(t \cos \theta - s \sin \theta, t \sin \theta + s \cos \theta) ds$$

along the single line L corresponding to the fixed values of θ and t. We repeat this process for every value of t and then change the angle θ and repeat again. In this way we obtain the integrals of f over every line L in the plane. We denote by $r_f(\theta, t)$ the integral

$$r_f(\theta, t) = \int_L f(x, y) ds.$$

The function $r_f(\theta, t)$ is called the *Radon transform* of f.

For fixed θ the function $r_f(\theta, t)$ is a function of the single real variable t; let $R_f(\theta, \omega)$ be its Fourier transform. Then

$$R_f(\theta, \omega) = \int r_f(\theta, t) e^{i\omega t} dt$$

$$= \int \int f(t \cos \theta - s \sin \theta, t \sin \theta + s \cos \theta) e^{i\omega t} ds dt$$

$$= \int \int f(x, y) e^{i\omega(x \cos \theta + y \sin \theta)} dx dy = F(\omega \cos \theta, \omega \sin \theta),$$

where $F(\omega\cos\theta, \omega\sin\theta)$ is the two-dimensional Fourier transform of the function $f(x,y)$, evaluated at the point $(\omega\cos\theta, \omega\sin\theta)$; this relationship is called the *Central Slice Theorem*. For fixed θ, as we change the value of ω, we obtain the values of the function F along the points of the line making the angle θ with the horizontal axis. As θ varies in $[0, \pi)$, we get all the values of the function F. Once we have F, we can obtain f using the formula for the two-dimensional inverse Fourier transform. We conclude that we are able to determine f from its line integrals.

The Fourier-transform inversion formula for two-dimensional functions tells us that the function $f(x,y)$ can be obtained as

$$f(x,y) = \frac{1}{4\pi^2} \int \int F(u,v)e^{-i(xu+yv)}\,dudv. \tag{26.1}$$

The *filtered backprojection* methods commonly used in the clinic are derived from different ways of calculating the double integral in Equation (26.1).

26.1.3   The Algebraic Approach

Although there is some flexibility in the mathematical description of the image reconstruction problem in transmission tomography, one popular approach is the algebraic formulation of the problem. In this formulation, the problem is to solve, at least approximately, a large system of linear equations, $Ax = b$.

The attenuation function is discretized, in the two-dimensional case, by imagining the body to consist of finitely many squares, or *pixels*, within which the function has a constant, but unknown, value. This value at the jth pixel is denoted x_j. In the three-dimensional formulation, the body is viewed as consisting of finitely many cubes, or *voxels*. The beam is sent through the body along various lines and both initial and final beam strength is measured. From that data we can calculate a discrete line integral along each line. For $i = 1, ..., I$ we denote by L_i the ith line segment through the body and by b_i its associated line integral. Denote by A_{ij} the length of the intersection of the jth pixel with L_i; therefore, A_{ij} is nonnegative. Most of the pixels do not intersect line L_i, so A is quite sparse. Then the data value b_i can be described, at least approximately, as

$$b_i = \sum_{j=1}^{J} A_{ij}x_j. \tag{26.2}$$

Both I, the number of lines, and J, the number of pixels or voxels, are quite large, although they certainly need not be equal, and they are typically unrelated.

The matrix A is large and rectangular. The system $Ax = b$ may or may not have exact solutions. We are always free to select J, the number of pixels, as large as we wish, limited only by computation costs. We may also have some choice as to the number I of lines, but within the constraints posed by the scanning machine and the desired duration and dosage of the scan. When the system is underdetermined $(J > I)$, there may be infinitely many exact solutions; in such cases we usually impose constraints and prior knowledge to select an appropriate solution. As we mentioned earlier, noise in the data, as well as error in our model of the physics of the scanning procedure, may make an exact solution undesirable, anyway. When the system is overdetermined $(J < I)$, we may seek a least-squares approximate solution, or some other approximate solution. We may have prior knowledge about the physics of the materials present in the body that can provide us with upper bounds for x_j, as well as information about body shape and structure that may tell where $x_j = 0$. Incorporating such information in the reconstruction algorithms can often lead to improved images [129].

26.2   Emission Tomography

In *single-photon emission computed tomography* (SPECT) and *positron emission tomography* (PET) the patient is injected with, or inhales, a chemical to which a radioactive substance has been attached. The recent book edited by Wernick and Aarsvold [153] describes the cutting edge of emission tomography. The particular chemicals used in emission tomography are designed to become concentrated in the particular region of the body under study. Once there, the radioactivity results in photons that travel through the body and, at least some of the time, are detected by the scanner. The function of interest is the actual concentration of the radioactive material at each spatial location within the region of interest. Learning what the concentrations are will tell us about the functioning of the body at the various spatial locations. Tumors may take up the chemical (and its radioactive passenger) more avidly than normal tissue, or less avidly, perhaps. Malfunctioning portions of the brain may not receive the normal amount of the chemical and will, therefore, exhibit an abnormal amount of radioactivity.

As in the transmission tomography case, this nonnegative function is discretized and represented as the vector x. The quantity b_i, the ith entry of the vector b, is the photon count at the ith detector; in coincidence-detection PET a detection is actually a nearly simultaneous detection of a photon at two different detectors. The entry A_{ij} of the matrix A is the probability that a photon emitted at the jth pixel or voxel will be detected at the ith detector.

In [140], Rockmore and Macovski suggest that, in the emission tomography case, one take a statistical view, in which the quantity x_j is the expected number of emissions at the jth pixel during the scanning time, so that the expected count at the ith detector is

$$E(b_i) = \sum_{j=1}^{J} A_{ij} x_j. \tag{26.3}$$

They further suggest that the problem of finding the x_j be viewed as a parameter-estimation problem, for which a maximum-likelihood technique might be helpful. These suggestions inspired work by Shepp and Vardi [143], Lange and Carson [116], Vardi, Shepp, and Kaufman [152], and others, and led to the expectation maximization maximum likelihood (EMML) method for reconstruction.

The system of equations $Ax = b$ is obtained by replacing the expected count, $E(b_i)$, with the actual count, b_i; obviously, an exact solution of the system is not needed in this case. As in the transmission case, we seek an approximate, and nonnegative, solution of $Ax = b$, where, once again, all the entries of the system are nonnegative.

26.2.1   Maximum-Likelihood Parameter Estimation

The measured data in tomography are values of random variables. The probabilities associated with these random variables are used in formulating the image reconstruction problem as one of solving a large system of linear equations. We can also use the stochastic model of the data to formulate the problem as a statistical parameter-estimation problem, which suggests the image be estimated using likelihood maximization. When formulated that way, the problem becomes a constrained optimization problem. The desired image can then be calculated using general-purpose iterative optimization algorithms, or iterative algorithms designed specifically to solve the particular problem.

26.3   Image Reconstruction in Tomography

Image reconstruction from tomographic data is an increasingly important area of applied numerical linear algebra, particularly for medical diagnosis. For in-depth discussion of these issues, the reader should consult the books by Herman [94, 99], Kak and Slaney [111], Natterer [131], Natterer and Wübbeling [132], and Wernick and Aarsvold [153]. In the algebraic approach, the problem is to solve, at least approximately, a large system of linear equations, $Ax = b$. The vector x is large because it is usually a vectorization of a discrete approximation of a function of two or three continuous spatial variables. The size of the system necessitates the use of iterative solution methods [118]. Because the entries of x usually represent intensity levels, of beam attenuation in transmission tomography, and of radionuclide concentration in emission tomography, we require x to be nonnegative; the physics of the situation may impose additional constraints on the entries of x. In practice, we often have prior knowledge about the function represented, in discrete form, by the vector x and we may wish to include this knowledge in the reconstruction. In tomography the entries of A and b are also nonnegative. Iterative algorithms tailored to find solutions to these special, constrained problems may out-perform general iterative solution methods [129]. To be medically useful in the clinic, the algorithms need to produce acceptable reconstructions early in the iterative process.

The *Fourier* approach to tomographic image reconstruction maintains, at least initially, the continuous model for the attenuation function. The data are taken to be line integrals through the attenuator, that is, values of its so-called *x-ray transform*, which, in the two-dimensional case, is the Radon transform. The Central Slice Theorem then relates the Radon-transform values to values of the Fourier transform of the attenuation function. Image reconstruction then becomes estimation of the (inverse) Fourier transform. In magnetic-resonance imaging (MRI), we again have the measured data related to the function we wish to image, the proton density function, by a Fourier relation.

In the transmission and emission tomography, the data are photon counts, so it is natural to adopt a statistical model and to convert the image reconstruction problem into a statistical parameter-estimation problem. The estimation can be done using maximum likelihood (ML) or maximum a posteriori (MAP) Bayesian methods, which then require iterative optimization algorithms.

27 | Intensity-Modulated Radiation Therapy

In [55] Censor et al. extend the CQ algorithm to solve what they call the *multiset split feasibility problem* (MSSFP). In the sequel [56] this extended CQ algorithm is used to determine dose intensities for *intensity-modulated radiation therapy* (IMRT) that satisfy both dose constraints and radiation-source constraints.

27.1   The Extended CQ Algorithm

For $n = 1, ..., N$, let C_n be a nonempty, closed convex subset of R^J. For $m = 1, ..., M$, let Q_m be a nonempty, closed convex subset of R^I. Let D be a real I by J matrix. The MSSFP is to find a member x of $C = \cap_{n=1}^N C_n$ for which $h = Dx$ is a member of $Q = \cap_{m=1}^M Q_m$. A somewhat more general problem is to find a minimizer of the proximity function

$$p(x) = \frac{1}{2} \sum_{n=1}^N \alpha_n \|P_{C_n} x - x\|_2^2 + \frac{1}{2} \sum_{m=1}^M \beta_m \|P_{Q_m} Dx - Dx\|_2^2,$$

with respect to the nonempty, closed convex set $\Omega \subseteq R^N$, where α_n and β_m are positive and

$$\sum_{n=1}^N \alpha_n + \sum_{m=1}^M \beta_m = 1.$$

They show that $\nabla p(x)$ is L-Lipschitz, for

$$L = \sum_{n=1}^N \alpha_n + \rho(D^T D) \sum_{m=1}^M \beta_m.$$

The algorithm given in [55] has the iterative step

$$x^{k+1} = P_\Omega\Big(x^k + s\Big(\sum_{n=1}^{N} \alpha_n(P_{C_n}x^k - x^k) + \sum_{m=1}^{M} \beta_m D^T(P_{Q_m}Dx^k - Dx^k)\Big)\Big),$$

for $0 < s < 2/L$. This algorithm converges to a minimizer of $p(x)$ over Ω, whenever such a minimizer exists, and to a solution, within Ω, of the MSSFP, whenever such solutions exist.

27.2 Intensity-Modulated Radiation Therapy

For $i = 1, ..., I$, and $j = 1, ..., J$, let $h_i \geq 0$ be the dose absorbed by the ith voxel of the patient's body, $x_j \geq 0$ be the intensity of the jth beamlet of radiation, and $D_{ij} \geq 0$ be the dose absorbed at the ith voxel due to a unit intensity of radiation at the jth beamlet. In intensity space, we have the obvious constraints that $x_j \geq 0$. In addition, there are *implementation constraints*; the available treatment machine will impose its own requirements, such as a limit on the difference in intensities between adjacent beamlets. In dosage space, there will be a lower bound on the dosage delivered to those regions designated as *planned target volumes* (PTV), and an upper bound on the dosage delivered to those regions designated as *organs at risk* (OAR).

27.3 Equivalent Uniform Dosage Functions

Suppose that S_t is either a PTV or an OAR, and suppose that S_t contains N_t voxels. For each dosage vector $h = (h_1, ..., h_I)^T$ define the *equivalent uniform dosage* (EUD) function $e_t(h)$ by

$$e_t(h) = \Big(\frac{1}{N_t}\sum_{i \in S_t}(h_i)^\alpha\Big)^{1/\alpha}, \tag{27.1}$$

where $0 < \alpha < 1$ if S_t is a PTV, and $\alpha > 1$ if S_t is an OAR. The function $e_t(h)$ is convex, for h nonnegative, when S_t is an OAR, and $-e_t(h)$ is convex, when S_t is a PTV. The constraints in dosage space take the form

$$e_t(h) \leq a_t,$$

when S_t is an OAR, and

$$-e_t(h) \leq b_t,$$

when S_t is a PTV. Therefore, we require that $h = Dx$ lie within the intersection of these convex sets. The constraint sets are convex sets of the form $\{x|f(x) \leq 0\}$, for particular convex functions f. Therefore, the cyclic subgradient projection (CSP) method is used to find the solution to the MSSFP.

28 | Magnetic-Resonance Imaging

The field of *magnetic-resonance imaging* (MRI) [97] is a relatively new and rapidly expanding area of medical diagnosis. In addition to being an important application, MRI provides us with a nice example of a remote-sensing problem involving image reconstruction from Fourier-transform data. Therefore, the Fourier-transform estimation and extrapolation techniques we have considered previously play a major role in this field.

28.1 An Overview of MRI

Protons have *spin*, which, for our purposes here, can be viewed as a charge distribution in the nucleus revolving around an axis. Associated with the resulting current is a *magnetic dipole moment* collinear with the axis of the spin. In elements with an odd number of protons, such as hydrogen, the nucleus itself will have a net magnetic moment. The objective in MRI is to determine the density of such elements in a volume of interest within the body. This is achieved by forcing the individual spinning nuclei to emit signals that, while too weak to be detected alone, are detectable in the aggregate. The signals are generated by the precession that results when the axes of the magnetic dipole moments are first aligned and then perturbed.

In much of MRI, it is the distribution of hydrogen in water molecules that is the object of interest, although the imaging of phosphorus to study energy transfer in biological processing is also important. There is ongoing work using tracers containing fluorine to target specific areas of the body and avoid background resonance.

28.2   Alignment

In the absence of an external magnetic field, the axes of these magnetic dipole moments have random orientation, dictated mainly by thermal effects. When an external magnetic field is introduced, it induces a small fraction, about one in 10^5, of the dipole moments to begin to align their axes with that of the external magnetic field. Only because the number of protons per unit of volume is so large do we get a significant number of moments aligned in this way. A strong external magnetic field, about $20,000$ times that of the earth's, is required to produce enough alignment to generate a detectable signal.

When the axes of the aligned magnetic dipole moments are perturbed, they begin to precess, like a spinning top, around the axis of the external magnetic field, at the *Larmor frequency*, which is proportional to the intensity of the external magnetic field. If the magnetic field intensity varies spatially, then so does the Larmor frequency. Each precessing magnetic dipole moment generates a signal; taken together, they contain information about the density of the element at the various locations within the body. As we shall see, when the external magnetic field is appropriately chosen, a Fourier relationship can be established between the information extracted from the received signal and this density function.

28.3   Slice Isolation

When the external magnetic field is the *static field* $B_0\mathbf{k}$, that is, the magnetic field has strength B_0 and axis $\mathbf{k} = (0,0,1)$, then the Larmor frequency is the same everywhere and equals $\omega_0 = \gamma B_0$, where γ is the gyromagnetic constant. If, instead, we impose an external magnetic field $(B_0 + G_z(z - z_0))\mathbf{k}$, for some constant G_z, then the Larmor frequency is ω_0 only within the plane $z = z_0$. This external field now includes a *gradient field*.

28.4   Tipping

When a magnetic dipole moment that is aligned with $\mathbf{k}$ is given a component in the x, y-plane, it begins to precess around the z-axis, with frequency equal to its Larmor frequency. To create this x, y-plane component, we apply a *radio-frequency* (RF) field

$$H_1(t)(\cos(\omega t)\mathbf{i} + \sin(\omega t)\mathbf{j}).$$

The function $H_1(t)$ typically lasts only for a short while, and the effect of imposing this RF field is to tip the aligned magnetic dipole moment axes away from the z-axis, initiating precession. Those dipole axes that tip most are those whose Larmor frequency is ω. Therefore, if we first isolate the slice $z = z_0$ and then choose $\omega = \omega_0$, we tip primarily those dipole axes within the plane $z = z_0$. The dipoles that have been tipped ninety degrees into the x, y-plane generate the strongest signal. How much tipping occurs also depends on $H_1(t)$, so it is common to select $H_1(t)$ to be constant over the time interval $[0, \tau]$, and zero elsewhere, with integral $\frac{\pi}{2\gamma}$. This $H_1(t)$ is called a $\frac{\pi}{2}$-pulse, and tips those axes with Larmor frequency ω_0 into the x, y-plane.

28.5   Imaging

The information we seek about the proton density function is contained within the received signal. By carefully adding gradient fields to the external field, we can make the Larmor frequency spatially varying, so that each frequency component of the received signal contains a piece of the information we seek. The proton density function is then obtained through Fourier transformations.

28.5.1   The Line-Integral Approach

Suppose that we have isolated the plane $z = z_0$ and tipped the aligned axes using a $\frac{\pi}{2}$-pulse. After the tipping has been completed, we introduce an external field $(B_0 + G_x x)\mathbf{k}$, so that now the Larmor frequency of dipoles within the plane $z = z_0$ is $\omega(x) = \omega_0 + \gamma G_x x$, which depends on the x-coordinate of the point. The result is that the component of the received signal associated with the frequency $\omega(x)$ is due solely to those dipoles having that x coordinate. Performing an FFT of the received signal gives us line integrals of the density function along lines in the x, y-plane having fixed x-coordinate.

More generally, if we introduce an external field $(B_0 + G_x x + G_y y)\mathbf{k}$, the Larmor frequency is constant at $\omega(x, y) = \omega_0 + \gamma(G_x x + G_y y) = \omega_0 + \gamma s$ along lines in the x, y-plane with equation

$$G_x x + G_y y = s.$$

Again performing an FFT on the received signal, we obtain the integral of the density function along these lines. In this way, we obtain the three-dimensional Radon transform of the desired density function. The Central Slice Theorem for this case tells us that we can obtain the Fourier transform of the density function by performing a one-dimensional Fourier transform with respect to the variable s. For each fixed (G_x, G_y) we obtain this Fourier transform along a ray through the origin. By varying the (G_x, G_y)

we get the entire Fourier transform. The desired density function is then obtained by Fourier inversion.

28.5.2 Phase Encoding

In the line-integral approach, the line-integral data is used to obtain values of the Fourier transform of the density function along lines through the origin in Fourier space. It would be more convenient to have Fourier-transform values on the points of a rectangular grid. We can obtain this by selecting the gradient fields to achieve *phase encoding*.

Suppose that, after the tipping has been performed, we impose the external field $(B_0+G_y y)\mathbf{k}$ for T seconds. The effect is to alter the precession frequency from ω_0 to $\omega(y) = \omega_0 + \gamma G_y y$. A harmonic $e^{i\omega_0 t}$ is changed to

$$e^{i\omega_0 t} e^{i\gamma G_y y t},$$

so that, after T seconds, we have

$$e^{i\omega_0 T} e^{i\gamma G_y y T}.$$

For $t \geq T$, the harmonic $e^{i\omega_0 t}$ returns, but now it is

$$e^{i\omega_0 t} e^{i\gamma G_y y T}.$$

The effect is to introduce a phase shift of $\gamma G_y y T$. Each point with the same y-coordinate has the same phase shift.

After time T, when this gradient field is turned off, we impose a second external field, $(B_0 + G_x x)\mathbf{k}$. Because this gradient field alters the Larmor frequencies, at times $t \geq T$ the harmonic $e^{i\omega_0 t} e^{i\gamma G_y y T}$ is transformed into

$$e^{i\omega_0 t} e^{i\gamma G_y y T} e^{i\gamma G_x x t}.$$

The received signal is now

$$S(t) = e^{i\omega_0 t} \int\int \rho(x,y) e^{i\gamma G_y y T} e^{i\gamma G_x x t} dx dy,$$

where $\rho(x, y)$ is the value of the proton density function at the point (x, y). Removing the $e^{i\omega_0 t}$ factor, we have

$$\int\int \rho(x,y) e^{i\gamma G_y y T} e^{i\gamma G_x x t} dx dy,$$

which is the Fourier transform of $\rho(x, y)$ at the point $(\gamma G_x t, \gamma G_y T)$. By selecting equi-spaced values of t and altering the G_y, we can get the Fourier transform values on a rectangular grid.

28.6   The General Formulation

The external magnetic field generated in the MRI scanner is generally described by

$$H(r,t) = (H_0 + \mathbf{G}(t) \cdot \mathbf{r})\mathbf{k} + H_1(t)(\cos(\omega t)\mathbf{i} + \sin(\omega t)\mathbf{j}).$$

The vectors $\mathbf{i}, \mathbf{j}$, and $\mathbf{k}$ are the unit vectors along the coordinate axes, and $\mathbf{r} = (x, y, z)$. The vector-valued function $\mathbf{G}(t) = (G_x(t), G_y(t), G_z(t))$ produces the *gradient field*

$$\mathbf{G}(t) \cdot \mathbf{r}.$$

The magnetic field component in the x, y plane is the radio frequency field.

If $\mathbf{G}(t) = 0$, then the Larmor frequency is ω_0 everywhere. Using $\omega = \omega_0$ in the RF field, with a $\frac{\pi}{2}$-pulse, will then tip the aligned axes into the x, y-plane and initiate precession. If $\mathbf{G}(t) = \theta$, for some direction vector θ, then the Larmor frequency is constant on planes $\theta \cdot \mathbf{r} = s$. Using an RF field with frequency $\omega = \gamma(H_0 + s)$ and a $\frac{\pi}{2}$-pulse will then tip the axes in this plane into the x, y-plane. The strength of the received signal will then be proportional to the integral, over this plane, of the proton density function. Therefore, the measured data will be values of the three-dimensional Radon transform of the proton density function, which is related to its three-dimensional Fourier transform by the Central Slice Theorem. Later, we shall consider two more widely used examples of $\mathbf{G}(t)$.

28.7   The Received Signal

We assume now that the function $H_1(t)$ is a *short $\frac{\pi}{2}$-pulse*, that is, it has constant value over a short time interval $[0, \tau]$ and has integral $\frac{\pi}{2\gamma}$. The received signal produced by the precessing magnetic dipole moments is approximately

$$S(t) = \int_{R^3} \rho(\mathbf{r}) \exp(-i\gamma(\int_0^t \mathbf{G}(s)ds) \cdot \mathbf{r}) \exp(-t/T_2)d\mathbf{r},$$

where $\rho(\mathbf{r})$ is the proton density function, and T_2 is the *transverse* or *spin-spin* relaxation time. The vector integral in the exponent is

$$\int_0^t \mathbf{G}(s)ds = (\int_0^t G_x(s)ds, \int_0^t G_y(s)ds, \int_0^t G_z(s)ds).$$

Now imagine approximating the function $G_x(s)$ over the interval $[0, t]$ by a step function that is constant over small subintervals, that is, $G_x(s)$

is approximately $G_x(n\Delta)$ for s in the interval $[n\Delta, (n+1)\Delta)$, with $n = 1, ..., N$ and $\Delta = \frac{t}{N}$. During the interval $[n\Delta, (n+1)\Delta)$, the presence of this gradient field component causes the phase to change by the amount $x\gamma G_x(n\Delta)\Delta$, so that by the time we reach $s = t$ the phase has changed by

$$x \sum_{n=1}^{N} G_x(n\Delta)\Delta,$$

which is approximately $x \int_0^t G_x(s)ds$.

28.7.1   An Example of $\mathbf{G}(t)$

Suppose now that $g > 0$ and θ is an arbitrary direction vector. Let

$$\mathbf{G}(t) = g\theta, \text{ for } \tau \leq t,$$

and $\mathbf{G}(t) = 0$ otherwise. Then the received signal $S(t)$ is

$$S(t) = \int_{R^3} \rho(\mathbf{r}) \exp(-i\gamma g(t - \tau)\theta \cdot \mathbf{r})d\mathbf{r}$$
$$= (2\pi)^{3/2}\hat{\rho}(\gamma g(t - \tau)\theta), \qquad (28.1)$$

for $\tau \leq t << T_2$, where $\hat{\rho}$ denotes the three-dimensional Fourier transform of the function $\rho(\mathbf{r})$.

From Equation (28.1) we see that, by selecting different direction vectors and by sampling the received signal $S(t)$ at various times, we can obtain values of the Fourier transform of ρ along lines through the origin in the Fourier domain, called k-space. If we had these values for all θ and for all t we would be able to determine $\rho(\mathbf{r})$ exactly. Instead, we have much the same problem as in transmission tomography: only finitely many θ and only finitely many samples of $S(t)$. Noise is also a problem, because the resonance signal is not strong, even though the external magnetic field is.

We may wish to avoid having to estimate the function $\rho(\mathbf{r})$ from finitely many noisy values of its Fourier transform. We can do this by selecting the gradient field $\mathbf{G}(t)$ differently.

28.7.2   Another Example of $\mathbf{G}(t)$

The vector-valued function $\mathbf{G}(t)$ can be written as

$$\mathbf{G}(t) = (G_1(t), G_2(t), G_3(t)).$$

Now we let

$$G_2(t) = g_2,$$

and

$$G_3(t) = g_3,$$

for $0 \leq t \leq \tau$, and zero otherwise, and

$$G_1(t) = g_1,$$

for $\tau \leq t$, and zero otherwise. This means that only $H_0\mathbf{k}$ and the RF field are present up to time τ, and then the RF field is shut off and the gradient field is turned on. Then, for $t \geq \tau$, we have

$$S(t) = (2\pi)^{3/2}\hat{M}_0(\gamma(t - \tau)g_1, \gamma\tau g_2, \gamma\tau g_3).$$

By selecting

$$t_n = n\Delta t + \tau, \text{ for } n = 1, ..., N,$$
$$g_{2k} = k\Delta g,$$
$$g_{3i} = i\Delta g,$$

for $i, k = -m, ..., m$, we have values of the Fourier transform, $\hat{M}_0$, on a Cartesian grid in three-dimensional k-space. The proton density function, ρ, can then be approximated using the fast Fourier transform.

29 | Hyperspectral Imaging

Hyperspectral image processing provides an excellent example of the need for estimating Fourier transform values from limited data. In this chapter we describe one novel approach, due to Mooney et al. [127]; the presentation here follows [19].

29.1 Spectral Component Dispersion

In this hyperspectral-imaging problem the electromagnetic energy reflected or emitted by a point, such as light reflected from a location on the earth's surface, is passed through a prism to separate the components as to their wavelengths. Due to the dispersion of the different frequency components caused by the prism, these components are recorded in the image plane not at a single spatial location, but at distinct points along a line. Since the received energy comes from a region of points, not a single point, what is received in the image plane is a superposition of different wavelength components associated with different points within the object. The first task is to reorganize the data so that each location in the image plane is associated with all the components of a single point of the object being imaged; this is a Fourier-transform estimation problem, which we can solve using band-limited extrapolation.

The points of the image plane are in one-to-one correspondence with points of the object. These spatial locations in the image plane and in the object are discretized into finite two-dimensional grids. Once we have reorganized the data we have, for each grid point in the image plane, a function of wavelength, describing the intensity of each component of the energy from the corresponding grid point on the object. Practical considerations limit the fineness of the grid in the image plane; the resulting discretization of the object is into pixels. In some applications, such as satellite imaging, a single pixel may cover an area several meters on a side. Achieving subpixel resolution is one goal of hyperspectral imaging; capturing other subtleties of the scene is another.

Within a single pixel of the object, there may well be a variety of object types, each reflecting or emitting energy differently. The data we now have corresponding to a single pixel are therefore a mixture of the energies associated with each of the subobjects within the pixel. With prior knowledge of the possible types and their reflective or emissive properties, we can separate the mixture to determine which object types are present within the pixel and to what extent. This mixture problem can be solved using the RBI-EMML method.

29.2   A Single Point Source

From an abstract perspective the problem is the following: F and f are a Fourier-transform pair, as are G and g; F and G have finite support. We measure G and want F; g determines some, but not all, of the values of f. We will have, of course, only finitely many measurements of G from which to estimate values of g. Having estimated finitely many values of g, we have the corresponding estimates of f. We apply band-limited extrapolation of these finitely many values of f to estimate F. In fact, once we have estimated values of F, we may not be finished; each value of F is a mixture whose individual components may be what we really want. For this unmixing step we use the RBI-EMML algorithm.

The region of the object that we wish to image is described by the two-dimensional spatial coordinate $\mathbf{x} = (x_1, x_2)$. For simplicity, we take these coordinates to be continuous, leaving until the end the issue of discretization. We shall also denote by $\mathbf{x}$ the point in the image plane corresponding to the point $\mathbf{x}$ on the object; the units of distance between two such points in one plane and their corresponding points in the other plane may, of course, be quite different. For each $\mathbf{x}$ we let $F(\mathbf{x}, \lambda)$ denote the intensity of the component at wavelength λ of the electromagnetic energy that is reflected from or emitted by location $\mathbf{x}$. We shall assume that $F(\mathbf{x}, \lambda) = 0$ for $(\mathbf{x}, \lambda)$ outside some bounded portion of three-dimensional space.

Consider, for a moment, the case in which the energy sensed by the imaging system comes from a single point $\mathbf{x}$. If the dispersion axis of the prism is oriented according to the unit vector $\mathbf{p}_\theta$, for some $\theta \in [0, 2\pi)$, then the component at wavelength λ of the energy from $\mathbf{x}$ on the object is recorded not at $\mathbf{x}$ in the image plane but at the point $\mathbf{x} + \mu(\lambda - \lambda_0)\mathbf{p}_\theta$. Here, $\mu > 0$ is a constant and λ_0 is the wavelength for which the component from point $\mathbf{x}$ of the object is recorded at $\mathbf{x}$ in the image plane.

29.3   Multiple Point Sources

Now imagine energy coming to the imaging system for all the points within the imaged region of the object. Let $G(\mathbf{x}, \theta)$ be the intensity of the energy received at location $\mathbf{x}$ in the image plane when the prism orientation is θ. It follows from the description of the sensing that

$$G(\mathbf{x}, \theta) = \int_{-\infty}^{+\infty} F(\mathbf{x} - \mu(\lambda - \lambda_0)\mathbf{p}_\theta, \lambda) d\lambda. \qquad (29.1)$$

The limits of integration are not really infinite due to the finiteness of the aperture and the focal plane of the imaging system. Our data will consist of finitely many values of $G(\mathbf{x}, \theta)$, as $\mathbf{x}$ varies over the grid points of the image plane and θ varies over some finite discretized set of angles.

We begin the image processing by taking the two-dimensional inverse Fourier transform of $G(\mathbf{x}, \theta)$ with respect to the spatial variable $\mathbf{x}$ to get

$$g(\mathbf{y}, \theta) = \frac{1}{(2\pi)^2} \int G(\mathbf{x}, \theta) \exp(-i\mathbf{x} \cdot \mathbf{y}) d\mathbf{x}. \qquad (29.2)$$

Inserting the expression for G in Equation (29.1) into Equation (29.2), we obtain

$$g(\mathbf{y}, \theta) = \exp(i\mu\lambda_0\mathbf{p}_\theta \cdot \mathbf{y}) \int \exp(-i\mu\lambda\mathbf{p}_\theta \cdot \mathbf{y}) f(\mathbf{y}, \lambda) d\lambda, \qquad (29.3)$$

where $f(\mathbf{y}, \lambda)$ is the two-dimensional inverse Fourier transform of $F(\mathbf{x}, \lambda)$ with respect to the spatial variable $\mathbf{x}$. Therefore,

$$g(\mathbf{y}, \theta) = \exp(i\mu\lambda_0\mathbf{p}_\theta \cdot \mathbf{y}) \mathcal{F}(\mathbf{y}, \gamma_\theta),$$

where $\mathcal{F}(\mathbf{y}, \gamma)$ denotes the three-dimensional inverse Fourier transform of $F(\mathbf{x}, \lambda)$ and $\gamma_\theta = \mu\mathbf{p}_\theta \cdot \mathbf{y}$. We see then that each value of $g(\mathbf{y}, \theta)$ that we estimate from our measurements provides us with a single estimated value of $\mathcal{F}$.

We use the measured values of $G(\mathbf{x}, \theta)$ to estimate values of $g(\mathbf{y}, \theta)$ guided by the discussion in our earlier chapter on discretization. Having obtained finitely many estimated values of $\mathcal{F}$, we use the support of the function $F(\mathbf{x}, \lambda)$ in three-dimensional space to perform a band-limited extrapolation estimate of the function F.

Alternatively, for each fixed $\mathbf{y}$ for which we have values of $g(\mathbf{y}, \theta)$ we use the PDFT to solve Equation (29.3), obtaining an estimate of $f(\mathbf{y}, \lambda)$ as a function of the continuous variable λ. Then, for each fixed λ, we again use the PDFT to estimate $F(\mathbf{x}, \lambda)$ from the values of $f(\mathbf{y}, \lambda)$ previously obtained.

29.4 Solving the Mixture Problem

Once we have the estimated function $F(\mathbf{x}, \lambda)$ on a finite grid in three-dimensional space, we can use the RBI-EMML method, as in [126], to solve the mixture problem and identify the individual object types contained within the single pixel denoted $\mathbf{x}$. For each fixed $\mathbf{x}$ corresponding to a pixel, denote by $\mathbf{b} = (b_1, ..., b_I)^T$ the column vector with entries $b_i = F(\mathbf{x}, \lambda_i)$, where λ_i, $i = 1, ..., I$ constitute a discretization of the wavelength space of those λ for which $F(\mathbf{x}, \lambda) > 0$. We assume that this energy intensity distribution vector $\mathbf{b}$ is a superposition of those vectors corresponding to a number of different object types; that is, we assume that

$$\mathbf{b} = \sum_{j=1}^{J} a_j \mathbf{q}_j, \qquad (29.4)$$

for some $a_j \geq 0$ and intensity distribution vectors $\mathbf{q}_j$, $j = 1, ..., J$. Each column vector $\mathbf{q}_j$ is a model for what $\mathbf{b}$ would be if there had been only one object type filling the entire pixel. These $\mathbf{q}_j$ are assumed to be known a priori. Our objective is to find the a_j.

With Q the I by J matrix whose jth column is $\mathbf{q}_j$ and $\mathbf{a}$ the column vector with entries a_j we write Equation (29.4) as $\mathbf{b} = Q\mathbf{a}$. Since the entries of Q are nonnegative, the entries of $\mathbf{b}$ are positive, and we seek a nonnegative solution a, we can use any of the entropy-based iterative algorithms discussed earlier. Because of its simplicity of form and speed of convergence our preference is the RBI-EMML algorithm. The recent master's thesis of E. Meidunas [126] discusses just such an application.

30 | Planewave Propagation

In this chapter we demonstrate how the Fourier transform arises naturally as we study the signals received in the farfield from an array of tranmitters or reflectors. We restrict our attention to single-frequency, or narrowband, signals.

30.1 Transmission and Remote Sensing

For pedagogical reasons, we shall discuss separately what we call the transmission and the remote-sensing problems, although the two problems are opposite sides of the same coin, in a sense. In the one-dimensional transmission problem, it is convenient to imagine the transmitters located at points $(x, 0)$ within a bounded interval $[-A, A]$ of the x-axis, and the measurements taken at points P lying on a circle of radius D, centered at the origin. The radius D is large, with respect to A. It may well be the case that no actual sensing is to be performed, but rather, we are simply interested in what the received signal pattern is at points P distant from the transmitters. Such would be the case, for example, if we were analyzing or constructing a transmission pattern of radio broadcasts. In the remote-sensing problem, in contrast, we imagine, in the one-dimensional case, that our sensors occupy a bounded interval of the x-axis, and the transmitters or reflectors are points of a circle whose radius is large, with respect to the size of the bounded interval. The actual size of the radius does not matter and we are interested in determining the amplitudes of the transmitted or reflected signals, as a function of angle only. Such is the case in astronomy, farfield sonar or radar, and the like. Both the transmission and remote-sensing problems illustrate the important role played by the Fourier transform.

295

30.2   The Transmission Problem

We identify two distinct transmission problems: the direct problem and the inverse problem. In the direct transmission problem, we wish to determine the farfield pattern, given the complex amplitudes of the transmitted signals. In the inverse transmission problem, the array of transmitters or reflectors is the object of interest; we are given, or we measure, the farfield pattern and wish to determine the amplitudes. For simplicity, we consider only single-frequency signals.

We suppose that each point x in the interval $[-A, A]$ transmits the signal $f(x)e^{i\omega t}$, where $f(x)$ is the complex amplitude of the signal and $\omega > 0$ is the common fixed frequency of the signals. Let $D > 0$ be large, with respect to A, and consider the signal received at each point P given in polar coordinates by $P = (D, \theta)$. The distance from $(x, 0)$ to P is approximately $D - x \cos \theta$, so that, at time t, the point P receives from $(x, 0)$ the signal $f(x)e^{i\omega(t-(D-x\cos\theta)/c)}$, where c is the propagation speed. Therefore, the combined signal received at P is

$$B(P, t) = e^{i\omega t}e^{-i\omega D/c} \int_{-A}^{A} f(x)e^{ix\frac{\omega \cos \theta}{c}}\, dx.$$

The integral term, which gives the farfield pattern of the transmission, is

$$F(\frac{\omega \cos \theta}{c}) = \int_{-A}^{A} f(x)e^{ix\frac{\omega \cos \theta}{c}}\, dx,$$

where $F(\gamma)$ is the Fourier transform of $f(x)$, given by

$$F(\gamma) = \int_{-A}^{A} f(x)e^{ix\gamma}\, dx.$$

How $F(\frac{\omega \cos \theta}{c})$ behaves, as a function of θ, as we change A and ω, is discussed in some detail in Chapter 12 of [48].

Consider, for example, the function $f(x) = 1$, for $|x| \le A$, and $f(x) = 0$, otherwise. The Fourier transform of $f(x)$ is

$$F(\gamma) = 2A\mathrm{sinc}(A\gamma),$$

where $\mathrm{sinc}(t)$ is defined to be

$$\mathrm{sinc}(t) = \frac{\sin(t)}{t},$$

for $t \ne 0$, and $\mathrm{sinc}(0) = 1$. Then $F(\frac{\omega \cos \theta}{c}) = 2A$ when $\cos \theta = 0$, so when $\theta = \frac{\pi}{2}$ and $\theta = \frac{3\pi}{2}$. We will have $F(\frac{\omega \cos \theta}{c}) = 0$ when $A\frac{\omega \cos \theta}{c} = \pi$, or

$\cos\theta = \frac{\pi c}{A\omega}$. Therefore, the transmission pattern has no nulls if $\frac{\pi c}{A\omega} > 1$. In order for the transmission pattern to have nulls, we need $A > \frac{\lambda}{2}$, where $\lambda = \frac{2\pi c}{\omega}$ is the wavelength. This rather counterintuitive fact, namely that we need more signals transmitted in order to receive less at certain locations, illustrates the phenomenon of destructive interference.

30.3   Reciprocity

For certain remote-sensing applications, such as sonar and radar array processing and astronomy, it is convenient to switch the roles of sender and receiver. Imagine that superimposed planewave fields are sensed at points within some bounded region of the interior of the sphere, having been transmitted or reflected from the points P on the surface of a sphere whose radius D is large with respect to the bounded region. The *reciprocity principle* tells us that the same mathematical relation holds between points P and $(x, 0)$, regardless of which is the sender and which the receiver. Consequently, the data obtained at the points $(x, 0)$ are then values of the inverse Fourier transform of the function describing the amplitude of the signal sent from each point P.

30.4   Remote Sensing

A basic problem in remote sensing is to determine the nature of a distant object by measuring signals transmitted by or reflected from that object. If the object of interest is sufficiently remote, that is, is in the *farfield*, the data we obtain by sampling the propagating spatio-temporal field is related, approximately, to what we want by *Fourier transformation*. The problem is then to estimate a function from finitely many (usually noisy) values of its *Fourier transform*. The application we consider here is a common one of remote-sensing of transmitted or reflected waves propagating from distant sources. Examples include optical imaging of planets and asteroids using reflected sunlight, radio-astronomy imaging of distant sources of radio waves, active and passive sonar, and radar imaging.

30.5   The Wave Equation

In many areas of remote sensing, what we measure are the fluctuations in time of an electromagnetic or acoustic field. Such fields are described mathematically as solutions of certain partial differential equations, such as the *wave equation*. A function $u(x, y, z, t)$ is said to satisfy the *three-dimensional wave equation* if

$$u_{tt} = c^2(u_{xx} + u_{yy} + u_{zz}) = c^2 \nabla^2 u,$$

where u_{tt} denotes the second partial derivative of u with respect to the time variable t twice and $c > 0$ is the (constant) speed of propagation. More complicated versions of the wave equation permit the speed of propagation c to vary with the spatial variables x, y, z, but we shall not consider that here.

We use the method of *separation of variables* at this point, to get some idea about the nature of solutions of the wave equation. Assume, for the moment, that the solution $u(t, x, y, z)$ has the simple form

$$u(t, x, y, z) = f(t)g(x, y, z).$$

Inserting this separated form into the wave equation, we get

$$f''(t)g(x, y, z) = c^2 f(t)\nabla^2 g(x, y, z)$$

or

$$f''(t)/f(t) = c^2 \nabla^2 g(x, y, z)/g(x, y, z).$$

The function on the left is independent of the spatial variables, while the one on the right is independent of the time variable; consequently, they must both equal the same constant, which we denote $-\omega^2$. From this we have two separate equations,

$$f''(t) + \omega^2 f(t) = 0, \tag{30.1}$$

$$\nabla^2 g(x, y, z) + \frac{\omega^2}{c^2} g(x, y, z) = 0. \tag{30.2}$$

Equation (30.2) is the *Helmholtz equation*.

Equation (30.1) has for its solutions the functions $f(t) = \cos(\omega t)$ and $\sin(\omega t)$, or, in complex form, the complex exponential functions $f(t) = e^{i\omega t}$ and $f(t) = e^{-i\omega t}$. Functions $u(t, x, y, z) = f(t)g(x, y, z)$ with such time dependence are called *time-harmonic* solutions.

30.6   Planewave Solutions

Suppose that, beginning at time $t = 0$, there is a localized disturbance. As time passes, that disturbance spreads out spherically. When the radius of the sphere is very large, the surface of the sphere appears planar, to an observer on that surface, who is said then to be in the *far field*. This motivates the study of solutions of the wave equation that are constant on planes; these are the so-called *planewave solutions*.

Let $\mathbf{s} = (x, y, z)$ and $u(\mathbf{s}, t) = u(x, y, z, t) = e^{i\omega t}e^{i\mathbf{k}\cdot\mathbf{s}}$. Then we can show that u satisfies the wave equation $u_{tt} = c^2\nabla^2 u$ for any real vector $\mathbf{k}$, so long as $||\mathbf{k}||^2 = \omega^2/c^2$. This solution is a planewave associated with frequency ω and *wavevector* $\mathbf{k}$; at any fixed time the function $u(\mathbf{s}, t)$ is constant on any plane in three-dimensional space having $\mathbf{k}$ as a normal vector.

In radar and sonar, the field $u(\mathbf{s}, t)$ being sampled is usually viewed as a discrete or continuous superposition of planewave solutions with various amplitudes, frequencies, and wavevectors. We sample the field at various spatial locations $\mathbf{s}$, for various times t. Here we simplify the situation a bit by assuming that all the planewave solutions are associated with the same frequency, ω. If not, we can perform an FFT on the functions of time received at each sensor location $\mathbf{s}$ and keep only the value associated with the desired frequency ω.

30.7 Superposition and the Fourier Transform

In the continuous superposition model, the field is

$$u(\mathbf{s}, t) = e^{i\omega t}\int F(\mathbf{k})e^{i\mathbf{k}\cdot\mathbf{s}}d\mathbf{k}.$$

Our measurements at the sensor locations $\mathbf{s}$ give us the values

$$f(\mathbf{s}) = \int F(\mathbf{k})e^{i\mathbf{k}\cdot\mathbf{s}}d\mathbf{k}. \tag{30.3}$$

The data are then Fourier transform values of the complex function $F(\mathbf{k})$; $F(\mathbf{k})$ is defined for all three-dimensional real vectors $\mathbf{k}$, but is zero, in theory, at least, for those $\mathbf{k}$ whose squared length $||\mathbf{k}||^2$ is not equal to ω^2/c^2. Our goal is then to estimate $F(\mathbf{k})$ from measured values of its Fourier transform. Since each $\mathbf{k}$ is a normal vector for its planewave field component, determining the value of $F(\mathbf{k})$ will tell us the strength of the planewave component coming from the direction $\mathbf{k}$.

30.7.1 The Spherical Model

We can imagine that the sources of the planewave fields are the points P that lie on the surface of a large sphere centered at the origin. For each P, the ray from the origin to P is parallel to some wavevector $\mathbf{k}$. The function $F(\mathbf{k})$ can then be viewed as a function $F(P)$ of the points P. Our measurements will be taken at points $\mathbf{s}$ inside this sphere. The radius of the sphere is assumed to be orders of magnitude larger than the distance between sensors. The situation is that of astronomical observation of the heavens using ground-based antennas. The sources of the optical or electromagnetic signals reaching the antennas are viewed as lying on a large sphere surrounding the earth. Distance to the sources is not considered now, and all we are interested in are the amplitudes $F(\mathbf{k})$ of the fields associated with each direction $\mathbf{k}$.

30.8   Sensor Arrays

In some applications the sensor locations are essentially arbitrary, while in others their locations are carefully chosen. Sometimes, the sensors are collinear, as they are in sonar, for towed arrays.

30.8.1   The Two-Dimensional Array

Suppose now that the sensors are in locations $\mathbf{s} = (x, y, 0)$, for various x and y; then we have a *planar array* of sensors. Then the dot product $\mathbf{s} \cdot \mathbf{k}$ that occurs in Equation (30.3) is

$$\mathbf{s} \cdot \mathbf{k} = xk_1 + yk_2;$$

we cannot *see* the third component, k_3. However, since we know the size of the vector $\mathbf{k}$, we can determine $|k_3|$. The only ambiguity that remains is that we cannot distinguish sources on the upper hemisphere from those on the lower one. In most cases, such as astronomy, it is obvious in which hemisphere the sources lie, so the ambiguity is resolved.

The function $F(\mathbf{k})$ can then be viewed as $F(k_1, k_2)$, a function of the two variables k_1 and k_2. Our measurements give us values of $f(x, y)$, the two-dimensional Fourier transform of $F(k_1, k_2)$. Because of the limitation $\|\mathbf{k}\| = \frac{\omega}{c}$, the function $F(k_1, k_2)$ has bounded support. Consequently, its Fourier transform cannot have bounded support. As a result, we can never have all the values of $f(x, y)$, and so we cannot hope to reconstruct $F(k_1, k_2)$ exactly, even for noise-free data.

30.8.2   The One-Dimensional Array

If the sensors are located at points $\mathbf{s}$ having the form $\mathbf{s} = (x, 0, 0)$, then we have a *line array* of sensors. The dot product in Equation (30.3) becomes

$$\mathbf{s} \cdot \mathbf{k} = xk_1.$$

Now the ambiguity is greater than in the planar array case. Once we have k_1, we know that

$$k_2^2 + k_3^2 = (\frac{\omega}{c})^2 - k_1^2,$$

which describes points P lying on a circle on the surface of the distant sphere, with the vector $(k_1, 0, 0)$ pointing at the center of the circle. It is said then that we have a *cone of ambiguity*. One way to resolve the situation is to assume $k_3 = 0$; then $|k_2|$ can be determined and we have remaining only the ambiguity involving the sign of k_2. Once again, in many applications, this remaining ambiguity can be resolved by other means.

Once we have resolved any ambiguity, we can view the function $F(\mathbf{k})$ as $F(k_1)$, a function of the single variable k_1. Our measurements give us values of $f(x)$, the Fourier transform of $F(k_1)$. As in the two-dimensional case, the restriction on the size of the vectors $\mathbf{k}$ means that the function $F(k_1)$ has bounded support. Consequently, its Fourier transform, $f(x)$, cannot have bounded support. Therefore, we shall never have all of $f(x)$, and so we cannot hope to reconstruct $F(k_1)$ exactly, even for noise-free data.

30.8.3 Limited Aperture

In both the one- and two-dimensional problems, the sensors will be placed within some bounded region, such as $|x| \leq A$, $|y| \leq B$ for the two-dimensional problem, or $|x| \leq A$ for the one-dimensional case. These bounded regions are the *apertures* of the arrays. The larger these apertures are, in units of the wavelength, the better the resolution of the reconstructions.

In digital array processing there are only finitely many sensors, which then places added limitations on our ability to reconstruct the field amplitude function $F(\mathbf{k})$.

30.9 The Remote-Sensing Problem

We shall begin our discussion of the remote-sensing problem by considering an extended object transmitting or reflecting a single-frequency, or *narrowband*, signal. The narrowband, extended-object case is a good place to begin, since a point object is simply a limiting case of an extended object, and broadband received signals can always be filtered to reduce their frequency band.

30.9.1   The Solar-Emission Problem

In [16] Bracewell discusses the *solar-emission* problem. In 1942, it was observed that radio-wave emissions in the one-meter wavelength range were arriving from the sun. Were they coming from the entire disk of the sun or were the sources more localized, in sunspots, for example? The problem then was to view each location on the sun's surface as a potential source of these radio waves and to determine the intensity of emission corresponding to each location.

For electromagnetic waves the propagation speed is the speed of light in a vacuum, which we shall take here to be $c = 3 \times 10^8$ meters per second. The wavelength λ for gamma rays is around one Angstrom, which is 10^{-10} meters; for x-rays it is about one millimicron, or 10^{-9} meters. The visible spectrum has wavelengths that are a little less than one micron, that is, 10^{-6} meters. Shortwave radio has a wavelength around one millimeter; microwaves have wavelengths between one centimeter and one meter. Broadcast radio has a λ running from about 10 meters to 1000 meters, while the so-called long radio waves can have wavelengths several thousand meters long.

The sun has an angular diameter of 30 minutes of arc, or one-half of a degree, when viewed from earth, but the needed resolution was more like 3 minutes of arc. As we shall see shortly, such resolution requires a radio telescope 1000 wavelengths across, which means a diameter of 1 kilometer at a wavelength of 1 meter; in 1942 the largest military radar antennas were less than 5 meters across. A solution was found, using the method of reconstructing an object from line-integral data, a technique that surfaced again in tomography. The problem here is inherently two-dimensional, but, for simplicity, we shall begin with the one-dimensional case.

30.10   Sampling

In the one-dimensional case, the signal received at the point $(x, 0, 0)$ is essentially the inverse Fourier transform $f(x)$ of the function $F(k_1)$; for notational simplicity, we write $k = k_1$. The function $F(k)$ is supported on a bounded interval $|k| \leq \frac{\omega}{c}$, so $f(x)$ cannot have bounded support. As we noted earlier, to determine $F(k)$ exactly, we would need measurements of $f(x)$ on an unbounded set. But, which unbounded set?

Because the function $F(k)$ is zero outside the interval $[-\frac{\omega}{c}, \frac{\omega}{c}]$, the function $f(x)$ is *band-limited*. The *Nyquist spacing* in the variable x is therefore

$$\Delta_x = \frac{\pi c}{\omega}.$$

The wavelength λ associated with the frequency ω is defined to be

$$\lambda = \frac{2\pi c}{\omega},$$

so that

$$\Delta_x = \frac{\lambda}{2}.$$

The significance of the Nyquist spacing comes from the *Shannon Sampling Theorem*, which says that if we have the values $f(m\Delta_x)$, for all integers m, then we have enough information to recover $F(k)$ exactly. In practice, of course, this is never the case.

30.11 The Limited-Aperture Problem

In the remote-sensing problem, our measurements at points $(x, 0, 0)$ in the farfield give us the values $f(x)$. Suppose now that we are able to take measurements only for limited values of x, say for $|x| \le A$; then $2A$ is the *aperture* of our antenna or array of sensors. We describe this by saying that we have available measurements of $f(x)h(x)$, where $h(x) = \chi_A(x) = 1$, for $|x| \le A$, and zero otherwise. So, in addition to describing blurring and low-pass filtering, the convolution-filter model can also be used to model the limited-aperture problem. As in the low-pass case, the limited-aperture problem can be attacked using extrapolation, but with the same sort of risks described for the low-pass case. A much different approach is to increase the aperture by physically moving the array of sensors, as in *synthetic aperture radar* (SAR).

Returning to the farfield remote-sensing model, if we have Fourier transform data only for $|x| \le A$, then we have $f(x)$ for $|x| \le A$. Using $h(x) = \chi_A(x)$ to describe the limited aperture of the system, the point-spread function is $H(\gamma) = 2A\mathrm{sinc}(\gamma A)$, the Fourier transform of $h(x)$. The first zeros of the numerator occur at $|\gamma| = \frac{\pi}{A}$, so the main lobe of the point-spread function has width $\frac{2\pi}{A}$. For this reason, the resolution of such a limited-aperture imaging system is said to be on the order of $\frac{1}{A}$. Since $|k| \le \frac{\omega}{c}$, we can write $k = \frac{\omega}{c}\cos\theta$, where θ denotes the angle between the positive x-axis and the vector $\mathbf{k} = (k_1, k_2, 0)$; that is, θ points in the direction of the point P associated with the wavevector $\mathbf{k}$. The resolution, as measured by the width of the main lobe of the point-spread function $H(\gamma)$, in units of k, is $\frac{2\pi}{A}$, but, the angular resolution will depend also on the frequency ω. Since $k = \frac{2\pi}{\lambda}\cos\theta$, a distance of one unit in k may correspond to a large change in θ when ω is small, but only to a relatively small change in θ when ω is large. For this reason, the aperture of the array is usually measured in units of the wavelength; an aperture of $A = 5$ meters may be acceptable if the frequency is high, so that the wavelength is small, but not if the radiation is in the one-meter-wavelength range.

30.12   Resolution

If $F(k) = \delta(k)$ and $h(x) = \chi_A(x)$ describes the aperture-limitation of the imaging system, then the point-spread function is $H(\gamma) = 2A\mathrm{sinc}(\gamma A)$. The maximum of $H(\gamma)$ still occurs at $\gamma = 0$, but the main lobe of $H(\gamma)$ extends from $-\frac{\pi}{A}$ to $\frac{\pi}{A}$; the point source has been spread out. If the point-source object shifts, so that $F(k) = \delta(k - a)$, then the reconstructed image of the object is $H(k - a)$, so the peak is still in the proper place. If we know a priori that the object is a single point source, but we do not know its location, the spreading of the point poses no problem; we simply look for the maximum in the reconstructed image. Problems arise when the object contains several point sources, or when we do not know a priori what we are looking at, or when the object contains no point sources, but is just a continuous distribution.

Suppose that $F(k) = \delta(k - a) + \delta(k - b)$; that is, the object consists of two point sources. Then Fourier transformation of the aperture-limited data leads to the reconstructed image

$$R(k) = 2A\Big(\mathrm{sinc}(A(k - a)) + \mathrm{sinc}(A(k - b))\Big).$$

If $|b - a|$ is large enough, $R(k)$ will have two distinct maxima, at approximately $k = a$ and $k = b$, respectively. For this to happen, we need π/A, the width of the main lobe of the function $\mathrm{sinc}(Ak)$, to be less than $|b - a|$. In other words, to resolve the two point sources a distance $|b - a|$ apart, we need $A \geq \pi/|b - a|$. However, if $|b - a|$ is too small, the distinct maxima merge into one, at $k = \frac{a+b}{2}$ and resolution will be lost. How small is too small will depend on both A and ω.

Suppose now that $F(k) = \delta(k - a)$, but we do not know a priori that the object is a single point source. We calculate

$$R(k) = H(k - a) = 2A\mathrm{sinc}(A(k - a))$$

and use this function as our reconstructed image of the object, for all k. What we see when we look at $R(k)$ for some $k = b \neq a$ is $R(b)$, which is the same thing we see when the point source is at $k = b$ and we look at $k = a$. Point-spreading is, therefore, more than a cosmetic problem. When the object is a point source at $k = a$, but we do not know a priori that it is a point source, the spreading of the point causes us to believe that the object function $F(k)$ is nonzero at values of k other than $k = a$. When we look at, say, $k = b$, we see a nonzero value that is caused by the presence of the point source at $k = a$.

Suppose now that the object function $F(k)$ contains no point sources, but it is simply an ordinary function of k. If the aperture A is very small, then the function $H(k)$ is nearly constant over the entire extent of the object. The convolution of $F(k)$ and $H(k)$ is essentially the integral of $F(k)$, so the reconstructed object is $R(k) = \int F(k)dk$, for all k.

Let's see what this means for the solar-emission problem discussed earlier.

30.12.1   The Solar-Emission Problem Revisited

The wavelength of the radiation is $\lambda = 1$ meter. Therefore, $\frac{\omega}{c} = 2\pi$, and k in the interval $[-2\pi, 2\pi]$ corresponds to the angle θ in $[0, \pi]$. The sun has an angular diameter of 30 minutes of arc, which is about 10^{-2} radians. Therefore, the sun subtends the angles θ in $[\frac{\pi}{2} - (0.5) \cdot 10^{-2}, \frac{\pi}{2} + (0.5) \cdot 10^{-2}]$, which corresponds roughly to the variable k in the interval $[-3 \cdot 10^{-2}, 3 \cdot 10^{-2}]$. Resolution of 3 minutes of arc means resolution in the variable k of $3 \cdot 10^{-3}$. If the aperture is $2A$, then to achieve this resolution, we need

$$\frac{\pi}{A} \leq 3 \cdot 10^{-3},$$

or

$$A \geq \frac{\pi}{3} \cdot 10^3$$

meters, or A not less than about 1000 meters.

The radio-wave signals emitted by the sun are focused, using a parabolic radio-telescope. The telescope is pointed at the center of the sun. Because the sun is a great distance from the earth and the subtended arc is small (30 minutes), the signals from each point on the sun's surface arrive at the parabola nearly head-on, that is, parallel to the line from the vertex to the focal point, and are reflected to the receiver located at the focal point of the parabola. The effect of the parabolic antenna is not to discriminate against signals coming from other directions, since there are none, but to effect a summation of the signals received at points $(x, 0, 0)$, for $|x| \leq A$, where $2A$ is the diameter of the parabola. When the aperture is large, the function $h(x)$ is nearly one for all x and the signal received at the focal point is essentially

$$\int f(x)dx = F(0);$$

we are now able to distinguish between $F(0)$ and other values $F(k)$. When the aperture is small, $h(x)$ is essentially $\delta(x)$ and the signal received at the focal point is essentially

$$\int f(x)\delta(x)dx = f(0) = \int F(k)dk;$$

now all we get is the contribution from all the k, superimposed, and all resolution is lost.

Since the solar emission problem is clearly two-dimensional, and we need 3 minutes of resolution in both dimensions, it would seem that we would need a circular antenna with a diameter of about one kilometer, or a rectangular antenna roughly one kilometer on a side. We shall return to this problem later, once when we discuss multidimensional Fourier transforms, and then again when we consider tomographic reconstruction of images from line integrals.

30.13   Discrete Data

A familiar topic in signal processing is the passage from functions of continuous variables to discrete sequences. This transition is achieved by *sampling*, that is, extracting values of the continuous-variable function at discrete points in its domain. Our example of farfield propagation can be used to explore some of the issues involved in sampling.

Imagine an infinite *uniform line array* of sensors formed by placing receivers at the points $(n\Delta, 0, 0)$, for some $\Delta > 0$ and all integers n. Then our data are the values $f(n\Delta)$. Because we defined $k = \frac{\omega}{c}\cos\theta$, it is clear that the function $F(k)$ is zero for k outside the interval $[-\frac{\omega}{c}, \frac{\omega}{c}]$.

Our discrete array of sensors cannot distinguish between the signal arriving from θ and a signal with the same amplitude, coming from an angle α with

$$\frac{\omega}{c}\cos\alpha = \frac{\omega}{c}\cos\theta + \frac{2\pi}{\Delta}m,$$

where m is an integer. To resolve this ambiguity, we select $\Delta > 0$ so that

$$-\frac{\omega}{c} + \frac{2\pi}{\Delta} \geq \frac{\omega}{c},$$

or

$$\Delta \leq \frac{\pi c}{\omega} = \frac{\lambda}{2}.$$

The sensor spacing $\Delta_s = \frac{\lambda}{2}$ is the Nyquist spacing.

In the sunspot example, the object function $F(k)$ is zero for k outside of an interval much smaller than $[-\frac{\omega}{c}, \frac{\omega}{c}]$. Knowing that $F(k) = 0$ for

$|k| > K$, for some $0 < K < \frac{\omega}{c}$, we can accept ambiguities that confuse θ with another angle that lies outside the angular diameter of the object. Consequently, we can redefine the Nyquist spacing to be

$$\Delta_s = \frac{\pi}{K}.$$

This tells us that when we are imaging a distant object with a small angular diameter, the Nyquist spacing is greater than $\frac{\lambda}{2}$. If our sensor spacing has been chosen to be $\frac{\lambda}{2}$, then we have *oversampled*. In the oversampled case, band-limited extrapolation methods can be used to improve resolution (see [48]).

30.13.1   Reconstruction from Samples

From the data gathered at our infinite array we have extracted the Fourier transform values $f(n\Delta)$, for all integers n. The obvious question is whether or not the data is sufficient to reconstruct $F(k)$. We know that, to avoid ambiguity, we must have $\Delta \leq \frac{\pi c}{\omega}$. The good news is that, provided this condition holds, $F(k)$ is uniquely determined by this data and formulas exist for reconstructing $F(k)$ from the data; this is the content of the Shannon Sampling Theorem. Of course, this is only of theoretical interest, since we never have infinite data. Nevertheless, a considerable amount of traditional signal-processing exposition makes use of this infinite-sequence model. The real problem, of course, is that our data is always finite.

30.14   The Finite-Data Problem

Suppose that we build a uniform line array of sensors by placing receivers at the points $(n\Delta, 0, 0)$, for some $\Delta > 0$ and $n = -N, ..., N$. Then our data are the values $f(n\Delta)$, for $n = -N, ..., N$. Suppose, as previously, that the object of interest, the function $F(k)$, is nonzero only for values of k in the interval $[-K, K]$, for some $0 < K < \frac{\omega}{c}$. Once again, we must have $\Delta \leq \frac{\pi c}{\omega}$ to avoid ambiguity; but this is not enough, now. The finite Fourier data is no longer sufficient to determine a unique $F(k)$. The best we can hope to do is to estimate the true $F(k)$, using both our measured Fourier data and whatever prior knowledge we may have about the function $F(k)$, such as where it is nonzero, if it consists of Dirac delta point sources, or if it is nonnegative. The data is also noisy, and that must be accounted for in the reconstruction process.

In certain applications, such as sonar array processing, the sensors are not necessarily arrayed at equal intervals along a line, or even at the grid points of a rectangle, but in an essentially arbitrary pattern in two, or even three, dimensions. In such cases, we have values of the Fourier transform of the object function, but at essentially arbitrary values of the variable. How best to reconstruct the object function in such cases is not obvious.

30.15   Functions of Several Variables

Fourier transformation applies, as well, to functions of several variables. As in the one-dimensional case, we can motivate the multidimensional Fourier transform using the farfield propagation model. As we noted earlier, the solar emission problem is inherently a two-dimensional problem.

30.15.1   Two-Dimensional Farfield Object

Assume that our sensors are located at points $\mathbf{s} = (x, y, 0)$ in the x,y-plane. As discussed previously, we assume that the function $F(\mathbf{k})$ can be viewed as a function $F(k_1, k_2)$. Since, in most applications, the distant object has a small angular diameter when viewed from a great distance—the sun's is only 30 minutes of arc—the function $F(k_1, k_2)$ will be supported on a small subset of vectors (k_1, k_2).

30.15.2   Limited Apertures in Two Dimensions

Suppose we have the values of the Fourier transform, $f(x, y)$, for $|x| \leq A$ and $|y| \leq A$. We describe this limited-data problem using the function $h(x, y)$ that is one for $|x| \leq A$, and $|y| \leq A$, and zero, otherwise. Then the point-spread function is the Fourier transform of this $h(x, y)$, given by

$$H(\alpha, \beta) = 4AB\text{sinc}(A\alpha)\text{sinc}(B\beta).$$

The resolution in the horizontal (x) direction is on the order of $\frac{1}{A}$, and $\frac{1}{B}$ in the vertical, where, as in the one-dimensional case, aperture is best measured in units of wavelength.

Suppose our aperture is circular, with radius A. Then we have Fourier transform values $f(x, y)$ for $\sqrt{x^2 + y^2} \leq A$. Let $h(x, y)$ equal one, for $\sqrt{x^2 + y^2} \leq A$, and zero, otherwise. Then the point-spread function of this limited-aperture system is the Fourier transform of $h(x, y)$, given by $H(\alpha, \beta) = \frac{2\pi A}{r} J_1(rA)$, with $r = \sqrt{\alpha^2 + \beta^2}$. The resolution of this system is roughly the distance from the origin to the first null of the function $J_1(rA)$, which means that $rA = 4$, roughly.

For the solar emission problem, this says that we would need a circular aperture with radius approximately one kilometer to achieve 3 minutes of arc resolution. But this holds only if the antenna is stationary; a moving antenna is different! The solar emission problem was solved by using a rectangular antenna with a large A, but a small B, and exploiting the rotation of the earth. The resolution is then good in the horizontal, but bad in the vertical, so that the imaging system discriminates well between two distinct vertical lines, but cannot resolve sources within the same vertical line. Because B is small, what we end up with is essentially the integral of the function $f(x, z)$ along each vertical line. By tilting the antenna, and waiting for the earth to rotate enough, we can get these integrals along any set of parallel lines. The problem then is to reconstruct $F(k_1, k_2)$ from such line integrals. This is also the main problem in tomography.

30.16 Broadband Signals

We have spent considerable time discussing the case of a distant point source or an extended object transmitting or reflecting a single-frequency signal. If the signal consists of many frequencies, the so-called broadband case, we can still analyze the received signals at the sensors in terms of time delays, but we cannot easily convert the delays to phase differences, and thereby make good use of the Fourier transform. One approach is to filter each received signal, to remove components at all but a single frequency, and then to proceed as previously discussed. In this way we can process one frequency at a time. The object now is described in terms of a function of both $\mathbf{k}$ and ω, with $F(\mathbf{k}, \omega)$ the complex amplitude associated with the wave vector $\mathbf{k}$ and the frequency ω. In the case of radar, the function $F(\mathbf{k}, \omega)$ tells us how the material at P reflects the radio waves at the various frequencies ω, and thereby gives information about the nature of the material making up the object near the point P.

There are times, of course, when we do not want to decompose a broadband signal into single-frequency components. A satellite reflecting a TV signal is a broadband point source. All we are interested in is receiving the broadband signal clearly, free of any other interfering sources. The direction of the satellite is known and the antenna is turned to face the satellite. Each location on the parabolic dish reflects the same signal. Because of its parabolic shape, the signals reflected off the dish and picked up at the focal point have exactly the same travel time from the satellite, so they combine coherently, to give us the desired TV signal.

31 | Inverse Problems and the Laplace Transform

In the farfield propagation examples considered previously, we found the measured data to be related to the desired object function by a Fourier transformation. The image reconstruction problem then became one of estimating a function from finitely many noisy values of its Fourier transform. In this chapter we consider two inverse problems involving the Laplace transform. The first example is taken from Twomey's book [151].

31.1 The Laplace Transform and the Ozone Layer

The Laplace transform of the function $f(x)$ defined for $0 \leq x < +\infty$ is the function

$$\mathcal{F}(s) = \int_0^{+\infty} f(x)e^{-sx}dx.$$

31.1.1 Scattering of Ultraviolet Radiation

The sun emits ultraviolet (UV) radiation that enters the Earth's atmosphere at an angle θ_0 that depends on the sun's position, and with intensity $I(0)$. Let the x-axis be vertical, with $x = 0$ at the top of the atmosphere and x increasing as we move down to the Earth's surface, at $x = X$. The intensity at x is given by

$$I(x) = I(0)e^{-kx/\cos\theta_0}.$$

Within the ozone layer, the amount of UV radiation scattered in the direction θ is given by

$$S(\theta, \theta_0)I(0)e^{-kx/\cos\theta_0}\Delta p,$$

311

where $S(\theta, \theta_0)$ is a known parameter, and Δp is the change in the pressure of the ozone within the infinitesimal layer $[x, x + \Delta x]$, and so is proportional to the concentration of ozone within that layer.

31.1.2   Measuring the Scattered Intensity

The radiation scattered at the angle θ then travels to the ground, a distance of $X - x$, weakened along the way, and reaches the ground with intensity

$$S(\theta, \theta_0)I(0)e^{-kx/\cos\theta_0}e^{-k(X-x)/\cos\theta}\Delta p.$$

The total scattered intensity at angle θ is then a superposition of the intensities due to scattering at each of the thin layers, and it is then

$$S(\theta, \theta_0)I(0)e^{-kX/\cos\theta_0}\int_0^X e^{-x\beta}dp,$$

where

$$\beta = k\left[\frac{1}{\cos\theta_0} - \frac{1}{\cos\theta}\right].$$

This superposition of intensity can then be written as

$$S(\theta, \theta_0)I(0)e^{-kX/\cos\theta_0}\int_0^X e^{-x\beta}p'(x)dx.$$

31.1.3   The Laplace Transform Data

Using integration by parts, we get

$$\int_0^X e^{-x\beta}p'(x)dx = p(X)e^{-\beta X} - p(0) + \beta\int_0^X e^{-\beta x}p(x)dx.$$

Since $p(0) = 0$ and $p(X)$ can be measured, our data is then the Laplace transform value

$$\int_0^{+\infty} e^{-\beta x}p(x)dx;$$

note that we can replace the upper limit X with $+\infty$ if we extend $p(x)$ as zero beyond $x = X$.

The variable β depends on the two angles θ and θ_0. We can alter θ as we measure and θ_0 changes as the sun moves relative to the earth. In this way we get values of the Laplace transform of $p(x)$ for various values of β. The problem then is to recover $p(x)$ from these values. Because the Laplace transform involves a smoothing of the function $p(x)$, recovering $p(x)$ from its Laplace transform is more ill-conditioned than is the Fourier transform inversion problem.

31.2 The Laplace Transform and Energy Spectral Estimation

In x-ray transmission tomography, x-ray beams are sent through the object and the drop in intensity is measured. These measurements are then used to estimate the distribution of attenuating material within the object. A typical x-ray beam contains components with different energy levels. Because components at different energy levels will be attenuated differently, it is important to know the relative contribution of each energy level to the entering beam. The energy spectrum is the function $f(E)$ that describes the intensity of the components at each energy level $E > 0$.

31.2.1 The Attenuation Coefficient Function

Each specific material, say aluminum, for example, is associated with attenuation coefficients, which is a function of energy, which we shall denote by $\mu(E)$. A beam with the single energy E passing through a thickness x of the material will be weakened by the factor $e^{-\mu(E)x}$. By passing the beam through various thicknesses x of aluminum and registering the intensity drops, one obtains values of the absorption function

$$R(x) = \int_0^\infty f(E)e^{-\mu(E)x}dE. \tag{31.1}$$

Using a change of variable, we can write $R(x)$ as a Laplace transform.

31.2.2 The Absorption Function as a Laplace Transform

For each material, the attenuation function $\mu(E)$ is a strictly decreasing function of E, so $\mu(E)$ has an inverse, which we denote by g; that is, $g(t) = E$, for $t = \mu(E)$. Equation (31.1) can then be rewritten as

$$R(x) = \int_0^\infty f(g(t))e^{-tx}g'(t)dt.$$

We see then that $R(x)$ is the Laplace transform of the function $r(t) = f(g(t))g'(t)$. Our measurements of the intensity drops provide values of $R(x)$, for various values of x, from which we must estimate the functions $r(t)$, and, ultimately, $f(E)$.

32 | Detection and Classification

In some applications of remote sensing, our goal is simply to see what is "out there"; in sonar mapping of the sea floor, the data are the acoustic signals as reflected from the bottom, from which the changes in depth can be inferred. Such problems are *estimation* problems.

In other applications, such as sonar target detection or medical diagnostic imaging, we are looking for certain things, evidence of a surface vessel or submarine, in the sonar case, or a tumor or other abnormality in the medical case. These are *detection* problems. In the sonar case, the data may be used directly in the detection task, or may be processed in some way, perhaps frequency-filtered, prior to being used for detection. In the medical case, or in synthetic-aperture radar (SAR), the data is usually used to construct an image, which is then used for the detection task. In estimation, the goal can be to determine how much of something is present; detection is then a special case, in which we want to decide if the amount present is zero or not.

The detection problem is also a special case of *discrimination*, in which the goal is to decide which of two possibilities is true; in detection the possibilities are simply the presence or absence of the sought-for signal.

More generally, in *classification* or *identification*, the objective is to decide, on the basis of measured data, which of several possibilities is true.

32.1 Estimation

We consider only estimates that are linear in the data, that is, estimates of the form

$$\hat{\gamma} = b^\dagger x = \sum_{n=1}^{N} \overline{b_n} x_n,$$

where $b^\dagger$ denotes the conjugate transpose of the vector $b = (b_1, ..., b_N)^T$. The vector b that we use will be the *best linear unbiased estimator* (BLUE) [48] for the particular estimation problem.

32.1.1   The Simplest Case: A Constant in Noise

We begin with the simplest case, estimating the value of a constant, given several instances of the constant in additive noise. Our data are $x_n = \gamma + q_n$, for $n = 1, ..., N$, where γ is the constant to be estimated, and the q_n are noises. For convenience, we write

$$x = \gamma u + q,$$

where $x = (x_1, ..., x_N)^T$, $q = (q_1, ..., q_N)^T$, $u = (1, ..., 1)^T$, the expected value of the random vector q is $E(q) = 0$, and the covariance matrix of q is $E(qq^T) = Q$. The BLUE employs the vector

$$b = \frac{1}{u^\dagger Q^{-1} u} Q^{-1} u.$$

The BLUE estimate of γ is

$$\hat{\gamma} = \frac{1}{u^\dagger Q^{-1} u} u^\dagger Q^{-1} x.$$

If $Q = \sigma^2 I$, for some $\sigma > 0$, with I the identity matrix, then the noise q is said to be *white*. In this case, the BLUE estimate of γ is simply the average of the x_n.

32.1.2   A Known Signal Vector in Noise

Generalizing somewhat, we consider the case in which the data vector x has the form

$$x = \gamma s + q,$$

where $s = (s_1, ..., s_N)^T$ is a known signal vector. The BLUE estimator is

$$b = \frac{1}{s^\dagger Q^{-1} s} Q^{-1} s$$

and the BLUE estimate of γ is now

$$\hat{\gamma} = \frac{1}{s^\dagger Q^{-1} s} s^\dagger Q^{-1} x.$$

In numerous applications of signal processing, the signal vectors take the form of sampled sinusoids; that is, $s = e_\theta$, with

$$e_\theta = \frac{1}{\sqrt{N}}(e^{-i\theta}, e^{-2i\theta}, ..., e^{-Ni\theta})^T,$$

where θ is a frequency in the interval $[0, 2\pi)$. If the noise is white, then the BLUE estimate of γ is

$$\hat{\gamma} = \frac{1}{\sqrt{N}} \sum_{n=1}^{N} x_n e^{in\theta},$$

which is the *discrete Fourier transform* (DFT) of the data, evaluated at the frequency θ.

32.1.3   Multiple Signals in Noise

Suppose now that the data values are

$$x_n = \sum_{m=1}^{M} \gamma_m s_n^m + q_n,$$

where the signal vectors $s^m = (s_1^m, ..., s_N^m)^T$ are known and we want to estimate the γ_m. We write this in matrix-vector notation as

$$x = Sc + q,$$

where S is the matrix with entries $S_{nm} = s_n^m$, and our goal is to find $c = (\gamma_1, ..., \gamma_M)^T$, the vector of coefficients. The BLUE estimate of the vector c is

$$\hat{c} = (S^\dagger Q^{-1} S)^{-1} S^\dagger Q^{-1} x,$$

assuming that the matrix $S^\dagger Q^{-1} S$ is invertible, in which case we must have $M \leq N$.

If the signals s^m are mutually orthogonal and have length one, then $S^\dagger S = I$; if, in addition, the noise is white, the BLUE estimate of c is $\hat{c} = S^\dagger x$, so that

$$\hat{c}_m = \sum_{n=1}^{N} x_n \overline{s_n^m}.$$

This case arises when the signals are $s^m = e_{\theta_m}$, for $\theta_m = 2\pi m/M$, for $m = 1, ..., M$, in which case the BLUE estimate of c_m is

$$\hat{c}_m = \frac{1}{\sqrt{N}} \sum_{n=1}^{N} x_n e^{2\pi imn/M},$$

the DFT of the data, evaluated at the frequency θ_m. Note that when the frequencies θ_m are not these, the matrix $S^\dagger S$ is not I, and the BLUE estimate is not obtained from the DFT of the data.

32.2 Detection

As we noted previously, the detection problem is a special case of estimation. Detecting the known signal s in noise is equivalent to deciding if the coefficient γ is zero or not. The procedure is to calculate $\hat{\gamma}$, the BLUE estimate of γ, and say that s has been detected if $|\hat{\gamma}|$ exceeds a certain threshold. In the case of multiple known signals, we calculate $\hat{c}$, the BLUE estimate of the coefficient vector c, and base our decisions on the magnitudes of each entry of $\hat{c}$.

32.2.1 Parameterized Signal

It is sometimes the case that we know that the signal s we seek to detect is a member of a parametrized family, $\{s_\theta | \theta \in \Theta\}$, of potential signal vectors, but we do not know the value of the parameter θ. For example, we may be trying to detect a sinusoidal signal, $s = e_\theta$, where θ is an unknown frequency in the interval $[0, 2\pi)$. In sonar direction-of-arrival estimation, we seek to detect a farfield point source of acoustic energy, but we do not know the direction of the source. The BLUE estimator can be extended to these cases, as well [48]. For each fixed value of the parameter θ, we estimate γ using the BLUE, obtaining the estimate

$$\hat{\gamma}(\theta) = \frac{1}{s_\theta^\dagger Q^{-1} s_\theta} s_\theta^\dagger Q^{-1} x,$$

which is then a function of θ. If the maximum of the magnitude of this function exceeds a specified threshold, then we may say that there is a signal present corresponding to that value of θ.

Another approach would be to extend the model of multiple signals to include a continuum of possibilities, replacing the finite sum with an integral. Then the model of the data becomes

$$x = \int_{\theta \in \Theta} \gamma(\theta) s_\theta \, d\theta + q.$$

Let S now denote the integral operator

$$S(\gamma) = \int_{\theta \in \Theta} \gamma(\theta) s_\theta \, d\theta$$

that transforms a function γ of the variable θ into a vector. The adjoint operator, $S^\dagger$, transforms any N-vector v into a function of θ, according to

$$S^\dagger(v)(\theta) = \sum_{n=1}^{N} v_n \overline{(s_\theta)_n} = s_\theta^\dagger v \,.$$

Consequently, $S^\dagger Q^{-1} S$ is the function of θ given by

$$g(\theta) = (S^\dagger Q^{-1} S)(\theta) = \sum_{n=1}^{N} \sum_{j=1}^{N} Q_{nj}^{-1} (s_\theta)_j \overline{(s_\theta)_n},$$

so

$$g(\theta) = s_\theta^\dagger Q^{-1} s_\theta.$$

The generalized BLUE estimate of $\gamma(\theta)$ is then

$$\hat{\gamma}(\theta) = \frac{1}{g(\theta)} \sum_{j=1}^{N} a_j \overline{(s_\theta)_j} = \frac{1}{g(\theta)} s_\theta^\dagger a \,,$$

where $x = Qa$ or

$$x_n = \sum_{j=1}^{N} a_j Q_{nj},$$

for $j = 1, ..., N$, and so $a = Q^{-1} x$. This is the same estimate we obtained in the previous paragraph. The only difference is that, in the first case, we assume that there is only one signal active, and apply the BLUE for each fixed θ, looking for the one most likely to be active. In the second case, we choose to view the data as a noisy superposition of a continuum of the s_θ, not just one. The resulting estimate of $\gamma(\theta)$ describes how each of the individual signal vectors s_θ contribute to the data vector x. Nevertheless, the calculations we perform are the same.

If the noise is white, we have $a_j = x_j$ for each j. The function $g(\theta)$ becomes

$$g(\theta) = \sum_{n=1}^{N} |(s_\theta)_n|^2,$$

which is simply the square of the length of the vector s_θ. If, in addition, the signal vectors all have length one, then the estimate of the function $\gamma(\theta)$ becomes

$$\hat{\gamma}(\theta) = \sum_{n=1}^{N} x_n \overline{(s_\theta)_n} = s_\theta^\dagger x.$$

Finally, if the signals are sinusoids $s_\theta = e_\theta$, then

$$\hat{\gamma}(\theta) = \frac{1}{\sqrt{N}} \sum_{n=1}^{N} x_n e^{in\theta},$$

again, the DFT of the data vector.

32.3 Discrimination

The problem now is to decide if the data is $x = s^1 + q$ or $x = s^2 + q$, where s^1 and s^2 are known vectors. This problem can be converted into a detection problem: Do we have $x - s^1 = q$ or $x - s^1 = s^2 - s^1 + q$? Then the BLUE involves the vector $Q^{-1}(s^2 - s^1)$ and the discrimination is made based on the quantity $(s^2 - s^1)^\dagger Q^{-1} x$. If this quantity is near enough to zero we say that the signal is s^1; otherwise, we say that it is s^2. The BLUE in this case is sometimes called the *Hotelling linear discriminant*, and a procedure that uses this method to perform medical diagnostics is called a *Hotelling observer*.

More generally, suppose we want to decide if a given vector x comes from class C_1 or from class C_2. If we can find a vector b such that $b^T x > a$ for every x that comes from C_1, and $b^T x < a$ for every x that comes from C_2, then the vector b is a linear discriminant for deciding between the classes C_1 and C_2.

32.3.1 Channelized Observers

The N by N matrix Q can be quite large, particularly when x and q are vectorizations of two-dimensional images. If, in additional, the matrix Q is obtained from K observed instances of the random vector q, then for Q to be invertible, we need $K \geq N$. To avoid these and other difficulties, the *channelized* Hotelling linear discriminant is often used. The idea here is to replace the data vector x with Ux for an appropriately chosen J by N matrix U, with J much smaller than N; the value $J = 3$ is used by Gifford et al. in [91], with the channels chosen to capture image information within selected frequency bands.

32.3.2 An Example of Discrimination

Suppose that there are two groups of students, the first group denoted G_1, the second G_2. The math SAT scores for the students in G_1 are always above 500, while their verbal scores are always below 500. For the students in G_2 the opposite is true; the math scores are below 500, the verbal above. For each student we create the two-dimensional vector $x = (x_1, x_2)^T$ of SAT scores, with x_1 the math score, x_2 the verbal score. Let $b = (1, -1)^T$. Then for every student in G_1 we have $b^T x > 0$, while for those in G_2, we have $b^T x < 0$. Therefore, the vector b provides a linear discriminant.

Suppose we have a third group, G_3, whose math scores and verbal scores are both below 500. To discriminate between members of G_1 and G_3 we can use the vector $b = (1, 0)^T$ and $a = 500$. To discriminate between the groups G_2 and G_3, we can use the vector $b = (0, 1)^T$ and $a = 500$.

Now suppose that we want to decide from which of the three groups the vector x comes; this is classification.

32.4   Classification

The classification problem is to determine to which of several classes of vectors a given vector x belongs. For simplicity, we assume all vectors are real. The simplest approach to solving this problem is to seek linear discriminant functions; that is, for each class we want to have a vector b with the property that $b^T x > 0$ if and only if x is in the class. If the vectors x are randomly distributed according to one of the parametrized family of probability density functions (pdfs) $p(x; \omega)$ and the ith class corresponds to the parameter value ω_i, then we can often determine the discriminant vectors b^i from these pdfs. In many cases, however, we do not have the pdf and the b^i must be estimated through a learning or training step before they are used on as yet unclassified data vectors. In the discussion that follows we focus on obtaining b for one class, suppressing the index i.

32.4.1   The Training Stage

In the training stage a candidate for b is tested on vectors whose class membership is known, say $\{x^1, ..., x^M\}$. First, we replace each vector x^m that is not in the class with its negative. Then we seek b such that $b^T x^m > 0$ for all m. With A the matrix whose mth row is $(x^m)^T$ we can write the problem as $Ab > 0$. If the b we obtain has some entries very close to zero it might not work well enough on actual data; it is often better, then, to take a vector ϵ with small positive entries and require $Ab \geq \epsilon$. When we have found b for each class we then have the machinery to perform the classification task.

There are several problems to be overcome, obviously. The main one is that there may not be a vector b for each class; the problem $Ab \geq \epsilon$ need not have a solution. In classification this is described by saying that the vectors x^m are not linearly separable [81]. The second problem is finding the b for each class; we need an algorithm to solve $Ab \geq \epsilon$.

One approach to designing an algorithm for finding b is the following: for arbitrary b let $f(b)$ be the number of the x^m misclassified by vector b. Then minimize $f(b)$ with respect to b. Alternatively, we can minimize the function $g(b)$ defined to be the sum of the values $-b^T x^m$, taken over all the x^m that are misclassified; the $g(b)$ has the advantage of being continuously valued. The batch Perceptron algorithm [81] uses gradient descent methods to minimize $g(b)$. Another approach is to use the Agmon-Motzkin-Schoenberg (AMS) Algorithm to solve the system of linear inequalities $Ab \geq \epsilon$ [48].

When the training set of vectors is linearly separable, the batch Perceptron and the AMS Algorithm converge to a solution, for each class. When the training vectors are not linearly separable there will be a class for which the problem $Ab \geq \epsilon$ will have no solution. Iterative algorithms in this case cannot converge to a solution. Instead, they may converge to an approximate solution or, as with the AMS algorithm, converge subsequentially to a limit cycle of more than one vector.

32.4.2   Our Example Again

We return to the example given earlier, involving the three groups of students and their SAT scores. To be consistent with the conventions of this section, we define $x = (x_1, x_2)^T$ differently now. Let x_1 be the math SAT score, minus 500, and x_2 be the verbal SAT score, minus 500. The vector $b = (1, 0)^T$ has the property that $b^T x > 0$ for each x coming from G_1, but $b^T x < 0$ for each x not coming from G_1. Similarly, the vector $b = (0, 1)^T$ has the property that $b^T x > 0$ for all x coming from G_2, while $b^T x < 0$ for all x not coming from G_2. However, there is no vector b with the property that $b^T x > 0$ for x coming from G_3, but $b^T x < 0$ for all x not coming from G_3; the group G_3 is not linearly separable from the others. Notice, however, that if we perform our classification sequentially, we can employ linear classifiers. First, we use the vector $b = (1, 0)^T$ to decide if the vector x comes from G_1 or not. If it does, fine; if not, then use vector $b = (0, 1)^T$ to decide if it comes from G_2 or G_3.

32.5   More Realistic Models

In many important estimation and detection problems, the signal vector s is not known precisely. In medical diagnostics, we may be trying to detect a lesion, and may know it when we see it, but may not be able to describe it using a single vector s, which now would be a vectorized image. Similarly, in discrimination or classification problems, we may have several examples of each type we wish to identify, but we will be unable to reduce these types to single representative vectors. We now have to derive an analog of the BLUE that is optimal with respect to the examples that have been presented for training. The linear procedure we seek will be one that has performed best, with respect to a training set of examples. The *Fisher linear discriminant* is an example of such a procedure.

32.5.1 The Fisher Linear Discriminant

Suppose that we have available for training K vectors $x^1, ..., x^K$ in R^N, with vectors $x^1, ..., x^J$ in the class A, and the remaining $K - J$ vectors in the class B. Let w be an arbitrary vector of length one, and for each k let $y_k = w^T x^k$ be the projected data. The numbers y_k, $k = 1, ..., J$, form the set Y_A, the remaining ones the set Y_B. Let

$$\mu_A = \frac{1}{J} \sum_{k=1}^{J} x^k,$$

$$\mu_B = \frac{1}{K - J} \sum_{k=J+1}^{K} x^k,$$

$$m_A = \frac{1}{J} \sum_{k=1}^{J} y_k = w^T \mu_A,$$

$$m_B = \frac{1}{K - J} \sum_{k=J+1}^{K} y_k = w^T \mu_B.$$

Let

$$\sigma_A^2 = \sum_{k=1}^{J} (y_k - m_A)^2,$$

$$\sigma_B^2 = \sum_{k=J+1}^{K} (y_k - m_B)^2.$$

The quantity $\sigma^2 = \sigma_A^2 + \sigma_B^2$ is the *total within-class scatter* of the projected data. Define the function $F(w)$ to be

$$F(w) = \frac{(m_A - m_B)^2}{\sigma^2}.$$

The *Fisher linear discriminant* is the vector w for which $F(w)$ achieves its maximum.

Define the scatter matrices S_A and S_B as follows:

$$S_A = \sum_{k=1}^{J} (x^k - \mu_A)(x^k - \mu_A)^T,$$

and

$$S_B = \sum_{k=J+1}^{K} (x^k - \mu_B)(x^k - \mu_B)^T.$$

Then

$$S_{\text{within}} = S_A + S_B$$

is the *within-class scatter matrix* and

$$S_{\text{between}} = (\mu_A - \mu_B)(\mu_A - \mu_B)^T$$

is the *between-class scatter matrix*. The function $F(w)$ can then be written as

$$F(w) = w^T S_{\text{between}} w / w^T S_{\text{within}} w.$$

The w for which $F(w)$ achieves its maximum value is then

$$w = S_{\text{within}}^{-1}(\mu_A - \mu_B).$$

This vector w is the Fisher linear discriminant. When a new data vector x is obtained, we decide to which of the two classes it belongs by calculating the dot product $w^T x$.

IX | Appendices

A | Bregman-Legendre Functions

In [10] Bauschke and Borwein show convincingly that the Bregman-Legendre functions provide the proper context for the discussion of Bregman projections onto closed convex sets. The summary here follows closely the discussion given in [10].

A.1 Essential Smoothness and Essential Strict Convexity

Following [139] we say that a closed proper convex function f is *essentially smooth* if int$D$ is not empty, f is differentiable on int$D$, and $x^n \in$ int$D$, with $x^n \to x \in$ bdry$D$, implies that $||\nabla f(x^n)|| \to +\infty$. Here int$D$ and bdry$D$ denote the interior and boundary of the set D. A closed proper convex function f is *essentially strictly convex* if f is strictly convex on every convex subset of dom ∂f.

The closed proper convex function f is essentially smooth if and only if the subdifferential $\partial f(x)$ is empty for $x \in$ bdry$D$ and is $\{\nabla f(x)\}$ for $x \in$ int$D$ (so f is differentiable on int$D$) if and only if the function f^* is essentially strictly convex.

Definition A.1. A closed proper convex function f is said to be a Legendre function if it is both essentially smooth and essentially strictly convex.

So f is Legendre if and only if its conjugate function is Legendre, in which case the gradient operator ∇f is a topological isomorphism with ∇f^* as its inverse. The gradient operator ∇f maps int dom f onto int dom f^*. If int dom $f^* = R^J$, then the range of ∇f is R^J and the equation $\nabla f(x) = y$ can be solved for every $y \in R^J$. In order for int dom $f^* = R^J$ it is necessary and sufficient that the Legendre function f be *super-coercive*, that is,

$$\lim_{||x|| \to +\infty} \frac{f(x)}{||x||} = +\infty.$$

If the effective domain of f is bounded, then f is super-coercive and its gradient operator is a mapping onto the space R^J.

A.2   Bregman Projections onto Closed Convex Sets

Let f be a closed proper convex function that is differentiable on the nonempty set int$D$. The corresponding *Bregman distance* $D_f(x, z)$ is defined for $x \in R^J$ and $z \in$ int$D$ by

$$D_f(x, z) = f(x) - f(z) - \langle \nabla f(z), x - z \rangle.$$

Note that $D_f(x, z) \geq 0$ always and that $D_f(x, z) = +\infty$ is possible. If f is essentially strictly convex, then $D_f(x, z) = 0$ implies that $x = z$.

Let K be a nonempty closed convex set with $K \cap$ int$D \neq \emptyset$. Pick $z \in$ int$D$. The *Bregman projection* of z onto K, with respect to f, is

$$P_K^f(z) = \operatorname{argmin}_{x \in K \cap D} D_f(x, z).$$

If f is essentially strictly convex, then $P_K^f(z)$ exists. If f is strictly convex on D, then $P_K^f(z)$ is unique. If f is Legendre, then $P_K^f(z)$ is uniquely defined and is in int$D$; this last condition is sometimes called *zone consistency*.

Example A.2. Let $J = 2$ and $f(x)$ be the function that is equal to one-half the norm squared on D, the nonnegative quadrant, $+\infty$ elsewhere. Let K be the set $K = \{(x_1, x_2) | x_1 + x_2 = 1\}$. The Bregman projection of $(2, 1)$ onto K is $(1, 0)$, which is not in int$D$. The function f is not essentially smooth, although it is essentially strictly convex. Its conjugate is the function f^* that is equal to one-half the norm squared on D and equal to zero elsewhere; it is essentially smooth, but not essentially strictly convex.

If f is Legendre, then $P_K^f(z)$ is the unique member of $K \cap$ int$D$ satisfying the inequality

$$\langle \nabla f(P_K^f(z)) - \nabla f(z), P_K^f(z) - c \rangle \geq 0,$$

for all $c \in K$. From this we obtain *Bregman's Inequality*:

$$D_f(c, z) \geq D_f(c, P_K^f(z)) + D_f(P_K^f(z), z), \tag{A.1}$$

for all $c \in K$.

A.3   Bregman-Legendre Functions

Following Bauschke and Borwein [10], we say that a Legendre function f is a *Bregman-Legendre* function if the following properties hold.

(1) For x in D and any $a > 0$ the set $\{z | D_f(x, z) \le a\}$ is bounded;

(2) If x is in D but not in int$D$, for each positive integer n, y^n is in int$D$ with $y^n \to y \in$ bd$D$ and if $\{D_f(x, y^n)\}$ remains bounded, then $D_f(y, y^n) \to 0$, so that $y \in D$;

(3) If x^n and y^n are in int$D$, with $x^n \to x$ and $y^n \to y$, where x and y are in D but not in int$D$, and if $D_f(x^n, y^n) \to 0$ then $x = y$.

Bauschke and Borwein then prove that Bregman's SGP method converges to a member of K provided that one of the following holds:

(1) f is Bregman-Legendre;

(2) $K \cap$ int$D \ne \emptyset$ and dom f^* is open;

(3) dom f and dom f^* are both open.

A.4 Useful Results about Bregman-Legendre Functions

The following results are proved in somewhat more generality in [10].

(1) If $y^n \in$ int dom f, and $y^n \to y \in$ int dom f, then $D_f(y, y^n) \to 0$.

(2) If x and $y^n \in$ int dom f, and $y^n \to y \in$ bdry dom f, then $D_f(x, y^n) \to +\infty$.

(3) If $x^n \in D$, $x^n \to x \in D$, $y^n \in$ int D, $y^n \to y \in D$, $\{x, y\} \cap$ int $D \ne \emptyset$, and $D_f(x^n, y^n) \to 0$, then $x = y$ and $y \in$ int D.

(4) If x and y are in D, but are not in int D, $y^n \in$ int D, $y^n \to y$, and $D_f(x, y^n) \to 0$, then $x = y$.

As a consequence of these results we have the following.

(5) If $\{D_f(x, y^n)\} \to 0$, for $y^n \in$ int D and $x \in R^J$, then $\{y^n\} \to x$.

Proof (of Result 5): Since $\{D_f(x, y^n)\}$ is eventually finite, we have $x \in D$. By Property 1 above it follows that the sequence $\{y^n\}$ is bounded; without loss of generality, we assume that $\{y^n\} \to y$, for some $y \in \overline{D}$. If x is in int D, then, by Result 2 above, we know that y is also in int D. Applying Result 3, with $x^n = x$, for all n, we conclude that $x = y$. If, on the other hand, x is in D, but not in int D, then y is in D, by Result 2. There are two cases to consider: (1) y is in int D; (2) y is not in int D. In case (1) we have $D_f(x, y^n) \to D_f(x, y) = 0$, from which it follows that $x = y$. In case (2) we apply Result 4 to conclude that $x = y$.

B | Bregman-Paracontractive Operators

In a previous chapter, we considered operators that are paracontractive, with respect to some norm. In this chapter, we extend that discussion to operators that are paracontractive, with respect to some Bregman distance. Our objective here is to examine the extent to which the EKN Theorem 5.21 and its consequences can be extended to the broader class of Bregman paracontractions. Typically, these operators are not defined on all of $\mathcal{X}$, but on a restricted subset, such as the nonnegative vectors, in the case of entropy.

B.1   Bregman Paracontractions

Let f be a closed proper convex function that is differentiable on the nonempty set $\text{int}D$. The corresponding *Bregman distance* $D_f(x, z)$ is defined for $x \in R^J$ and $z \in \text{int}D$ by

$$D_f(x, z) = f(x) - f(z) - \langle \nabla f(z), x - z \rangle,$$

where $D = \{x \mid f(x) < +\infty\}$ is the essential domain of f. When the domain of f is not all of R^J, we define $f(x) = +\infty$, for x outside its domain. Note that $D_f(x, z) \geq 0$ always and that $D_f(x, z) = +\infty$ is possible. If f is essentially strictly convex, then $D_f(x, z) = 0$ implies that $x = z$.

Let C be a nonempty closed convex set with $C \cap \text{int}D \neq \emptyset$. Pick $z \in \text{int}D$. The *Bregman projection* of z onto C, with respect to f, is

$$P_C^f(z) = \text{argmin}_{x \in C \cap D} D_f(x, z).$$

If f is essentially strictly convex, then $P_C^f(z)$ exists. If f is strictly convex on D then $P_C^f(z)$ is unique. We assume that f is Legendre, so that $P_C^f(z)$

is uniquely defined and is in $\text{int}D$; this last condition is sometimes called *zone consistency*.

We shall make much use of *Bregman's Inequality* (A.1):

$$D_f(c, z) \geq D_f(c, P_C^f z) + D_f(P_C^f z, z). \tag{B.1}$$

Definition B.1. A continuous operator $T : \text{int}D \to \text{int}D$ is called a *Bregman paracontraction* (BPC) if, for every fixed point z of T, and for every x, we have

$$D_f(z, Tx) < D_f(z, x),$$

unless $Tx = x$.

In order for the Bregman distances $D_f(z, x)$ and $D_f(z, Tx)$ to be defined, it is necessary that $\nabla f(x)$ and $\nabla f(Tx)$ be defined, and so we need to restrict the domain and range of T in the manner above. This can sometimes pose a problem, when the iterative sequence $\{x^{k+1} = Tx^k\}$ converges to a point on the boundary of the domain of f. This happens, for example, in the EMML and SMART methods, in which each x^k is a positive vector, but the limit can have entries that are zero. One way around this problem is to extend the notion of a fixed point: say that z is an *asymptotic fixed point* of T if (z, z) is in the closure of the graph of T, that is, (z, z) is the limit of points of the form (x, Tx). Theorems for iterative methods involving Bregman paracontractions can then be formulated to involve convergence to an asymptotic fixed point [41]. In our discussion here, however, we shall not consider this more general situation.

B.1.1 Entropic Projections

As an example of a Bregman distance and Bregman paracontractions, consider the function $g(t) = t \log(t) - t$, with $g(0) = 0$, and the associated Bregman-Legendre function

$$f(x) = \sum_{j=1}^{J} g(x_j),$$

defined for vectors x in the nonnegative cone R_+^J. The corresponding Bregman distance is the Kullback-Leibler, or cross-entropy, distance

$$D_f(x, z) = f(x) - f(z) - \langle \nabla f(z), x - z \rangle = KL(x, z).$$

For any nonempty, closed, convex set C, the *entropic projection* operator P_C^e is defined such that $P_C^e z$ is the member x of $C \cap R_+^J$ for which $KL(x, z)$ is minimized.

Theorem B.2. *The operator $T = P_C^e$ is BPC, with respect to the cross-entropy distance.*

Proof: The fixed points of $T = P_C^e$ are the vectors c in $C \cap R_+^J$. From Bregman's Inequality (B.1) we have

$$D_f(c, x) - D_f(c, P_C^e x) \geq D_f(P_C^e x, x) \geq 0,$$

with equality if and only if $D_f(P_C^e x, x) = 0$, in which case $Tx = x$. $\qquad \square$

B.1.2 Weighted Entropic Projections

Generally, we cannot exhibit the entropic projection onto a closed, convex set C in closed form. When we consider the EMML and SMART algorithms, we shall focus on nonnegative systems $Ax = b$, in which the entries of A are nonnegative, those of b are positive, and we seek a nonnegative solution. For each $i = 1, ..., I$, let

$$H_i = \{x \geq 0 | (Ax)_i = b_i\}.$$

We cannot write the entropic projection of z onto H_i in closed form, but, for each positive vector z, the member of H_i that minimizes the weighted cross-entropy,

$$\sum_{j=1}^{J} A_{ij} KL(x_j, z_j) \qquad\qquad (B.2)$$

is

$$x_j = (Q_i^e z)_j = z_j \frac{b_i}{(Az)_i}.$$

Lemma B.3. *The operator Q_i^e is BPC, with respect to the Bregman distance in Equation (B.2).*

Proof: For each x in H_i,

$$\sum_{j=1}^{J} A_{ij} KL(x_j, z_j) - \sum_{j=1}^{J} A_{ij} KL(x_j, (Q_i^e z)_j) = KL(b_i, (Az)_i). \qquad \square$$

With $\sum_{i=1}^{I} A_{ij} = 1$, for each j, the iterative step of the EMML algorithm can be written as $x^{k+1} = Tx^k$, for

$$(Tx)_j = \sum_{i=1}^{I} A_{ij} (Q_i^e x)_j,$$

and that of the SMART is $x^{k+1} = Tx^k$, for

$$(Tx)_j = \prod_{i=1}^{I}[(Q_i^e x)_j]^{A_{ij}}.$$

It follows from the theory of these two algorithms that, in both cases, T is BPC, with respect to the cross-entropy distance.

B.2   Extending the EKN Theorem

Now we present a generalization of the EKN Theorem.

Theorem B.4. *For $i = 1, ..., I$, let T_i be BPC, for the Bregman distance D_f. Let $F = \cap_{i=1}^{I}\text{Fix}(T_i)$ be nonempty. Let $i(k) = k(\text{mod } I) + 1$ and $x^{k+1} = T_{i(k)}x^k$. Then the sequence $\{x^k\}$ converges to a member of F.*

Proof: Let z be a member of F. We know that

$$D_f(z, x^k) - D_f(z, x^{k+1}) \geq 0,$$

so that the sequence $\{D_f(z, x^k\}$ is decreasing, with limit $d \geq 0$. Then the sequence $\{x_k\}$ is bounded; select a cluster point, x^*. Then T_1x^* is also a cluster point, so we have

$$D_f(z, x) - D_f(z, T_1x) = 0,$$

from which we conclude that $T_1x = x$. Similarly, $T_2T_1x^* = T_2x^*$ is a cluster point, and $T_2x^* = x^*$. Continuing in this manner, we show that x^* is in F. Then $\{D_f(x^*, x^k)\} \to 0$, so that $\{x^k\} \to x^*$. □

We have the following generalization of Corollary 5.25:

Corollary B.5. *For $i = 1, ..., I$, let T_i be BPC, for the Bregman distance D_f. Let $F = \cap_{i=1}^{I}\text{Fix}(T_i)$ be nonempty. Let $T = T_I T_{I-1} \cdots T_2 T_1$. Then the sequence $\{T^k x_0\}$ converges to a member of F.*

Proof: Let z be in F. Since $D_f(z, T_i x) \leq D_f(z, x)$, for each i, it follows that

$$D_f(z, x) - D_f(z, Tx) \geq 0.$$

If equality holds, then

$$D_f(z, (T_I T_{I-1} \cdots T_1)x) = D_f(z, (T_{I-1} \cdots T_1)x) \ldots$$
$$= D_f(z, T_1 x) = D_f(z, x),$$

from which we can conclude that $T_i x = x$, for each i. Therefore, $Tx = x$, and T is BPC. □

Corollary B.6. *If F is not empty, then $F = \text{Fix}(T)$.*

B.3 Multiple Bregman Distances

We saw earlier that both the EMML and the SMART algorithms involve Bregman projections with respect to distances that vary with the sets $C_i = H_i$. This suggests that Theorem B.4 could be extended to include continuous operators T_i that are BPC, with respect to Bregman distances D_{f_i} that vary with i. However, there is a counter-example in [46] that shows that the sequence $\{x^{k+1} = T_{i(k)}x^k\}$ need not converge to a fixed point of T. The problem is that we need some Bregman distance D_h that is independent of i, with $\{D_h(z, x^k\}$ decreasing. The result we present now is closely related to the MSGP algorithm.

B.3.1 Assumptions and Notation

We make the following assumptions throughout this section. The function h is super-coercive and Bregman-Legendre with essential domain $D = \operatorname{dom} h$. For $i = 1, 2, ..., I$ the function f_i is also Bregman-Legendre, with $D \subseteq \operatorname{dom} f_i$, so that $\operatorname{int} D \subseteq \operatorname{int} \operatorname{dom} f_i$. For all $x \in \operatorname{dom} h$ and $z \in \operatorname{int} \operatorname{dom} h$ we have $D_h(x, z) \geq D_{f_i}(x, z)$, for each i.

B.3.2 The Algorithm

The *multidistance* extension of Theorem B.4 concerns the algorithm with the following iterative step:

$$x^{k+1} = \nabla h^{-1}\Big(\nabla h(x^k) - \nabla f_{i(k)}(x^k) + \nabla f_{i(k)}(T_{i(k)}(x^k))\Big). \tag{B.3}$$

B.3.3 A Preliminary Result

For each $k = 0, 1, ...$ define the function $G^k(\cdot) : \operatorname{dom} h \to [0, +\infty)$ by

$$G^k(x) = D_h(x, x^k) - D_{f_{i(k)}}(x, x^k) + D_{f_{i(k)}}(x, T_{i(k)}(x^k)).$$

The next proposition provides a useful identity, which can be viewed as an analogue of Pythagoras' theorem. The proof is not difficult and we omit it.

Proposition B.7. *For each $x \in \operatorname{dom} h$, each $k = 0, 1, ...,$ and x^{k+1} given by Equation (B.3) we have*

$$G^k(x) = G^k(x^{k+1}) + D_h(x, x^{k+1}). \tag{B.4}$$

Consequently, x^{k+1} is the unique minimizer of the function $G^k(\cdot)$.

This identity (B.4) is the key ingredient in the proof of convergence of the algorithm.

B.3.4 Convergence of the Algorithm

We shall prove the following convergence theorem:

Theorem B.8. *Let F be nonempty. Let $x^0 \in \text{int dom } h$ be arbitrary. Any sequence x^k obtained from the iterative scheme given by Equation (B.3) converges to $x^\infty \in F \cap \text{dom } h$.*

Proof: Let z be in F. Then it can be shown that

$$D_h(z, x^k) - D_h(z, x^{k+1}) = G^k(x^{k+1}) + D_{f_i}(z, x^k) - D_{f_i}(z, T_{i(k)}x^k).$$

Therefore, the sequence $\{D_h(z, x^k)\}$ is decreasing, and the nonnegative sequences $\{G^k(x^{k+1})\}$ and $\{D_{f_i}(z, x^k) - D_{f_i}(z, T_{i(k)}x^k)\}$ converge to zero. The sequence $\{x^{mI}\}$ is then bounded and we can select a subsequence $\{x^{m_n I}\}$ with limit point $x^{*,0}$. Since the sequence $\{x^{m_n I+1}\}$ is bounded, it has a subsequence with limit $x^{*,1}$. But, since

$$D_{f_1}(z, x^{m_n I}) - D_{f_1}(z, x^{m_n I+1}) \to 0,$$

we conclude that $T_1 x^{*,0} = x^{*,0}$. Continuing in this way, we eventually establish that $T_i x^{*,0} = x^{*,0}$, for each i. So, $x^{*,0}$ is in F. Using $x^{*,0}$ in place of z, we find that $\{D_h(x^{*,0}, x^k)\}$ is decreasing; but a subsequence converges to zero, so the entire sequence converges to zero, and $\{x^k\} \to x^{*,0}$.   □

C | The Fourier Transform

The Fourier Transform plays an important role in several of the applications we consider in this book. For the convenience of the reader, we review here the basic properties of the Fourier transform.

C.1   Fourier-Transform Pairs

Let $f(x)$ be defined for the real variable x in $(-\infty, \infty)$. The *Fourier transform* of $f(x)$ is the function of the real variable γ given by

$$F(\gamma) = \int_{-\infty}^{\infty} f(x)e^{i\gamma x}dx. \tag{C.1}$$

Precisely how we interpret the infinite integrals that arise in the discussion of the Fourier transform will depend on the properties of the function $f(x)$. A detailed treatment of this issue, which is beyond the scope of this book, can be found in almost any text on the Fourier transform (see, for example, [89]).

C.1.1   Reconstructing from Fourier-Transform Data

Our goal is often to reconstruct the function $f(x)$ from measurements of its Fourier transform $F(\gamma)$. But, how?

If we have $F(\gamma)$ for all real γ, then we can recover the function $f(x)$ using the *Fourier Inversion Formula*:

$$f(x) = \frac{1}{2\pi} \int_{-\infty}^{\infty} F(\gamma)e^{-i\gamma x}d\gamma. \tag{C.2}$$

The functions $f(x)$ and $F(\gamma)$ are called a *Fourier-transform pair*. Once again, the proper interpretation of Equation (C.2) will depend on the properties of the functions involved. If both $f(x)$ and $F(\gamma)$ are measurable and absolutely integrable, then both functions are continuous. To illustrate some of the issues involved, we consider the functions in the Schwartz class [89]

C.1.2 Functions in the Schwartz class

A function $f(x)$ is said to be in the *Schwartz class*, or to be a *Schwartz function*, if $f(x)$ is infinitely differentiable and

$$|x|^m f^{(n)}(x) \to 0$$

as x goes to $-\infty$ and $+\infty$. Here $f^{(n)}(x)$ denotes the nth derivative of $f(x)$. An example of a Schwartz function is $f(x) = e^{-x^2}$, with Fourier transform $F(\gamma) = \sqrt{\pi}e^{-\gamma^2/4}$. The following proposition tells us that Schwartz functions are absolutely integrable on the real line, and so the Fourier transform is well defined.

Proposition C.1. *If $f(x)$ is a Schwartz function, then*

$$\int_{-\infty}^{\infty} |f(x)|dx < +\infty.$$

Proof: There is a constant $M > 0$ such that $|x|^2|f(x)| \le 1$, for $|x| \ge M$. Then

$$\int_{-\infty}^{\infty} |f(x)|dx \le \int_{-M}^{M} |f(x)|dx + \int_{|x|\ge M} |x|^{-2}dx < +\infty.$$

If $f(x)$ is a Schwartz function, then so is its Fourier transform. To prove the Fourier Inversion Formula it is sufficient to show that

$$f(0) = \int_{-\infty}^{\infty} F(\gamma)d\gamma/2\pi.$$

Write

$$f(x) = f(0)e^{-x^2} + (f(x) - f(0)e^{-x^2}) = f(0)e^{-x^2} + g(x). \qquad \text{(C.3)}$$

Then $g(0) = 0$, so $g(x) = xh(x)$, where $h(x) = g(x)/x$ is also a Schwartz function. Then the Fourier transform of $g(x)$ is the derivative of the Fourier transform of $h(x)$; that is,

$$G(\gamma) = H'(\gamma).$$

The function $H(\gamma)$ is a Schwartz function, so it goes to zero at the infinities. Computing the Fourier transform of both sides of Equation (C.3), we obtain

$$F(\gamma) = f(0)\sqrt{\pi}e^{-\gamma^2/4} + H'(\gamma).$$

Therefore,

$$\int_{-\infty}^{\infty} F(\gamma)d\gamma = 2\pi f(0) + H(+\infty) - H(-\infty) = 2\pi f(0).$$

To prove the Fourier Inversion Formula, we let $K(\gamma) = F(\gamma)e^{-ix_0\gamma}$, for fixed x_0. Then the inverse Fourier transform of $K(\gamma)$ is $k(x) = f(x+x_0)$, and therefore

$$\int_{-\infty}^{\infty} K(\gamma)d\gamma = 2\pi k(0) = 2\pi f(x_0).$$

In the next subsection we consider a discontinuous $f(x)$.

C.1.3 An Example

Consider the function $f(x) = \frac{1}{2A}$, for $|x| \leq A$, and $f(x) = 0$, otherwise. The Fourier transform of this $f(x)$ is

$$F(\gamma) = \frac{\sin(A\gamma)}{A\gamma},$$

for all real $\gamma \neq 0$, and $F(0) = 1$. Note that $F(\gamma)$ is nonzero throughout the real line, except for isolated zeros, but that it goes to zero as we go to the infinities. This is typical behavior. Notice also that the smaller the A, the slower $F(\gamma)$ dies out; the first zeros of $F(\gamma)$ are at $|\gamma| = \frac{\pi}{A}$, so the main lobe widens as A goes to zero. The function $f(x)$ is not continuous, so its Fourier transform cannot be absolutely integrable. In this case, the Fourier Inversion Formula must be interpreted as involving convergence in the L^2 norm.

C.1.4 The Issue of Units

When we write $\cos \pi = -1$, it is with the understanding that π is a measure of angle, in radians; the function cos will always have an independent variable in units of radians. By extension, the same is true of the complex exponential functions. Therefore, when we write $e^{ix\gamma}$, we understand the product $x\gamma$ to be in units of radians. If x is measured in seconds, then γ is in units of radians per second; if x is in meters, then γ is in units of radians per meter. When x is in seconds, we sometimes use the variable $\frac{\gamma}{2\pi}$; since 2π is then in units of radians per cycle, the variable $\frac{\gamma}{2\pi}$ is in units of cycles per second, or Hertz. When we sample $f(x)$ at values of x spaced Δ apart, the Δ is in units of x-units per sample, and the reciprocal, $\frac{1}{\Delta}$, which is called the *sampling frequency*, is in units of samples per x-units. If x is in seconds, then Δ is in units of seconds per sample, and $\frac{1}{\Delta}$ is in units of samples per second.

C.2   The Dirac Delta

Consider what happens in the limit, as $A \to 0$. Then we have an infinitely high point source at $x = 0$; we denote this by $\delta(x)$, the *Dirac delta*. The Fourier transform approaches the constant function with value 1, for all γ; the Fourier transform of $f(x) = \delta(x)$ is the constant function $F(\gamma) = 1$, for all γ. The Dirac delta $\delta(x)$ has the *sifting property*:

$$\int h(x)\delta(x)dx = h(0),$$

for each function $h(x)$ that is continuous at $x = 0$.

Because the Fourier transform of $\delta(x)$ is the function $F(\gamma) = 1$, the Fourier Inversion Formula tells us that

$$\delta(x) = \frac{1}{2\pi} \int_{-\infty}^{\infty} e^{-i\gamma x} d\gamma. \tag{C.4}$$

Obviously, this integral cannot be understood in the usual way. The integral in Equation (C.4) is a symbolic way of saying that

$$\int h(x)\left(\frac{1}{2\pi} \int_{-\infty}^{\infty} e^{-i\gamma x} d\gamma\right)dx = \int h(x)\delta(x)dx = h(0), \tag{C.5}$$

for all $h(x)$ that are continuous at $x = 0$; that is, the integral in Equation (C.4) has the sifting property, so it acts like $\delta(x)$. Interchanging the order of integration in Equation (C.5), we obtain

$$\int h(x)\left(\frac{1}{2\pi} \int_{-\infty}^{\infty} e^{-i\gamma x} d\gamma\right)dx = \frac{1}{2\pi} \int_{-\infty}^{\infty} \left(\int h(x)e^{-i\gamma x} dx\right)d\gamma$$

$$= \frac{1}{2\pi} \int_{-\infty}^{\infty} H(-\gamma)d\gamma$$

$$= \frac{1}{2\pi} \int_{-\infty}^{\infty} H(\gamma)d\gamma = h(0).$$

We shall return to the Dirac delta when we consider farfield point sources.

It may seem paradoxical that when A is larger, its Fourier transform dies off more quickly. The Fourier transform $F(\gamma)$ goes to zero faster for larger A because of destructive interference. Because of differences in their complex phases, the magnitude of the sum of the signals received from various parts of the object is much smaller than we might expect, especially when A is large. For smaller A the signals received at a sensor are much more *in phase* with one another, and so the magnitude of the sum remains large. A more quantitative statement of this phenomenon is provided by the *uncertainty principle* (see [47]).

C.3   Practical Limitations

In actual remote-sensing problems, antennas cannot be of infinite extent. In digital signal processing, moreover, there are only finitely many sensors. We never measure the entire Fourier transform $F(\gamma)$, but, at best, just part of it; in the direct transmission problem we measure $F(\gamma)$ only for $\gamma = k$, with $|k| \leq \frac{\omega}{c}$. In fact, the data we are able to measure are almost never exact values of $F(\gamma)$, but rather, values of some distorted or blurred version. To describe such situations, we usually resort to *convolution-filter* models.

C.3.1   Convolution Filtering

Imagine that what we measure are not values of $F(\gamma)$, but of $F(\gamma)H(\gamma)$, where $H(\gamma)$ is a function that describes the limitations and distorting effects of the measuring process, including any blurring due to the medium through which the signals have passed, such as refraction of light as it passes through the atmosphere. If we apply the Fourier Inversion Formula to $F(\gamma)H(\gamma)$, instead of to $F(\gamma)$, we get

$$ g(x) = \frac{1}{2\pi} \int F(\gamma)H(\gamma)e^{-i\gamma x}dx. \qquad (C.6) $$

The function $g(x)$ that results is $g(x) = (f * h)(x)$, the *convolution* of the functions $f(x)$ and $h(x)$, with the latter given by

$$ h(x) = \frac{1}{2\pi} \int H(\gamma)e^{-i\gamma x}dx. $$

Note that, if $f(x) = \delta(x)$, then $g(x) = h(x)$; that is, our reconstruction of the object from distorted data is the function $h(x)$ itself. For that reason, the function $h(x)$ is called the *point-spread function* of the imaging system.

Convolution filtering refers to the process of converting any given function, say $f(x)$, into a different function, say $g(x)$, by convolving $f(x)$ with a fixed function $h(x)$. Since this process can be achieved by multiplying $F(\gamma)$ by $H(\gamma)$ and then inverse Fourier transforming, such convolution filters are studied in terms of the properties of the function $H(\gamma)$, known in this context as the *system transfer function*, or the *optical transfer function* (OTF); when γ is a frequency, rather than a spatial frequency, $H(\gamma)$ is called the *frequency-response function* of the filter. The magnitude of $H(\gamma)$, $|H(\gamma)|$, is called the *modulation transfer function* (MTF). The study of convolution filters is a major part of signal processing. Such filters provide both reasonable models for the degradation signals suffer and useful tools for reconstruction.

Let us rewrite Equation (C.6), replacing $F(\gamma)$ and $H(\gamma)$ with their definitions, as given by Equation (C.1). Then we have

$$g(x) = \int (\int f(t)e^{i\gamma t}dt)(\int h(s)e^{i\gamma s}ds)e^{-i\gamma x}d\gamma.$$

Interchanging the order of integration, we get

$$g(x) = \int \int f(t)h(s)(\int e^{i\gamma(t+s-x)}d\gamma)dsdt.$$

Now using Equation (C.4) to replace the inner integral with $\delta(t + s - x)$, the next integral becomes

$$\int h(s)\delta(t + s - x)ds = h(x - t).$$

Finally, we have

$$g(x) = \int f(t)h(x - t)dt;$$

this is the definition of the convolution of the functions f and h.

C.3.2   Low-Pass Filtering

A major problem in image reconstruction is the removal of blurring, which is often modelled using the notion of convolution filtering. In the one-dimensional case, we describe blurring by saying that we have available measurements not of $F(\gamma)$, but of $F(\gamma)H(\gamma)$, where $H(\gamma)$ is the frequency-response function describing the blurring. If we know the nature of the blurring, then we know $H(\gamma)$, at least to some degree of precision. We can try to remove the blurring by taking measurements of $F(\gamma)H(\gamma)$, dividing these numbers by the value of $H(\gamma)$, and then inverse Fourier transform-ing. The problem is that our measurements are always noisy, and typical functions $H(\gamma)$ have many zeros and small values, making division by $H(\gamma)$ dangerous, except where the values of $H(\gamma)$ are not too small. These values of γ tend to be the smaller ones, centered around zero, so that we end up with estimates of $F(\gamma)$ itself only for the smaller values of γ. The result is a *low-pass filtering* of the object $f(x)$.

To investigate such low-pass filtering, we suppose that $H(\gamma) = 1$, for $|\gamma| \leq \Gamma$, and is zero, otherwise. Then the filter is called the ideal Γ-lowpass filter. In the farfield propagation model, the variable x is spatial, and the variable γ is spatial frequency, related to how the function $f(x)$ changes spatially, as we move x. Rapid changes in $f(x)$ are associated with values of $F(\gamma)$ for large γ. For the case in which the variable x is time, the variable γ

becomes frequency, and the effect of the low-pass filter on $f(x)$ is to remove its higher-frequency components.

One effect of low-pass filtering in image processing is to smooth out the more rapidly changing features of an image. This can be useful if these features are simply unwanted oscillations, but if they are important detail, the smoothing presents a problem. Restoring such wanted detail is often viewed as removing the unwanted effects of the low-pass filtering; in other words, we try to recapture the missing high-spatial-frequency values that have been zeroed out. Such an approach to image restoration is called *frequency-domain extrapolation*. How can we hope to recover these missing spatial frequencies, when they could have been anything? To have some chance of estimating these missing values we need to have some prior information about the image being reconstructed.

C.4 Two-Dimensional Fourier Transforms

More generally, we consider a function $f(x, z)$ of two real variables. Its Fourier transformation is

$$F(\alpha, \beta) = \int \int f(x, z) e^{i(x\alpha + z\beta)} dx dz.$$

For example, suppose that $f(x, z) = 1$ for $\sqrt{x^2 + z^2} \leq R$, and zero, otherwise. Then we have

$$F(\alpha, \beta) = \int_{-\pi}^{\pi} \int_0^R e^{-i(\alpha r \cos\theta + \beta r \sin\theta)} r dr d\theta.$$

In polar coordinates, with $\alpha = \rho \cos\phi$ and $\beta = \rho \sin\phi$, we have

$$F(\rho, \phi) = \int_0^R \int_{-\pi}^{\pi} e^{ir\rho \cos(\theta - \phi)} d\theta dr.$$

The inner integral is well known;

$$\int_{-\pi}^{\pi} e^{ir\rho \cos(\theta - \phi)} d\theta = 2\pi J_0(r\rho),$$

where J_0 denotes the 0th order Bessel function. Using the identity

$$\int_0^z t^n J_{n-1}(t) dt = z^n J_n(z),$$

we have

$$F(\rho, \phi) = \frac{2\pi R}{\rho} J_1(\rho R).$$

Notice that, since $f(x, z)$ is a radial function, that is, dependent only on the distance from $(0,0)$ to (x, z), its Fourier transform is also radial.

The first positive zero of $J_1(t)$ is around $t = 4$, so when we measure F at various locations and find $F(\rho, \phi) = 0$ for a particular (ρ, ϕ), we can estimate $R \approx 4/\rho$. So, even when a distant spherical object, like a star, is too far away to be imaged well, we can sometimes estimate its size by finding where the intensity of the received signal is zero [113].

C.4.1 Two-Dimensional Fourier Inversion

Just as in the one-dimensional case, the Fourier transformation that produced $F(\alpha, \beta)$ can be inverted to recover the original $f(x, y)$. The Fourier Inversion Formula in this case is

$$f(x, y) = \frac{1}{4\pi^2} \int \int F(\alpha, \beta) e^{-i(\alpha x + \beta y)} d\alpha d\beta.$$

It is important to note that this procedure can be viewed as two one-dimensional Fourier inversions: first, we invert $F(\alpha, \beta)$, as a function of, say, β only, to get the function of α and y

$$g(\alpha, y) = \frac{1}{2\pi} \int F(\alpha, \beta) e^{-i\beta y} d\beta;$$

second, we invert $g(\alpha, y)$, as a function of α, to get

$$f(x, y) = \frac{1}{2\pi} \int g(\alpha, y) e^{-i\alpha x} d\alpha.$$

If we write the functions $f(x, y)$ and $F(\alpha, \beta)$ in polar coordinates, we obtain alternative ways to implement the two-dimensional Fourier inversion. We consider these other ways when we discuss the tomography problem of reconstructing a function $f(x, y)$ from line-integral data.

D | The EM Algorithm

The so-called *EM algorithm*, discussed by Dempster, Laird, and Rubin [75], is a general framework for deriving iterative methods for maximum likelihood parameter estimation. The book by McLachnan and Krishnan [125] is a good source for the history of this general method. There is a problem with the way the EM algorithm is usually described in the literature. That description is fine for the case of discrete random vectors, but needs to be modified to apply to continuous ones. We consider some of these issues in this chapter. We begin with the usual formulation of the EM algorithm, as it applies to the discrete case.

D.1   The Discrete Case

We denote by Z a random vector, taking values in R^N, by $h : R^N \rightarrow R^I$ a function from R^N to R^I, with $N > I$, and $Y = h(Z)$ the corresponding random vector taking values in R^I. The random vector Z has probability function $f(z; x)$, where x is a parameter in the parameter space $\mathcal{X}$. The probability function associated with Y is then

$$g(y; x) = \sum_{z \in h^{-1}(y)} f(z; x) \leq 1.$$

The random vector Y is usually called the *incomplete data*, and Z the *complete data*. The EM algorithm is typically used when maximizing $f(z; x)$ is easier than maximizing $g(y; x)$, but we have only y, an instance of Y, and not a value of Z.

The conditional probability function for Z, given $Y = y$ and x, is

$$b(z; y, x) = f(z; x)/g(y; x),$$

for $z \in h^{-1}(y)$, and $b(z; y, x) = 0$, otherwise. The *E-step* of the EM algorithm is to calculate the conditional expected value of the random variable $\log f(Z; x)$, given y and the current estimate x_k of x:

$$Q(x; x_k) = E(\log f(Z; x)|y, x_k) = \sum_{z \in h^{-1}(y)} b(z; y, x_k) \log f(z; x).$$

The *M-step* is to select x_{k+1} as a maximizer of $Q(x; x_k)$. Denote by $H(x; x_k)$ the conditional expected value of the random variable $\log b(Z; y, x)$, given y and x_k:

$$H(x; x_k) = \sum_{z \in h^{-1}(y)} b(z; y, x_k) \log b(z; y, x).$$

Then, for all $x \in \mathcal{X}$, we have

$$Q(x; x_k) = H(x; x_k) + L(x),$$

for $L(x) = \log g(y; x)$.

For positive scalars a and b, let $KL(a, b)$ denote the Kullback-Leibler distance

$$KL(a, b) = a \log \frac{a}{b} + b - a.$$

Also let $KL(a, 0) = +\infty$ and $KL(0, b) = b$. Extend the KL distance component-wise to vectors with nonnegative entries. It follows from the inequality $\log t \leq t - 1$ that $KL(a, b) \geq 0$ and $KL(a, b) = 0$ if and only if $a = b$. Then we have

$$Q(x; x_k) = -KL(b(\cdot; y, x_k), f(\cdot; x)),$$

and

$$H(x_k; x_k) = H(x; x_k) + KL(b(\cdot; y, x_k), b(\cdot; y, x)),$$

where

$$KL(b(\cdot; y, x_k), b(\cdot; y, x)) = \sum_z KL(b(z; y, x_k), b(z; y, x)) \geq 0.$$

Therefore,

$$
\begin{aligned}
L(x_k) = Q(x_k; x_k) - H(x_k; x_k) &\leq Q(x_{k+1}; x_k) - H(x_k; x_k) \\
&= Q(x_{k+1}; x_k) - H(x_{k+1}; x_k) - KL(b(x_k), b(x_{k+1})) \\
&= L(x_{k+1}) - KL(b(x_k), b(x_{k+1})).
\end{aligned}
$$

The sequence $\{L(x_k)\}$ is increasing and nonpositive, so convergent. The sequence $\{KL(b(x_k), b(x_{k+1}))\}$ converges to zero.

In the discrete case, the EM algorithm is an *alternating minimization* method. The function $KL(b(\cdot; y, x_k), f(\cdot; x))$ is minimized by the choice $x = x_{k+1}$, and the function $KL(b(\cdot; y, x), f(\cdot; x_{k+1}))$ is minimized by the choice $x = x_{k+1}$. Therefore, the EM algorithm can be viewed as the result of alternately minimizing $KL(b(\cdot; y, u), f(\cdot; v))$, first with respect to the variable u, and then with respect to the variable v.

Without further assumptions, we can say no more; see [155]. We would like to conclude that the sequence $\{x_k\}$ converges to a maximizer of $L(x)$, but we have no metric on the parameter space $\mathcal{X}$. We need an identity that relates the nonnegative quantity

$$KL(b(\cdot; y, x_k), f(\cdot; x)) - KL(b(\cdot; y, x_k), f(\cdot; x_{k+1}))$$

to the difference, in parameter space, between x and x_{k+1}. For example, for the EMML algorithm in the Poisson mixture case, we have

$$KL(x_{k+1}, x) = KL(b(\cdot; y, x_k), f(\cdot; x)) - KL(b(\cdot; y, x_k), f(\cdot; x_{k+1})).$$

D.2 The Continuous Case

The usual approach to the EM algorithm in this case is to mimic the discrete case. A problem arises when we try to define $g(y; x)$ as

$$g(y; x) = \int_{z \in h^{-1}(y)} f(z; x) dz;$$

the set $h^{-1}(y)$ typically has measure zero in R^N. We need a different approach.

Suppose that there is a second function $c : R^N \to R^{N-I}$ such that the function $G(z) = G(h(z), c(z)) = (y, w)$ has inverse $H(y, w) = z$. Then, given y, let $W(y) = \{w = c(z) | y = h(z)\}$. Then, with $J(y, w)$ the Jacobian, the pdf of the random vector Y is

$$g(y; x) = \int_{W(y)} f(H(y, w); x) J(y, w) dw,$$

and the pdf for the random vector $W = c(Z)$ is

$$b(H(y, w); y, x) = f(H(y, w); x) J(y, w) / g(y; x),$$

for $w \in W(y)$. Given y, and having found x_k, we minimize

$$KL(b(H(y, w); x_k), f(H(y, w); x)),$$

with respect to x, to get x_{k+1}.

D.2.1   An Example

Suppose that Z_1 and Z_2 are independent and uniformly distributed on the interval $[0, x]$, where $x > 0$ is an unknown parameter. Let $Y = Z_1 + Z_2$. Then

$$g(y; x) = y/x^2,$$

for $0 \leq y \leq x$, and

$$g(y; x) = (2x - y)/x^2,$$

for $x \leq y \leq 2x$. Given y, the maximum likelihood estimate of x is y. The pdf for the random vector $Z = (Z_1, Z_2)$ is

$$f(z_1, z_2; x) = \frac{1}{x^2} \chi_{[0,x]}(z_1) \chi_{[0,x]}(z_2).$$

The conditional pdf of Z, given y and x_k, is

$$b(z_1, z_2; y, x_k) = \frac{1}{y} \chi_{[0,x_k]}(z_1) \chi_{[0,x_k]}(z_2),$$

for $0 \leq y \leq x_k$, and for $x_k \leq y \leq 2x_k$ it is

$$b(z_1, z_2; y, x_k) = \frac{1}{2x_k - y} \chi_{[0,x_k]}(z_1) \chi_{[0,x_k]}(z_2).$$

Suppose that $c(z) = c(z_1, z_2) = z_2$ and $W = c(Z)$. Then $W(y) = [0, y]$ and the conditional pdf of W, given y and x_k, is $b(y - w, w; y, x_k)$. If we choose $x_0 \geq y$, then $x_1 = y$, which is the ML estimator. But, if we choose x_0 in the interval $[\frac{y}{2}, y]$, then $x_1 = x_0$ and the EM iteration stagnates. Note that the function $L(x) = \log g(y; x)$ is continuous, but not differentiable. It is concave for x in the interval $[\frac{y}{2}, y]$ and convex for $x \geq y$.

E | Using Prior Knowledge in Remote Sensing

The problem is to reconstruct a (possibly complex-valued) function f : $R^D \to C$ from finitely many measurements g_n, $n = 1, ..., N$, pertaining to f. The function $f(r)$ represents the physical object of interest, such as the spatial distribution of acoustic energy in sonar, the distribution of x-ray-attenuating material in transmission tomography, the distribution of radionuclide in emission tomography, the sources of reflected radio waves in radar, and so on. Often the reconstruction, or estimate, of the function f takes the form of an image in two or three dimensions; for that reason, we also speak of the problem as one of *image reconstruction*. The data are obtained through measurements. Because there are only finitely many measurements, the problem is highly underdetermined and even noise-free data are insufficient to specify a unique solution.

E.1 The Optimization Approach

One way to solve such under-determined problems is to replace $f(r)$ with a vector in C^N and to use the data to determine the N entries of this vector. An alternative method is to model $f(r)$ as a member of a family of linear combinations of N preselected basis functions of the multivariable r. Then the data is used to determine the coefficients. This approach offers the user the opportunity to incorporate prior information about $f(r)$ in the choice of the basis functions. Such finite-parameter models for $f(r)$ can be obtained through the use of the minimum-norm estimation procedure, as we shall see. More generally, we can associate a *cost* with each data-consistent function of r, and then minimize the cost over all the potential solutions to the problem. Using a norm as a cost function is one way to proceed, but there are others. These optimization problems can often be solved only through the use of discretization and iterative algorithms.

E.2   Introduction to Hilbert Space

In many applications the data are related linearly to f. To model the operator that transforms f into the data vector, we need to select an ambient space containing f. Typically, we choose a Hilbert space. The selection of the inner product provides an opportunity to incorporate prior knowledge about f into the reconstruction. The inner product induces a norm and our reconstruction is that function, consistent with the data, for which this norm is minimized. We shall illustrate the method using Fourier-transform data and prior knowledge about the support of f and about its overall shape.

Our problem, then, is to estimate a (possibly complex-valued) function $f(r)$ of D real variables $r = (r_1, ..., r_D)$ from finitely many measurements, g_n, $n = 1, ..., N$. We shall assume, in this chapter, that these measurements take the form

$$g_n = \int_S f(r)\overline{h_n(r)}dr, \tag{E.1}$$

where S denotes the support of the function $f(r)$, which, in most cases, is a bounded set. For the purpose of estimating, or reconstructing, $f(r)$, it is convenient to view Equation (E.1) in the context of a Hilbert space, and to write

$$g_n = \langle f, h_n \rangle,$$

where the usual Hilbert space inner product is defined by

$$\langle f, h \rangle_2 = \int_S f(r)\overline{h(r)}dr,$$

for functions $f(r)$ and $h(r)$ supported on the set S. Of course, for these integrals to be defined, the functions must satisfy certain additional properties, but a more complete discussion of these issues is outside the scope of this chapter. The Hilbert space so defined, denoted $L^2(S)$, consists (essentially) of all functions $f(r)$ for which the norm

$$\|f\|_2 = \sqrt{\int_S |f(r)|^2 dr} \tag{E.2}$$

is finite.

E.2.1   Minimum-Norm Solutions

Our estimation problem is highly under-determined; there are infinitely many functions in $L^2(S)$ that are consistent with the data and might be the right answer. Such under-determined problems are often solved by acting conservatively, and selecting as the estimate that function consistent with the data that has the smallest norm. At the same time, however, we often have some prior information about f that we would like to incorporate in the estimate. One way to achieve both of these goals is to select the norm to incorporate prior information about f, and then to take as the estimate of f the function consistent with the data, for which the chosen norm is minimized.

The data vector $g = (g_1, ..., g_N)^T$ is in C^N and the linear operator $\mathcal{H}$ from $L^2(S)$ to C^N takes f to g; so we write $g = \mathcal{H}f$. Associated with the mapping $\mathcal{H}$ is its adjoint operator, $\mathcal{H}^\dagger$, going from C^N to $L^2(S)$ and given, for each vector $a = (a_1, ..., a_N)^T$, by

$$\mathcal{H}^\dagger a(r) = a_1 h_1(r) + ... + a_N h_N(r). \tag{E.3}$$

The operator from C^N to C^N defined by $\mathcal{H}\mathcal{H}^\dagger$ corresponds to an N by N matrix, which we shall also denote by $\mathcal{H}\mathcal{H}^\dagger$. If the functions $h_n(r)$ are linearly independent, then this matrix is positive-definite, therefore invertible.

Given the data vector g, we can solve the system of linear equations

$$g = \mathcal{H}\mathcal{H}^\dagger a \tag{E.4}$$

for the vector a. Then the function

$$\hat{f}(r) = \mathcal{H}^\dagger a(r) \tag{E.5}$$

is consistent with the measured data and is the function in $L^2(S)$ with the smallest norm for which this is true. The function $w(r) = f(r) - \hat{f}(r)$ has the property $\mathcal{H}w = 0$. It is easy to see that

$$||f||_2^2 = ||\hat{f}||_2^2 + ||w||_2^2.$$

The estimate $\hat{f}(r)$ is the *minimum-norm solution*, with respect to the norm defined in Equation (E.2). If we change the norm on $L^2(S)$, or, equivalently, the inner product, then the minimum-norm solution will change.

For any continuous linear operator T on $L^2(S)$, the adjoint operator, denoted $T^\dagger$, is defined by

$$\langle Tf, h \rangle_2 = \langle f, T^\dagger h \rangle_2.$$

The adjoint operator will change when we change the inner product.

E.3   A Class of Inner Products

Let T be a continuous, linear, and invertible operator on $L^2(S)$. Define the T inner product to be

$$\langle f, h \rangle_T = \langle T^{-1} f, T^{-1} h \rangle_2.$$

We can then use this inner product to define the problem to be solved. We now say that

$$g_n = \langle f, t^n \rangle_T,$$

for known functions $t^n(r)$. Using the definition of the T inner product, we find that

$$g_n = \langle f, h^n \rangle_2 = \langle T f, T h^n \rangle_T.$$

The adjoint operator for T, with respect to the T-norm, is denoted T^*, and is defined by

$$\langle T f, h \rangle_T = \langle f, T^* h \rangle_T.$$

Therefore,

$$g_n = \langle f, T^* T h^n \rangle_T.$$

Lemma E.1. *We have $T^* T = T T^\dagger$.*

Consequently, we have

$$g_n = \langle f, T T^\dagger h^n \rangle_T.$$

E.4   Minimum-T-Norm Solutions

The function $\tilde{f}$ consistent with the data and having the smallest T-norm has the algebraic form

$$\hat{f} = \sum_{m=1}^{N} a_m T T^\dagger h^m. \tag{E.6}$$

Applying the T-inner product to both sides of Equation (E.6), we get

$$g_n = \langle \hat{f}, T T^\dagger h^n \rangle_T$$

$$= \sum_{m=1}^{N} a_m \langle T T^\dagger h^m, T T^\dagger h^n \rangle_T.$$

Therefore,

$$g_n = \sum_{m=1}^{N} a_m \langle T^\dagger h^m, T^\dagger h^n \rangle_2. \tag{E.7}$$

We solve this system for the a_m and insert them into Equation (E.6) to get our reconstruction. The Gram matrix that appears in Equation (E.7) is positive-definite, but is often ill conditioned; increasing the main diagonal by a percent or so usually is sufficient regularization.

E.5   The Case of Fourier-Transform Data

To illustrate these minimum-T-norm solutions, we consider the case in which the data are values of the Fourier transform of f. Specifically, suppose that

$$g_n = \int_S f(x)e^{-i\omega_n x}dx,$$

for arbitrary values ω_n.

E.5.1   The $L^2(-\pi, \pi)$ Case

Assume that $f(x) = 0$, for $|x| > \pi$. The minimum-2-norm solution has the form

$$\hat{f}(x) = \sum_{m=1}^{N} a_m e^{i\omega_m x},$$

with

$$g_n = \sum_{m=1}^{N} a_m \int_{-\pi}^{\pi} e^{i(\omega_m - \omega_n)x}dx.$$

For the equi-spaced values $\omega_n = n$ we find that $a_m = g_m$ and the minimum-norm solution is

$$\hat{f}(x) = \sum_{n=1}^{N} g_n e^{inx}.$$

E.5.2   The Over-Sampled Case

Suppose that $f(x) = 0$ for $|x| > A$, where $0 < A < \pi$. Then we use $L^2(-A, A)$ as the Hilbert space. For equi-spaced data at $\omega_n = n$, we have

$$g_n = \int_{-\pi}^{\pi} f(x)\chi_A(x)e^{-inx} dx,$$

so that the minimum-norm solution has the form

$$\hat{f}(x) = \chi_A(x) \sum_{m=1}^{N} a_m e^{imx},$$

with

$$g_n = 2 \sum_{m=1}^{N} a_m \frac{\sin A(m-n)}{m-n}.$$

The minimum-norm solution is support-limited to $[-A, A]$ and consistent with the Fourier-transform data.

E.5.3   Using a Prior Estimate of f

Suppose that $f(x) = 0$ for $|x| > \pi$ again, and that $p(x)$ satisfies

$$0 < \epsilon \leq p(x) \leq E < +\infty,$$

for all x in $[-\pi, \pi]$. Define the operator T by $(Tf)(x) = \sqrt{p(x)}f(x)$. The T-norm is then

$$\langle f, h \rangle_T = \int_{-\pi}^{\pi} f(x)\overline{h(x)}p(x)^{-1} dx.$$

It follows that

$$g_n = \int_{-\pi}^{\pi} f(x)p(x)e^{-inx}p(x)^{-1} dx,$$

so that the minimum T-norm solution is

$$\hat{f}(x) = \sum_{m=1}^{N} a_m p(x)e^{imx} = p(x) \sum_{m=1}^{N} a_m e^{imx}, \tag{E.8}$$

where

$$g_n = \sum_{m=1}^{N} a_m \int_{-\pi}^{\pi} p(x)e^{i(m-n)x} dx.$$

If we have prior knowledge about the support of f, or some idea of its shape, we can incorporate that prior knowledge into the reconstruction through the choice of $p(x)$.

The reconstruction in Equation (E.8) was presented in [25], where it was called the PDFT method. The PDFT was based on a noniterative version of the Gerchberg-Papoulis bandlimited extrapolation procedure, discussed earlier in [24]. The PDFT was then applied to image reconstruction problems in [26]. An application of the PDFT was presented in [29]. In [28] we extended the PDFT to a nonlinear version, the indirect PDFT (IPDFT), that generalizes Burg's maximum entropy spectrum estimation method. The PDFT was applied to the phase problem in [31] and in [32] both the PDFT and IPDFT were examined in the context of Wiener filter approximation. More recent work on these topics is discussed in the book [48].

F | Optimization in Remote Sensing

Once again, the basic problem is to reconstruct or estimate a (possibly complex-valued) function $f_0(r)$ of several real variables, from finitely many measurements pertaining to $f_0(r)$. As previously, we shall assume that the measurements g_n take the form

$$g_n = \int_S f_0(r)\overline{h_n(r)}dr,$$

for $n = 1, ..., N$. The problem is highly under-determined; there are infinitely many functions consistent with the data. One approach to solving such problems is to select a cost function $C(f) \geq 0$ and minimize $C(f)$ over all functions $f(r)$ consistent with the measured data. As we saw previously, cost functions that are Hilbert-space norms are reasonable choices. How we might select the cost function is the subject of this chapter.

F.1   The General Form of the Cost Function

We shall consider cost functions of the form

$$C(f) = \int_S F(f(r), p(r))dr,$$

where $p(r)$ is a fixed prior estimate of the true $f(r)$ and $F(y, z) \geq 0$ is to be determined. Such cost functions are viewed as measures of distance between the functions $f(r)$ and $p(r)$. Therefore, we also write

$$D(f, p) = \int_S F(f(r), p(r))dr.$$

Our goal is to impose reasonable conditions on these distances $D(f, p)$ sufficiently restrictive to eliminate all but a small class of suitable distances.

357

F.2   The Conditions

In order for $D(f, p)$ to be viewed as a distance measure, we want $D(f, f) = 0$ for all appropriate f. Therefore, we require

Axiom F.1. $F(y, y) = 0$, for all suitable y.

We also want $D(f, p) \geq D(p, p)$ for all appropriate f and p, so we require

Axiom F.2. $F_y(y, y) = 0$, for all suitable y.

To make $D(f, p)$ strictly convex in f we impose

Axiom F.3. $F_{y,y}(y, z) > 0$, for all suitable y and z.

Given $p(r)$ and the data, we find our estimate by minimizing $D(f, p)$ over all appropriate $f(r)$ consistent with the data. The Lagrangian is then

$$L(f, \lambda) = D(f, p) + \sum_{n=1}^{N} \lambda_n \left(g_n - \int_S f(r) \overline{h_n(r)} dr \right).$$

Taking the first partial derivative of $L(f, \lambda)$ with respect to f gives the Euler equation

$$F_y(f(r), p(r)) = \sum_{n=1}^{N} \lambda_n h_n(r). \qquad \text{(F.1)}$$

Given the data, we must find the λ_n for which the resulting $f(r)$ is consistent with the data.

As we vary the values of g_n, the values of the λ_n will change also. The functions $t(r)$ satisfying

$$F_y(t(r), p(r)) = \sum_{n=1}^{N} \lambda_n h_n(r),$$

for some choice of the λ_n, will form the family denoted $\mathcal{T}$. We see from Equation (F.1) that our optimal f is a member of $\mathcal{T}$. The functions consistent with the data we denote by $\mathcal{Q}$. We seek those functions $F(y, z)$ for which Axiom 4 holds:

Axiom F.4. In all cases, the member of $\mathcal{T}$ that minimizes $D(f_0, t)$ is the function $f(r)$ in $\mathcal{Q}$ that minimizes $D(f, p)$.

Our goal is to find an estimate $f(r)$ that is close to the true $f_0(r)$. We are relying on data consistency to provide such an estimate. At the very least, we hope that data consistency produces the best approximation of $f_0(r)$ within $\mathcal{T}$. This will depend on our choice of the cost function. Axiom 4 says that, among all the functions in $\mathcal{T}$, the one that is closest to the true $f_0(r)$ is the one that is consistent with the data.

In [109] it was shown that the functions $F(y, z)$ that satisfy these four axioms must also have the property

$$F_{z,y,y}(y, z) = 0,$$

for all suitable y and z. It follows that there is a strictly convex function $H(y)$ such that

$$F(y, z) = H(y) - H(z) - H'(z)(y - z). \tag{F.2}$$

As we saw in our discussion of Bregman-Legendre functions, the Bregman distances have the form in Equation (F.2).

If $\hat{f}(r)$ is the member of $\mathcal{Q}$ that minimizes $D(f, p)$, then

$$D(f, p) = D(f, \hat{f}) + D(\hat{f}, p).$$

There are many F that fit this description. If we impose one more axiom, we can reduce the choice significantly.

Axiom F.5. Let $\hat{f}$ minimize $D(f, p)$ over f in $\mathcal{Q}$. Then, for any suitable constant c, $\hat{f}$ also minimizes $D(f, cp)$, over f in $\mathcal{Q}$.

Axiom F.5'. Let $\hat{f}$ minimize $D(f, p)$ over f in $\mathcal{Q}$. Then, for any suitable constant c, $c\hat{f}$ minimizes $D(f, p)$, over f consistent with the data cg_n.

If the function F satisfies either of these two additional axioms, for all appropriate choices of p, then F is a positive multiple of the Kullback-Leibler distance, that is,

$$F(y, z) = c^2 [y \log \frac{y}{z} + z - y],$$

for $y > 0$ and $z > 0$.

Bibliography

[1] S. Agmon. "The Relaxation Method for Linear Inequalities." *Canadian Journal of Mathematics* 6 (1954), 382–392.

[2] A. Anderson and A. Kak. "Simultaneous Algebraic Reconstruction Technique (SART): A Superior Implementation of the ART Algorithm." *Ultrasonic Imaging* 6 (1984), 81–94.

[3] J.-P. Aubin. *Optima and Equilibria: An Introduction to Nonlinear Analysis.* New York: Springer-Verlag, 1993.

[4] O. Axelsson. *Iterative Solution Methods.* Cambridge, UK: Cambridge University Press, 1994.

[5] J.-B. Baillon, R. E. Bruck, and S. Reich. "On the Asymptotic Behavior of Nonexpansive Mappings and Semigroups in Banach Spaces." *Houston Journal of Mathematics* 4 (1978), 1–9.

[6] J.-B. Baillon and G. Haddad. "Quelques proprietes des operateurs angle-bornes et n-cycliquement monotones." *Israel J. of Mathematics* 26 (1977), 137–150.

[7] H. Bauschke. "The Approximation of Fixed Points of Compositions of Nonexpansive Mappings in Hilbert Space." *Journal of Mathematical Analysis and Applications* 202 (1996), 150–159.

[8] H. Bauschke. "Projection Algorithms: Results and Open Problems." In *Inherently Parallel Algorithms in Feasibility and Optimization and their Applications*, edited by D. Butnariu, Y. Censor, and S. Reich, pp. 11–22, Studies in Computational Mathematics 8. Amsterdam: Elsevier Science, 2001.

[9] H. Bauschke and J. Borwein. "On Projection Algorithms for Solving Convex Feasibility Problems." *SIAM Review* 38:3 (1996), 367–426.

[10] H. Bauschke and J. Borwein. "Legendre Functions and the Method of Random Bregman Projections." *Journal of Convex Analysis* 4 (1997), 27–67.

[11] H. Bauschke, J. Borwein, and A. Lewis. "The Method of Cyclic Projections for Closed Convex Sets in Hilbert Space." In *Contemporary Mathematics: Recent Developments in Optimization Theory and Nonlinear Analysis*, Vol. 204, pp. 1–38. Providence, RI: American Mathematical Society, 1997.

[12] H. Bauschke and A. Lewis. "Dykstra's Algorithm with Bregman Projections: A Convergence Proof." *Optimization* 48 (2000), 409–427.

[13] M. Bertero and P. Boccacci. *Introduction to Inverse Problems in Imaging.* Bristol, UK: Institute of Physics Publishing, 1998.

[14] D. P. Bertsekas. "A New Class of Incremental Gradient Methods for Least Squares Problems." *SIAM Journal on Optimization* 7 (1997), 913–926.

[15] J. Borwein and A. Lewis. *Convex Analysis and Nonlinear Optimization.* Canadian Mathematical Society Books in Mathematics. New York: Springer-Verlag, 2000.

[16] R. C. Bracewell. "Image Reconstruction in Radio Astronomy." In *Image Reconstruction from Projections*, edited by G. T. Herman, pp. 81–104, Topics in Applied Physics 32. Berlin: Springer-Verlag, 1979.

[17] L. M. Bregman. "The Relaxation Method of Finding the Common Point of Convex Sets and Its Application to the Solution of Problems in Convex Programming." *USSR Computational Mathematics and Mathematical Physics* 7 (1967), 200–217.

[18] L. Bregman, Y. Censor, and S. Reich. "Dykstra's Algorithm as the Nonlinear Extension of Bregman's Optimization Method." *Journal of Convex Analysis* 6:2 (1999), 319–333.

[19] A. Brodzik and J. Mooney. "Convex Projections Algorithm for Restoration of Limited-Angle Chromotomographic Images." *Journal of the Optical Society of America A* 16:2 (1999), 246–257.

[20] J. Browne and A. DePierro. "A Row-Action Alternative to the EM Algorithm for Maximizing Likelihoods in Emission Tomography." *IEEE Transactions in Medical Imaging* 15 (1996), 687–699.

[21] R. E. Bruck and S. Reich. "Nonexpansive Projections and Resolvents of Accretive Operators in Banach Spaces." *Houston Journal of Mathematics* 3 (1977), 459–470.

[22] R. L. Burden and J. D. Faires. *Numerical Analysis.* Boston: PWS-Kent, 1993.

[23] E. Burger and M. Starbird. *Coincidences, Chaos, and All That Math Jazz.* New York: W W Norton, 2006.

[24] C. Byrne and R. Fitzgerald. "A Unifying Model for Spectrum Estimation." In *Proceedings of the RADC Workshop on Spectrum Estimation*, Griffiss AFB, Rome, NY, 1979.

[25] C. Byrne and R. Fitzgerald. "Reconstruction from Partial Information, with Applications to Tomography." *SIAM J. Applied Math.* 42:4 (1982), 933–940.

[26] C. Byrne, R. Fitzgerald, M. Fiddy, T. Hall, and A. Darling. "Image Restoration and Resolution Enhancement." *J. Opt. Soc. Amer.* 73 (1983), 1481–1487.

[27] C. Byrne and D. Wells. "Limit of Continuous and Discrete Finite-Band Gerchberg Iterative Spectrum Extrapolation." *Optics Letters* 8:10 (1983), 526–527.

[28] C. Byrne and R. Fitzgerald. "Spectral Estimators that Extend the Maximum Entropy and Maximum Likelihood Methods." *SIAM J. Applied Math.* 44:2 (1984), 425–442.

[29] C. Byrne, B. M. Levine, and J. C. Dainty. "Stable Estimation of the Probability Density Function of Intensity from Photon Frequency Counts." *JOSA Communications* 1:11 (1984), 1132–1135.

[30] C. Byrne and D. Wells. "Optimality of Certain Iterative and Non-Iterative Data Extrapolation Procedures." *Journal of Mathematical Analysis and Applications* 111:1 (1985), 26–34.

[31] C. Byrne and M. Fiddy. "Estimation of Continuous Object Distributions from Fourier Magnitude Measurements." *JOSA A* 4 (1987), 412–417.

[32] C. Byrne and M. Fiddy. "Images as Power Spectra; Reconstruction as Wiener Filter Approximation." *Inverse Problems* 4 (1988), 399–409.

[33] C. Byrne, D. Haughton, and T. Jiang. "High-Resolution Inversion of the Discrete Poisson and Binomial Transformations." *Inverse Problems* 9 (1993), 39–56.

[34] C. Byrne. "Iterative Image Reconstruction Algorithms Based on Cross-Entropy Minimization." *IEEE Transactions on Image Processing* IP-2 (1993), 96–103.

[35] C. Byrne. "Erratum and Addendum to 'Iterative Image Reconstruction Algorithms Based on Cross-Entropy Minimization.'" *IEEE Transactions on Image Processing* IP-4 (1995), 225–226.

[36] C. Byrne. "Iterative Reconstruction Algorithms Based on Cross-Entropy Minimization." In *Image Models (and Their Speech Model Cousins)*, edited by S.E. Levinson and L. Shepp, pp. 1–11, IMA Volumes in Mathematics and Its Applications 80. New York: Springer-Verlag, 1996.

[37] C. Byrne. "Block-Iterative Methods for Image Reconstruction from Projections." *IEEE Transactions on Image Processing* IP-5 (1996), 792–794.

[38] C. Byrne. "Convergent Block-Iterative Algorithms for Image Reconstruction from Inconsistent Data." *IEEE Transactions on Image Processing* IP-6 (1997), 1296–1304.

[39] C. Byrne. "Accelerating the EMML Algorithm and Related Iterative Algorithms by Rescaled Block-Iterative (RBI) Methods." *IEEE Transactions on Image Processing* IP-7 (1998), 100–109.

[40] C. Byrne. "Iterative Deconvolution and Deblurring with Constraints." *Inverse Problems* 14 (1998), 1455–1467.

[41] C. Byrne. "Iterative Projection onto Convex Sets Using Multiple Bregman Distances." *Inverse Problems* 15 (1999), 1295–1313.

[42] C. Byrne. "Block-Iterative Interior Point Optimization Methods for Image Reconstruction from Limited Data." *Inverse Problems* 16 (2000), 1405–1419.

[43] C. Byrne. "Bregman-Legendre Multidistance Projection Algorithms for Convex Feasibility and Optimization." In *Inherently Parallel Algorithms in Feasibility and Optimization and their Applications*, edited by D. Butnariu, Y. Censor, and S. Reich, pp. 87–100, Studies in Computational Mathematics 8. Amsterdam: Elsevier Science, 2001.

[44] C. Byrne. "Likelihood Maximization for List-Mode Emission Tomographic Image Reconstruction." *IEEE Transactions on Medical Imaging* 20:10 (2001), 1084–1092.

[45] C. Byrne. "Iterative Oblique Projection onto Convex Sets and the Split Feasibility Problem." *Inverse Problems* 18 (2002), 441–453.

[46] C. Byrne. "A Unified Treatment of Some Iterative Algorithms in Signal Processing and Image Reconstruction." *Inverse Problems* 20 (2004), 103–120.

[47] C. Byrne. "Choosing Parameters in Block-Iterative or Ordered-Subset Reconstruction Algorithms." *IEEE Transactions on Image Processing* 14:3 (2005), 321–327.

[48] C. Byrne. *Signal Processing: A Mathematical Approach.* Wellesley, MA: A K Peters, Ltd., 2005.

[49] C. Byrne. "Feedback in Iterative Algorithms." Unpublished lecture notes, 2005.

[50] C. Byrne and S. Ward. "Estimating the Largest Singular Value of a Sparse Matrix." Unpublished report, 2005.

[51] C. Byrne and Y. Censor. "Proximity Function Minimization Using Multiple Bregman Projections, with Applications to Split Feasibility and Kullback-Leibler Distance Minimization." *Annals of Operations Research* 105 (2001), 77–98.

[52] Y. Censor. "Row-Action Methods for Huge and Sparse Systems and their Applications." *SIAM Review* 23 (1981), 444–464.

[53] Y. Censor, P. P. B. Eggermont, and D. Gordon. "Strong Underrelaxation in Kaczmarz's Method for Inconsistent Systems." *Numerische Mathematik* 41 (1983), 83–92.

[54] Y. Censor and T. Elfving. "A Multiprojection Algorithm Using Bregman Projections in a Product Space." *Numerical Algorithms* 8 (1994), 221–239.

[55] Y. Censor, T. Elfving, N. Kopf, and T. Bortfeld. "The Multiple-Sets Split Feasibility Problem and Its Application for Inverse Problems." *Inverse Problems* 21 (2005), 2071–2084.

[56] Y. Censor, T. Bortfeld, B. Martin, and A. Trofimov. "A Unified Approach for Inversion Problems in Intensity-Modulated Radiation Therapy." *Physics in Medicine and Biology* 51 (2006), 2353–2365.

[57] Y. Censor and S. Reich. "Iterations of Paracontractions and Firmly Nonexpansive Operators with Applications to Feasibility and Optimization", *Optimization* 37 (1996), 323–339.

[58] Y. Censor and S. Reich. "The Dykstra Algorithm for Bregman Projections." *Communications in Applied Analysis* 2 (1998), 323–339.

[59] Y. Censor and J. Segman. "On Block-Iterative Maximization." *Journal of Information and Optimization Sciences* 8 (1987), 275–291.

[60] Y. Censor, and S. A. Zenios. "Proximal Minimization Algorithm with *D*-Functions" *Journal of Optimization Theory and Applications* 73:3 (1992), 451–464.

[61] Y. Censor and S. A. Zenios. *Parallel Optimization: Theory, Algorithms and Applications.* New York: Oxford University Press, 1997.

[62] J.-H. Chang, J. M. M. Anderson, and J. R. Votaw. "Regularized Image Reconstruction Algorithms for Positron Emission Tomography." *IEEE Transactions on Medical Imaging* 23:9 (2004), 1165–1175.

[63] W. Cheney and A. Goldstein. "Proximity Maps for Convex Sets." *Proc. Am. Math. Soc.* 10 (1959), 448–450.

[64] G. Cimmino. "Calcolo approssimato per soluzioni die sistemi di equazioni lineari." *La Ricerca Scientifica XVI, Series II, Anno IX* 1 (1938), 326–333.

[65] P. Combettes. "The Foundations of Set Theoretic Estimation." *Proceedings of the IEEE* 81:2 (1993), 182–208.

[66] P. Combettes. "The Convex Feasibility Problem in Image Recovery." *Advances in Imaging and Electron Physics* 95 (1996), 155–270.

[67] P. Combettes. "Fejér Monotonicity in Convex Optimization." In *Encyclopedia of Optimization*, edited by C. A. Floudas and P. M. Pardalos, Boston: Kluwer Academic Publishers, 2000.

[68] P. Combettes and J. Trussell. "Method of Successive Projections for Finding a Common Point of Sets in a Metric Space." *Journal of Optimization Theory and Applications* 67:3 (1990), 487–507.

[69] P. Combettes and V. Wajs. "Signal Recovery by Proximal Forward-Backward Splitting." *Multiscale Modeling and Simulation* 4:4 (2005), 1168–1200.

[70] I. Csiszár and G. Tusnády. "Information Geometry and Alternating Minimization Procedures." *Statistics and Decisions* Supp. 1 (1984), 205–237.

[71] I. Csiszár. "A Geometric Interpretation of Darroch and Ratcliff's Generalized Iterative Scaling." *The Annals of Statistics* 17:3 (1989), 1409–1413.

[72] I. Csiszár. "Why Least Squares and Maximum Entropy? An Axiomatic Approach to Inference for Linear Inverse Problems." *The Annals of Statistics* 19:4 (1991), 2032–2066.

[73] J. Darroch and D. Ratcliff. "Generalized Iterative Scaling for Log-Linear Models." *Annals of Mathematical Statistics* 43 (1972), 1470–1480.

[74] A. Dax. "The Convergence of Linear Stationary Iterative Processes for Solving Singular Unstructured Systems of Linear Equations." *SIAM Review* 32 (1990), 611–635.

[75] A. P. Dempster, N. M. Laird, and D. B. Rubin. "Maximum Likelihood from Incomplete Data via the EM Algorithm." *Journal of the Royal Statistical Society, Series B* 37 (1977), 1–38.

[76] A. De Pierro. "A Modified Expectation Maximization Algorithm for Penalized Likelihood Estimation in Emission Tomography." *IEEE Transactions on Medical Imaging* 14 (1995), 132–137.

[77] A. De Pierro, and A. Iusem. "On the Asymptotic Behavior of Some Alternate Smoothing Series Expansion Iterative Methods." *Linear Algebra and its Applications* 130 (1990), 3–24.

[78] A. De Pierro, and M. Yamaguchi. "Fast EM-Like Methods for Maximum 'a posteriori' Estimates in Emission Tomography." *Transactions on Medical Imaging* 20:4 (2001), 280–288.

[79] F. Deutsch and I. Yamada. "Minimizing Certain Convex Functions over the Intersection of the Fixed Point Sets of Nonexpansive Mappings." *Numerical Functional Analysis and Optimization* 19 (1998), 33–56.

[80] R. Devaney. *An Introduction to Chaotic Dynamical Systems*. Reading, MA: Addison-Wesley, 1989.

[81] R. Duda, P. Hart, and D. Stork. *Pattern Classification*. New York: Wiley, 2001.

[82] J. Dugundji. *Topology*. Boston: Allyn and Bacon, Inc., 1970.

[83] R. Dykstra. "An Algorithm for Restricted Least Squares Regression." *J. Amer. Statist. Assoc.* 78:384 (1983), 837–842.

[84] P. P. B. Eggermont, G. T. Herman, and A. Lent. "Iterative Algorithms for Large Partitioned Linear Systems, with Applications to Image Reconstruction." *Linear Algebra and Its Applications* 40 (1981), 37–67.

[85] L. Elsner, L. Koltracht, and M. Neumann. "Convergence of Sequential and Asynchronous Nonlinear Paracontractions." *Numerische Mathematik* 62 (1992), 305–319.

[86] T. Farncombe. "Functional Dynamic SPECT Imaging Using a Single Slow Camera Rotation." PhD diss., University of British Columbia, 2000.

[87] J. Fessler, E. Ficaro, N. Clinthorne, and K. Lange. "Grouped-Coordinate Ascent Algorithms for Penalized-Likelihood Transmission Image Reconstruction." *IEEE Transactions on Medical Imaging* 16:2 (1997), 166–175.

[88] W. Fleming. *Functions of Several Variables*. Reading, MA: Addison-Wesley, 1965.

[89] C. Gasquet and F. Witomski. *Fourier Analysis and Applications*. Berlin: Springer-Verlag, 1998.

[90] S. Geman and D. Geman. "Stochastic Relaxation, Gibbs Distributions and the Bayesian Restoration of Images." *IEEE Transactions on Pattern Analysis and Machine Intelligence* PAMI-6 (1984), 721–741.

[91] H. Gifford, M. King, D. de Vries, and E. Soares. "Channelized Hotelling and Human Observer Correlation for Lesion Detection in Hepatic SPECT Imaging" *Journal of Nuclear Medicine* 41:3 (2000), 514–521.

[92] K. Goebel and S. Reich. *Uniform Convexity, Hyperbolic Geometry, and Nonexpansive Mappings.* New York: Dekker, 1984.

[93] E. Golshtein and N. Tretyakov. *Modified Lagrangians and Monotone Maps in Optimization.* New York: Wiley, 1996.

[94] R. Gordon, R. Bender, and G. T. Herman. "Algebraic Reconstruction Techniques (ART) for Three-Dimensional Electron Microscopy and X-Ray Photography." *J. Theoret. Biol.* 29 (1970), 471–481.

[95] P. Green. "Bayesian Reconstructions from Emission Tomography Data Using a Modified EM Algorithm." *IEEE Transactions on Medical Imaging* 9 (1990), 84–93.

[96] L. G. Gubin, B. T. Polyak, and E. V. Raik. "The Method of Projections for Finding the Common Point of Convex Sets." *USSR Computational Mathematics and Mathematical Physics* 7 (1967), 1–24.

[97] E. Haacke, R. Brown, M. Thompson, and R. Venkatesan. *Magnetic Resonance Imaging.* New York: Wiley-Liss, 1999.

[98] T. Hebert and R. Leahy. "A Generalized EM Algorithm for 3-D Bayesian Reconstruction from Poisson Data Using Gibbs Priors." *IEEE Transactions on Medical Imaging* 8 (1989), 194–202.

[99] G. T. Herman, editor. "Image Reconstruction from Projections." In *Topics in Applied Physics,* Vol. 32. Berlin: Springer-Verlag, 1979.

[100] G. T. Herman and F. Natterer, editors. "Mathematical Aspects of Computerized Tomography." In *Lecture Notes in Medical Informatics,* Vol. 8. Berlin: Springer-Verlag, 1981.

[101] G. T. Herman, Y. Censor, D. Gordon, and R. Lewitt. "Comment on 'A Statistical Model for Positron Emission Tomography.' " *Journal of the American Statistical Association* 80 (1985), 22–25.

[102] G. T. Herman and L. Meyer. "Algebraic Reconstruction Techniques Can Be Made Computationally Efficient." *IEEE Transactions on Medical Imaging* 12 (1993), 600–609.

[103] G. T. Herman. Conversation with the author, Philadelphia, PA, July, 1999.

[104] C. Hildreth. "A Quadratic Programming Procedure." *Naval Research Logistics Quarterly* 4 (1957), 79–85. Erratum, ibid., p. 361.

[105] S. Holte, P. Schmidlin, A. Linden, G. Rosenqvist, and L. Eriksson. "Iterative Image Reconstruction for Positron Emission Tomography: A Study of Convergence and Quantitation Problems." *IEEE Transactions on Nuclear Science* 37 (1990), 629–635.

[106] H. M. Hudson, B. Hutton, and R. S. Larkin. "Accelerated EM Reconstruction Using Ordered Subsets." *Journal of Nuclear Medicine* 33 (1992), 960.

[107] H. M. Hudson and R. S. Larkin. "Accelerated Image Reconstruction Using Ordered Subsets of Projection Data." *IEEE Transactions on Medical Imaging* 13 (1994), 601–609.

[108] B. Hutton, A. Kyme, Y. Lau, D. Skerrett, and R. Fulton. "A Hybrid 3-D Reconstruction/Registration Algorithm for Correction of Head Motion in Emission Tomography." *IEEE Transactions on Nuclear Science* 49:1 (2002), 188–194.

[109] L. Jones and C. Byrne "General Entropy Criteria for Inverse Problems, with Applications to Data Compression, Pattern Classification, and Cluster Analysis." *IEEE Transactions on Information Theory* 36:1 (1990), 23–30.

[110] S. Kaczmarz. "Angenäherte Auflösung von Systemen linearer Gleichungen." *Bulletin de l'Academie Polonaise des Sciences et Lettres* A35 (1937), 355–357.

[111] A. Kak and M. Slaney. *Principles of Computerized Tomographic Imaging.* Philadelphia, PA: SIAM, 2001.

[112] L. Koltracht and P. Lancaster. "Constraining Strategies for Linear Iterative Processes." *IMA J. Numer. Anal.* 10 (1990), 555–567.

[113] T. Körner. *Fourier Analysis.* Cambridge, UK: Cambridge University Press, 1988.

[114] S. Kullback and R. Leibler. "On Information and Sufficiency." *Annals of Mathematical Statistics* 22 (1951), 79–86.

[115] L. Landweber. "An Iterative Formula for Fredholm Integral Equations of the First Kind." *Amer. J. of Math.* 73 (1951), 615–624.

[116] K. Lange and R. Carson. "EM Reconstruction Algorithms for Emission and Transmission Tomography." *Journal of Computer Assisted Tomography* 8 (1984), 306–316.

[117] K. Lange, M. Bahn, and R. Little. "A Theoretical Study of Some Maximum Likelihood Algorithms for Emission and Transmission Tomography." *IEEE Trans. Med. Imag.* TMI-6:2 (1987), 106–114.

[118] R. Leahy and C. Byrne "Guest Editorial: Recent Development in Iterative Image Reconstruction for PET and SPECT." *IEEE Trans. Med. Imag.* 19 (2000), 257–260.

[119] R. Leahy, T. Hebert, and R. Lee. "Applications of Markov Random Field Models in Medical Imaging." *Progress in Clinical and Biological Research* 363 (1991), 1–14.

[120] A. Lent and Y. Censor. "Extensions of Hildreth's Row-Action Method for Quadratic Programming." *SIAM Journal on Control and Optimization* 18 (1980), 444–454.

[121] E. Levitan and G. T. Herman. "A Maximum a posteriori Probability Expectation Maximization Algorithm for Image Reconstruction in Emission Tomography." *IEEE Transactions on Medical Imaging* 6 (1987), 185–192.

[122] T. Li and J. A. Yorke. "Period Three Implies Chaos." *American Mathematics Monthly* 82 (1975), 985–992.

[123] D. Luenberger. *Optimization by Vector Space Methods*. New York: Wiley, 1969.

[124] W. Mann. "Mean Value Methods in Iteration." *Proc. Amer. Math. Soc.* 4 (1953), 506–510.

[125] G. J. McLachlan and T. Krishnan. *The EM Algorithm and Extensions*. New York: Wiley, 1997.

[126] E. Meidunas. "Re-scaled Block Iterative Expectation Maximization Maximum Likelihood (RBI-EMML) Abundance Estimation and Sub-pixel Material Identification in Hyperspectral Imagery." Master's thesis, University of Massachusetts Lowell, 2001.

[127] J. Mooney, V. Vickers, M. An, and A. Brodzik. "High-Throughput Hyperspectral Infrared Camera." *Journal of the Optical Society of America, A* 14:11 (1997), 2951–2961.

[128] T. Motzkin and I. Schoenberg. "The Relaxation Method for Linear Inequalities." *Canadian Journal of Mathematics* 6 (1954), 393–404.

[129] M. Narayanan, C. Byrne, and M. King. "An Interior Point Iterative Maximum-Likelihood Reconstruction Algorithm Incorporating Upper and Lower Bounds, with Application to SPECT Transmission Imaging." *IEEE Transactions on Medical Imaging* TMI-20:4 (2001), 342–353.

[130] S. Nash and A. Sofer. *Linear and Nonlinear Programming*. New York: McGraw-Hill, 1996.

[131] F. Natterer. *Mathematics of Computed Tomography*. New York: Wiley, 1986.

[132] F. Natterer and F. Wübbeling. *Mathematical Methods in Image Reconstruction*. Philadelphia: SIAM, 2001.

[133] J. Ortega and W. Rheinboldt. *Iterative Solution of Nonlinear Equations in Several Variables*, Classics in Applied Mathematics, 30. Philadelphia: SIAM, 2000

[134] A. Peressini, F. Sullivan, and J. Uhl. *The Mathematics of Nonlinear Programming*. Berlin: Springer-Verlag, 1988.

[135] P. Pretorius, M. King, T.-S. Pan, D. deVries, S. Glick, and C. Byrne "Reducing the Influence of the Partial Volume Effect on SPECT Activity Quantitation with 3D Modelling of Spatial Resolution in Iterative Reconstruction." *Phys.Med. Biol.* 43 (1998),407–420.

[136] S. Reich. "Weak Convergence Theorems for Nonexpansive Mappings in Banach Spaces." *Journal of Mathematical Analysis and Applications* 67 (1979), 274–276.

[137] S. Reich. "Strong Convergence Theorems for Resolvents of Accretive Operators in Banach Spaces." *Journal of Mathematical Analysis and Applications* 75 (1980), 287–292.

[138] S. Reich. "A Weak Convergence Theorem for the Alternating Method with Bregman Distances." In *Theory and Applications of Nonlinear Operators*, edited by A. Kartsatos, pp. 313–319. New York: Dekker, 1996.

[139] R. Rockafellar. *Convex Analysis*. Princeton, NJ: Princeton University Press, 1970.

[140] A. Rockmore and A. Macovski. "A Maximum Likelihood Approach to Emission Image Reconstruction from Projections." *IEEE Transactions on Nuclear Science* NS-23 (1976), 1428–1432.

[141] P. Schmidlin. "Iterative Separation of Sections in Tomographic Scintigrams." *Nucl. Med.* 15:1 (1972).

[142] M. Schroeder. *Fractals, Chaos, Power Laws*. New York: W. H. Freeman, 1991.

[143] L. Shepp and Y. Vardi. "Maximum Likelihood Reconstruction for Emission Tomography." *IEEE Transactions on Medical Imaging* MI-1 (1982), 113–122.

[144] M. Shieh, C. Byrne, and M. Fiddy. "Image Reconstruction: A Unifying Model for Resolution Enhancement and Data Extrapolation: Tutorial." *Journal of the Optical Society of America, A* 23:2 (2006), 258–266.

[145] M. Shieh, C. Byrne, M. Testorf, and M. Fiddy. "Iterative Image Reconstruction Using Prior Knowledge." *Journal of the Optical Society of America, A* 23:6 (2006), 1292–1300.

[146] M. Shieh and C. Byrne. "Image Reconstruction from Limited Fourier Data." *Journal of the Optical Society of America, A* 23:11 (2006), 2732–2736.

[147] E. Soares, C. Byrne, S. Glick, R. Appledorn, and M. King. "Implementation and Evaluation of an Analytic Solution to the Photon Attenuation and Nonstationary Resolution Reconstruction Problem in SPECT." *IEEE Transactions on Nuclear Science* 40:4 (1993), 1231–1237.

[148] H. Stark and Y. Yang. *Vector Space Projections: A Numerical Approach to Signal and Image Processing, Neural Nets and Optics*. New York: Wiley, 1998.

[149] K. Tanabe. "Projection Method for Solving a Singular System of Linear Equations and its Applications." *Numer. Math.* 17 (1971), 203–214.

[150] M. Teboulle. "Entropic Proximal Mappings with Applications to Nonlinear Programming." *Mathematics of Operations Research* 17:3 (1992), 670–690.

[151] S. Twomey. *Introduction to the Mathematics of Inversion in Remote Sensing and Indirect Measurement*. New York: Dover Publications, 1996.

[152] Y. Vardi, L. A. Shepp, and L. Kaufman. "A Statistical Model for Positron Emission Tomography." *Journal of the American Statistical Association* 80 (1985), 8–20.

[153] M. Wernick and J. Aarsvold, editors. *Emission Tomography: The Fundamentals of PET and SPECT*. San Diego: Elsevier Academic Press, 2004.

[154] G. A. Wright. "Magnetic Resonance Imaging." *IEEE Signal Processing Magazine* 14:1 (1997), 56–66.

[155] C. F. Wu. "On the Convergence Properties of the EM Algorithm." *Annals of Statistics* 11 (1983), 95–103.

[156] Q. Yang. "The Relaxed CQ Algorithm Solving the Split Feasibility Problem." *Inverse Problems* 20 (2004), 1261–1266.

[157] D. C. Youla. "Mathematical Theory of Image Restoration by the Method of Convex Projections." In *Image Recovery: Theory and Applications*, edited by H. Stark, pp. 29–78. Orlando, FL: Academic Press, 1987.

[158] D. C. Youla. "Generalized Image Restoration by the Method of Alternating Projections." *IEEE Transactions on Circuits and Systems* CAS-25:9 (1978), 694–702.

[159] R. Young. *Excursions in Calculus: An Interplay of the Continuous and Discrete.* Dolciani Mathematical Expositions Number 13. Washington, DC: The Mathematical Association of America, 1992.

Index

Printed in the United States
by Baker & Taylor Publisher Services